Geophysics and Environmental Physics

The series *Geophysics and Environmental Physics* (GEP), reports new developments in geophysics research and teaching with a high quality and the explicit aim to summarize and communicate current knowledge in an accessible way. Books published in this series are conceived as bridging material between graduate textbooks and the forefront of research.

The research fields covered by the series are:

- Atmospheric Physics, Environmental Physics, Biogeophysics and Physics of Climate;
- Earth Magnetism and Rock Physics, Seismology and Physics of Volcanoes;
- Geodynamics and Tectonics;
- Physical Oceanography and Hydrology;
- Surface Processes;
- Geodesy;
- Instrumentation and Data Systems;
- Nonlinear Processes in Geophysics;
- Planetary and Solar System Physics.

Proposal should be sent to a member of the Editorial Board, or directly to the Responsible Editor at Springer:

Dr Lisa Scalone - lisa.scalone@springernature.com

Charles-Antoine Guérin

High-Frequency Radars for the Oceanographer

Basic Principles

Springer

Charles-Antoine Guérin
Mediterranean Institute of Oceanography
(MIO, Aix-Marseille Univ, Univ Toulon,
CNRS, IRD)
Université de Toulon
Toulon, France

ISSN 2948-2186 ISSN 2948-2194 (electronic)
Geophysics and Environmental Physics
ISBN 978-3-031-95641-6 ISBN 978-3-031-95642-3 (eBook)
https://doi.org/10.1007/978-3-031-95642-3

This Springer imprint is published by the registered company Springer Nature Switzerland AG
The registered company address is: Gewerbestrasse 11, 6330 Cham, Switzerland

Preface

This book has been inspired by about 20 years of teaching waves, electromagnetism, remote sensing and signal processing theory at the university of Toulon, and as many years of research in radar remote sensing, the last decade being devoted specifically to oceanographic high frequency radars. This is a multidisciplinary field of research, the understanding of which requires many concepts in different fields of physics and engineering. The difficulty in training students in this field is to find complete and accessible textbooks to rely on. It seems to me that most of the existing books and review papers are aimed at an already specialised audience and are not easy to start with if you are not already familiar with the field.

I was always embarrassed when a Ph.D. student asked for simple references to understand the theory and applications of high-frequency radar. I ended up having to explain everything from scratch and start all over again with each new student. At some point, I decided to write a book myself to put things down on paper and make the transfer of knowledge easier. This has been a two-year process during which I have had the pleasure of returning to the fundamental, old and slow science, alongside my hectic life of a 21st century researcher with urgent and daily "things to do". In particular, I went through a series of brilliant papers written more than half a century ago and realised how much had already been done in an age when data and calculations were processed without computers but with extreme intelligence. In this book I have tried to pay tribute to these pioneers and to recall their main results in a unified and modern approach.

I have also tried to make the theory accessible to anyone who has some understanding of waves and physics and who knows how to calculate a Fourier transform. Some of my colleagues oceanographers have had a bad experience with electromagnetism in their undergraduate studies and I hope that the second part of the book dealing with radio waves propagation will help to change their minds.

Toulon, France
April 2025

Charles-Antoine Guérin

Acknowledgments My sincere thanks go to my wife, Anne Sentenac, who read my drafts mercilessly with a phycisist eye and forced me to clarify and rewrite many chapters. I am also grateful to my colleagues Philippe Forget and Stephan Grilli, who read and checked some parts of the manuscript, and to Alice de Saint Jacob, who accepted to be my "beta tester" student and helped me to correct many errors and inconsistencies. I would also like to thank Dylan Dumas for providing some nice figures with the RADAR PROCESSING MIO software, and Anne-Claire Bennis for providing some data from the Alderney race radars. I would also like to thank Baptiste Domps, Cedric Chavanne, Eric Gill, Roberto Gomez, Weimin Huang, Marc Saillard, Victor Shrira and Lucy Wyatt for useful discussions on various topics covered in this book. A special thanks goes to Joel T. Johnson (Ohio State University) for sharing documents and codes on ground wave propagation.

Competing Interests The author has no competing interests to declare that are relevant to the content of this manuscript.

Contents

Acronyms

ADC	Analog-to-Digital Converter
ADCP	Acoustic Doppler Current Profiler
BF	Beam-Forming
BLR	Bragg Line Ratio
CW	Continuous Wave
DF	Direction Finding
DoA	Direction of Arrival
EM	Electromagnetic
FOBR	First-order Bragg Region
FFT	Fast Fourier Transform
FMCW	Frequency Modulated Continuous Waveform
FWHM	Full Width at Half Maximum
GDOP	Geometrical Dilution of Precision
HFR	High-Frequency Radar
IFMCW	Interrupted Frequency Modulated Continuous Waveform
MUSIC	Multiple Signal Classification
NRCS	Normalized Radar Cross-Section
PRF	Pulse Repetition Frequency
RX	Radar receiver station
SNR	Signal-to-Noise Ratio
SSP	Signal Subspace Projection
SVD	Singular Value Decomposition
TX	Radar transmitter station
UWLS	Unweighted Least-Square
WLS	Weighted Least-Square

1 Introduction

Observing the oceans in coastal areas is essential for a number of environmental challenges such as climate and marine ecosystem monitoring or coastal hazard mitigation. It involves the measurement of key geophysical parameters that describe the dynamics of the sea surface, namely winds, waves and currents. Among the many remote sensing instruments that have been developed for this purpose, High-Frequency Radars (HFR) have the unique ability to provide synoptic and quasi-instantaneous maps of surface currents and, to a lesser extent, ocean wave parameters. They are deployed along the coasts and observe the sea at grazing incidence in the HF radio frequency band (3–30 MHz carrier frequency, 10–100 m radar wavelength). In this range of frequency and below, transmitted electromagnetic (EM) waves can be guided along the ocean surface well beyond the horizon following the Earth curvature, a phenomenon known as "ground wave" propagation. This remarkable property of radio waves at certain frequencies has been known since Marconi's first experiments more than a century ago and was used to establish the first long-distance wireless communications. HFR take advantage of the ground wave mechanism to observe the sea surface over long distances, typically between 30 and 200 km offshore, depending on the radar frequency, with a spatial resolution of a few square kilometers. Due to these very efficient propagation characteristics and an adapted electromagnetic transmitted waveform, they consume little power and operate around the clock. They have a typical observation time of one hour, allowing processing of the data in near-real-time. They are "coherent on receive", which means they can measure the phase shift of the backscattered echo and are therefore able to produce the ocean Doppler spectra which are used to characterize motion at the sea surface. Last but not least, they allow an unambiguous and accurate interpretation of the oceanic parameters since the physical mechanisms at play are fully described by an analytical electromagnetic and hydrodynamic theory.

C.-A. Guérin, *High-Frequency Radars for the Oceanographer*, Geophysics and Environmental Physics, https://doi.org/10.1007/978-3-031-95642-3_1

HFR have been used in oceanography for half a century to monitor ocean currents. They have long been installed and maintained for research and development purposes. In the last decade their expansion has been intensified with the emergence of many civil and environmental applications and a more operational management. The number of HFR stations in the world has increased considerably (more than 400 in 2019, see [223]) and the community of users has grown significantly, both in the academic world and in the industrial and governmental sectors. The difficulty with the use of oceanographic HFR is that they are at the interface of many disciplines: electronics and radar, radio wave propagation and antenna design, electromagnetic scattering theory, hydrodynamics and oceanography, signal processing...While the pioneering scientists of the 1970s were able to deal with all these aspects at the same time, the advances of modern research and technology have made things so complex that it is impossible for most users today to possess these multidisciplinary skills. Today, the student or researcher who tries to be more than a naive user of these instruments and who wants to take a critical look at what they produce needs to delve into a variety of literature and communities in order to find the relevant information. The aim of this book is to provide in one single place a synthesis of what it is needed to understand the basic principles of HFR in its various aspects without having to be a specialist in any of them. We have voluntarily limited ourselves to the main accepted principles, thereby renouncing to provide the most up-to-date survey of this rapidly growing field of research. The latest technical developments for specialists, the operational aspects, and the state of the regional networks or the data management policy can be found in several reviews [88,90,91,110,135,140,188,191,213,221,223,229,294]. As HFR is also a very competitive industrial sector we would like to point out that we did not intend to compare the performances of the different radar systems and manufacturers (because it depends so much on how they are used) but only compare them in their principles. We would also like to point out to the reader, who might be more interested in oceanographic applications, that we have only considered the stage of producing "raw'" surface currents with an HFR; we have not dealt with the post-processing stage, which makes it possible to filter, clean up and interpolate imperfect data in order to transform them into a higher-level geophysical product.

The first part of this book is entitled "From Antenna Voltage to Surface Currents"; it essentially describes the principles and techniques that allow us to interpret the received radar signal in terms of ocean currents. It requires no background in electromagnetism other than some elementary notions of waves and is therefore intended to appeal to a very wide community of users. The second part of the book, "From sea surface to antenna voltage", is aimed more specifically at those who wish to understand the physical mechanisms by which electromagnetic waves are propagated over and scattered from the sea surface. These two fundamental theoretical sides of HFR, namely antenna voltage processing and sea surface scattering, are reconciled in the third part "From principles to reality". We confront these general physical and mathematical principles with real data, based on our own HFR network in Toulon and share our everyday experience of using these instruments.

Chapter 2 describes the basic principles of radio-oceanography. The history of oceanographic HFR began with Crombie's discovery in 1955 that the backscattering of HF radio waves from the sea surface is essentially caused by the surface gravity waves whose wavelength is half that of the transmitted electromagnetic wavelength. This could be seen in the backscattered Doppler spectrum, which had a sharp peak at a predictable value called the "Bragg frequency". It was later realized that the remaining continuous sideband spectrum surrounding the Bragg peak was related to the ocean wave spectrum and could be therefore used to measure the sea state. At the same time, it was also observed that the Bragg peak was shifted from its theoretical value in the presence of a surface current, so that measuring this shift could be used to estimate the latter. By 1974, with the pioneering works of Crombie, Hasselmann, Stewart and Joy, and Barrick (as summarized in the old reviews [19,241,242]), the complete physical theory underpinning oceanographic HFR observation had been established. We review this golden age and attempt to present in a simple and intuitive manner the physical principles at the heart of HFR ocean remote sensing.

Chapter 3 describes the various processing steps that are applied to the radar signal before any geophysical information can be extracted from it. Very soon after the physical foundations were laid down the question arose as to how the measurement of surface currents and other sea state parameters could be made operational. This involved designing a suitable radar architecture and developing the appropriate processing of the radar signal to enable surface current mapping. A key step in this process is the ability to discriminate between the "directions of arrival" of the different sea patches contributing simultaneously to the sea echo, a problem known as "direction finding". Overcoming these questions led to the development of the "Coastal Ocean Dynamics Application Radar" (CODAR) by Barrick [22], which was the first operational HFR system for measuring the currents. Other systems were developed in the 1980s and 1990s in several other countries: the "Ocean Surface Current Radar" (OSCR) in the UK [112], the Okinawa Radio Observatory HFR in Japan [159], the "Ocean State Monitoring and Analyzing Radar" (OSMAR) in China [136] or the "Wellen Radar" (WERA) in Germany [110], to name but a few. All these systems share most of the long signal processing chain needed to convert the raw electrical voltage at the output of the receiving antennas into a nice surface current map.

Chapters 4 and 5 deal specifically with the techniques of direction finding, which are at the heart of surface current mapping. Some radar systems, such as the WERA or the Okinawa HFR, have been designed with phased array antennas, a configuration that allows azimuthal discrimination to be performed using the "beam forming" technique. Other radar systems, such as the CODAR or OSMAR, have developed compact antenna systems using "high-resolution" direction finding methods. While the first method is fairly standard, the second is more complex to understand and has been given a chapter of its own.

In Chap. 6 we recall the basic concepts of electromagnetic radiation in vacuum and review some elementary properties of antennas. In the presence of a flat interface separating the air from a conducting medium such as seawater, the polarization of transmitted waves plays an important role in their reflection properties.

This is addressed in Chap. 7 where we introduce the canonical horizontal/vertical polarization basis and the associated Fresnel coefficients, leading to a simple justification for the exclusive use of vertical polarization for HF ocean remote sensing. This leads to the central concept of "ground waves", which have been known and used long before the time of oceanographic radars. We return in Chap. 8 to the famous century-old Sommerfeld problem for the radiation of a dipole antenna over a conducting medium and try to derive a simple and self-contained solution for the electric field attenuation with the Norton factor.

In Chap. 9 we introduce the scattering of electromagnetic waves from the sea surface that follows the propagation problem. In the context of HFR the electromagnetic wavelength is much larger than the water wave height so that the description of the scattering process falls within the scope of the analytical perturbation theory. We adopt a general point of view to introduce the corresponding scattering amplitude and specialize on the grazing incidence where the main results of Barrick are recovered in Chap. 10 with a fully analytical description of the second-order Doppler cross-section in both monostatic and bistatic configurations.

In Chap. 11 we illustrate the properties of the radar signal with a set of experimental data drawn from our daily work with HFR at the University of Toulon. We show how the abstract concepts around the Doppler spectrum materialize in the handling of these data.

We end up in Chap. 12 with the derivation of the main geophysical products of HFR: surface currents, wind, waves, and tsunamis. Again, we limit ourselves to the main principles and techniques, taking into account the practical limitations of the instruments.

In order to avoid long technical digressions in the main course of this story, we have postponed some technical material to the Appendix, namely: a synthesis of the wave spectral models (Appendix A); some classical results of electromagnetic theory which are invoked in the derivation of the scattered field above the sea surface (Appendices B, C, and D); some complementary calculations on the Zenneck and Norton ground waves (Appendices E and F); explicit formulae for the second-order scattering amplitude (Appendix G); calculation methods for the second-order Doppler spectra and derived products (Appendices I and J).

Part I

From Antenna Voltage to Surface Currents

Physical Principles of Radio-Oceanography

2

In this introductory chapter, we give an overview of the physical principles that lie at the heart of the use of HFR for oceanography, so that the reader has a simple understanding of the main concepts. These will be treated with greater rigor and detail in the rest of the book.

2.1 Waves and Currents

Oceanographic HFR are active remote sensing instruments that send decametric radio waves (10–100 m) from the coast towards the sea surface. Their main purpose is to extract information on the sea state and the currents from the echo of ocean waves, which are a dynamical deformation of the surface mostly caused by the action of wind. Radio waves impinging at the air-water interface are mainly sensitive to the surface deformations of the same scale, which correspond to decametric *gravity waves* (Fig. 2.1). We refer to the classical textbooks on ocean waves (e.g. [134, 150]) for the description and properties of the latter and only recall the main concepts which are useful to HFR theory. Even though the ocean wave field has in general a random and irregular aspect, it is convenient to represent it as a superposition of perfectly sinusoïdal components. Denoting $\eta(\mathbf{r}, t)$ the vertical disturbance of the surface at the horizontal position $\mathbf{r} = (x, y)$ and the time t, these individual waves are of the form:

$$\eta(\mathbf{r}, t) = a \cos(\mathbf{k} \cdot \mathbf{r} - \omega t), \tag{2.1.1}$$

where a is called the *amplitude*, $\mathbf{k}$ is the *spatial wave vector* and ω is the (circular) *wave frequency*. Such deformation translates with time in the direction of the wave vector at the *phase velocity* $c(k) = \omega/k$. The *wavelength* $\lambda = 2\pi/k$ is the spatial period along the propagation direction and the time period $T = 2\pi/\omega$ is the time

C.-A. Guérin, *High-Frequency Radars for the Oceanographer*, Geophysics and Environmental Physics, https://doi.org/10.1007/978-3-031-95642-3_2

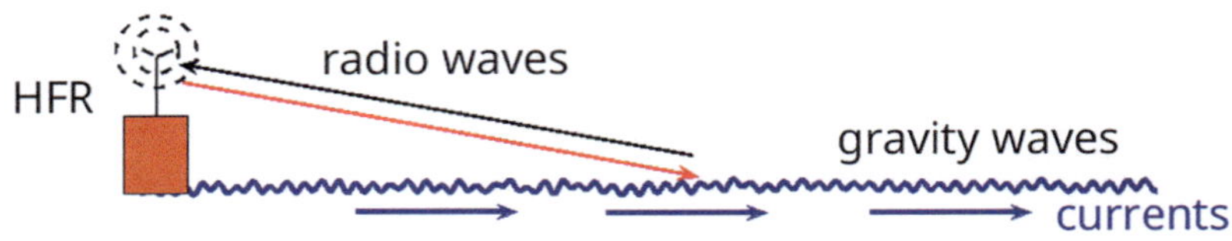

Fig. 2.1 HFR probe the sea surface with radio waves and measure the echo of gravity waves

period at the same position. The wavevector $\mathbf{k}$ and the wave frequency ω are not independent variables. There exists a *dispersion relation* that relates them to each other through a (possibly multi-valued) functional dependence $\omega = \omega(\mathbf{k})$. In the usual small-amplitude approximation of *linear wave theory*, gravity waves in deep-water satisfy the dispersion relation:

$$\omega = \sqrt{gk}, \tag{2.1.2}$$

where $g = 9.81$ m.s^{-2} is the Earth gravity. This implies a phase velocity:

$$c(k) = \sqrt{\frac{g}{k}}, \tag{2.1.3}$$

which is on the order of 4–12 m/s for decametric wavelengths. The deep-water condition is considered to hold as soon as the water depth is larger than at least half the wavelength, an assumption which is satisfied in most HFR applications. In addition to ocean waves, which transport energy but not matter, there are the *ocean currents*, which are a mass transport in the ocean volume associated to a local speed vector $\mathbf{U}(x, y, z)$. The interaction of ocean waves and currents is a complex topic and it is most often assumed that they add up linearly, in which case the surface current has a mere translational effect on gravity waves. In presence of a uniform current $\mathbf{U}$, the gravity-wave linear dispersion relation is modified to:

$$\omega = \sqrt{gk} + \mathbf{k} \cdot \mathbf{U} \tag{2.1.4}$$

This implies that the wave celerity is increased or decreased by the projected current speed along its propagation direction:

$$c(\mathbf{k}) = \sqrt{\frac{g}{k}} + \widehat{\mathbf{k}} \cdot \mathbf{U}, \tag{2.1.5}$$

where $\widehat{\mathbf{k}} = \mathbf{k}/k$ is the unit vector along $\mathbf{k}$. The current speed is most often on the order of $1-2$ m/s or less and is in general smaller than the celerity of decametric gravity waves, except for very strong tidal currents in specific places.

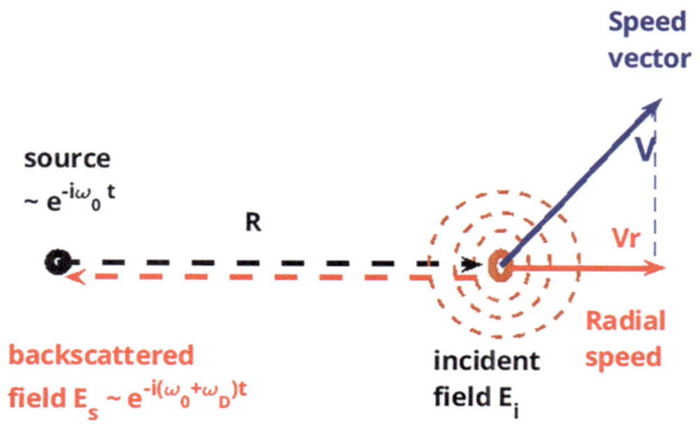

Fig. 2.2 The Doppler frequency shift only involves the radial velocity of the moving object, i.e. the projection of its speed onto the source-object axis

2.2 The Doppler Spectrum

The main geophysical product of oceanographic HFR is the ocean current, which is seen through the modified celerity of ocean waves. The principle at the heart of this measurement is the Doppler effect.[1] This well-known phenomenon, which is experienced in everyday life with acoustic or electromagnetic waves, refers to the change in the apparent frequency of the waves backscattered by a moving target when illuminated by an incident monochromatic wave. Consider a monochromatic source at radian frequency ω_0 radiating an electric field in the air towards a target at a distance R. In the far field, that is at a distance which is much larger than both the electromagnetic wavelength and the size of the source, it is possible to assimilate the incident field on the object to a spherical wave of the form (ignoring the amplitude factor and the vectorial nature of the field):

$$E_i(t) \sim \mathrm{e}^{\mathrm{i}(K_0 R - \omega_0 t)} \tag{2.2.6}$$

where $K_0 = \omega_0 / c_0$ is the electromagnetic wavenumber and c_0 the celerity of light in vacuum. According to a foundamental principle of wave scattering (known as "Huygens principle" in Physical Optics), the field scattered by the illuminated object can be seen as a secondary spherical wave whose complex amplitude is proportional

[1] Christian Doppler (1803–1853) was an Austrian physicist. He postulated the effect that bears his name in a 1842 report ("Über das farbige Licht der Doppelsterne und einiger anderer Gestirne des Himmels") in the Royal Society of Bohemian Sciences. A first experimental proof was given 3 years later by the Dutch Physicist Christoph Buys-Ballot by having musicians play the same tone with the trumpet on a moving train.

to the incident excitation. The backscattered field at the transmitter can thus be written as:

$$E_s(t) \sim \mathrm{e}^{\mathrm{i}(2K_0R-\omega_0t)} \tag{2.2.7}$$

Whenever the source and the illuminated object are in relative motion, their mutual distance $R(t)$ varies with time. If we assume that the object is moving with a constant velocity vector $\mathbf{V}$ then its relative distance to the source varies with the *radial velocity* V_r, which is the projection of the velocity vector onto the oriented source-target axis (Fig. 2.2):

$$R(t) = R_0 + V_rt \tag{2.2.8}$$

It follows that the phase $2K_0R_0 - \omega_0t(1 - 2V_r/c_0)$ of the backscattered signal has an additional dependence on time, resulting in an apparent frequency $-\omega_0(1 - 2V_r/c_0)$. The frequency shift with respect to the transmitted signal is called the *Doppler frequency shift*:

$$\omega = -\frac{2V_r}{c_0}\omega_0 \tag{2.2.9}$$

The minus sign indicates that an approaching object ($V_r < 0$) induces a positive shift and vice versa for a receding object. When using the ordinar frequencies $f = \omega/(2\pi)$ and $f_0 = \omega_0/(2\pi)$ we have the alternative formula:

$$f = -\frac{2V_r}{\lambda_0}, \tag{2.2.10}$$

where $\lambda_0 = c_0/f_0$ is the electromagnetic wavelength. This is the well-known formula of the *non-relativistic Doppler effect*; in its derivation one neglects the travel time of light from the source to the scatterer. The general formula, which is valid for arbitrary speeds, has a corrective relativistic term that involves the ratio V_r/c_0 (see e.g. [143]) but is of no use in the oceanographic context because the involved velocities are much smaller than the speed of light.

When sending radio waves to the ocean surface it is obvious that the motion of the waves will induce Doppler shifts in the backscattered signal. However, the infinite number of ocean waves with different velocities produces a continuum of possible Doppler frequency shifts in the sea echo. Furthermore, due to the random nature of waves at the sea surface, the backscattered electric field is better modeled with a random process than with a deterministic function. The power associated to each Doppler shift is described by a statistical function called the *Doppler spectrum* . This quantity is proportional to the *power spectrum* (or *spectral density*) of the complex time series associated to the backscattered electric field $E_s(t)$. It is defined as the Fourier transform:

$$\sigma(\omega) = \widehat{\gamma}(\omega) = \int_{-\infty}^{+\infty} \mathrm{e}^{-\mathrm{i}\omega t}\gamma(t)dt, \tag{2.2.11}$$

where

$$\gamma(t) = \alpha\langle E_s(t'+t)E_s^*(t')\rangle, \tag{2.2.12}$$

is proportional to the field autocorrelation function and the asterisk in E_s^* stands for the complex conjugate. In this definition the brackets $\langle\cdot\rangle$ indicate an ensemble average over the realizations of the random process $E_s(t)$, which is implicitly assumed to be stationary (at least for the duration of the measurement) in the sense that the field correlations between any two instants t' and $t'+t$ depends only on the time lag t. The factor α is a normalization constant whose value has no importance at this stage (see Eq. (9.4.28) for the exact definition). This electrical voltage undergoes some primary processing before it can be used. It is first demodulated from the carrier radar frequency and then sorted in range and direction. These operations, which will be discussed in detail in Chap. 3, lead in the end to a complex time series that carries the physical information associated with a patch of the ocean surface.

To better understand what a Doppler spectrum means, it is instructive to consider the case of a finite number N of independent targets moving with their own radial velocities with respect to the radar. Since the electromagnetic waves reflected back from the different objects add up at the receiver, the total complex backscattered field can be written (after demodulation by the radar frequency to remove the trivial $e^{-i\omega_0 t}$ factor) as:

$$E_s(t) \sim A_1 e^{-i(\omega_1 t-\varphi_1)} + \cdots + A_N e^{-i(\omega_N t-\varphi_N)}, \tag{2.2.13}$$

where $\omega_1, \ldots, \omega_N$ are the Doppler frequency shifts caused by the motion of the various targets, the $A_1, \ldots, A_N$ are positive coefficients quantifying the strength of each echo (depending on the distance, size, orientation, etc. of each object), and φ_1, ..., φ_N are constant phases related to the initial positions of the targets. The cross-product that is needed to form the autocorrelation function (2.2.12) is therefore:

$$E_s(t'+t)E_s^*(t') \sim \sum_{n=1}^{N} |A_n|^2 e^{i\omega_n t} + \sum_{m\neq n} A_n A_m^* e^{i(\omega_m-\omega_n)t'} e^{i(\varphi_m-\varphi_n)} e^{-i\omega_n t} \tag{2.2.14}$$

When taking the ensemble average, which is actually obtained by considering the successive measurements in time t', the oscillatory terms in the second summation cancel out, leaving only the first summation. The resulting Doppler spectrum is a sum of Dirac functions located at the Doppler frequencies:

$$\sigma(\omega) \sim |A_1|^2 \delta(\omega-\omega_1) + \cdots |A_N|^2 \delta(\omega-\omega_N) \tag{2.2.15}$$

Thus, these discrete frequency components indicate the various Doppler shifts caused by the different radial speeds and the magnitude of the Doppler spectrum indicates the power of those echos. Of course, this picture with Dirac functions holds only for a discrete set of scattering objects. In the case of a continuous scattering medium, such as the ocean surface, the Doppler spectrum takes the form of a continuous spectral density.

Since the autocorrelation function of the radar signal is complex, its Fourier transform does not enjoy the hermitian symmetry, which implies that the Doppler spectrum is generally not the same on the positive and negative frequency side, $\sigma(-\omega) \neq \sigma(\omega)$, and thus makes the difference between the objects that are approaching (positive Doppler shift) or receding (negative Doppler shifts) from the radar. The Doppler spectrum can be defined in either radian frequency (ω) or ordinar frequency unit $f = \omega/(2\pi)$ and the two units are used interchangeably in the HFR literature. The two definitions differ only by the factor 2π (i.e. $\sigma(f) = 2\pi\sigma(\omega)$) and, following the common usage, we will plot the Doppler spectra most often as a function of the ordinary frequency f. Due to its large dynamic range, it is in general expressed in *decibels* (dB), which is a base 10 logarithmic scale:

$$\sigma_{dB} = 10\log_{10}(\sigma) \tag{2.2.16}$$

The dB scale is very useful because it allows to compare very large and very small signals and to convert any multiplicative factor into a simple addition. For example, an increase of 3 dB is a factor of 2 and an increase of 10 dB is a factor of 10. The dynamic range of experimental ocean Doppler spectra is the difference between the highest measurable value (without distortion) and the noise floor. It is typically on the order of 60–80 dB, depending on the instrument.

2.3 The Bragg Wave

In 1955 Crombie[2] published a one-page paper in Nature [54], in which he reported for the first time a Doppler spectrum of the sea echo with a HFR. He found a pronounced peak at a given frequency in the Doppler spectrum, irrespective of wind and sea conditions. In view of the simple formula (2.2.15), this would be caused by many targets moving at a particular radial speed. He interpreted this peak as the constructive interference of the signals reflected by the crests of one ocean wave, namely that with a wavelength equal to half the radar wavelength:

$$\lambda_B = \frac{\lambda_0}{2} \tag{2.3.17}$$

The ocean wave with half the radar wavelength is called the *Bragg wave* and its wavenumber $k_B = 2\pi/\lambda_B = 2K_0$ is the *Bragg wavenumber*.[3] For HF radio frequencies, it is a decametric wave whose celerity is governed by the deep water gravity

[2] Douglass D. Crombie (1924–2011) was a New Zealander physicist and engineer. He started his very long career in New Zealand and continued in the US in the 60 s with the Institute of Telecommunication Science at Boulder and later with the Aerospace Corporation. He made an important contribution to the propagation of radio waves in the open sea and in the ionosphere and co-authored 63 papers on the subject. His most famous article is his 1977 Nature paper which has been cited over 1000 times.

[3] Sir William Henry Bragg (1862–1942) was a british physicist and chemist. He is well known for his work on X-ray crystallography for which he obtained the Nobel prize in 1915. He had nothing to do with oceanography but ironically left an important mark in this discipline because the mechanism of diffraction by a periodic structure is the same for water waves.

wave dispersion relation (2.1.3). By applying the Doppler shift formula (2.2.10) to the velocity of the Bragg wave, Crombie obtained the following formula for the observed Doppler shift:

$$f_B = \sqrt{\frac{g}{\pi \lambda_0}}, \tag{2.3.18}$$

which he found in very close agreement with the experimental value, thereby justifying his interpretation. This specific frequency is called the *Bragg frequency* and depends solely on the radar wavelength. The corresponding peak in the Doppler spectrum is called the *Bragg peak* . Unlike the Bragg frequency, the amplitude of the Bragg peak depends on the ocean wave height and consequently on the wind speed and direction as wind is the main driving parameter for the growth of ocean waves. Crombie's argument to identify the Bragg resonance relies on a simple physical-optics argument that is heuristic but gives the correct dominant contribution of the backscattered power, as confirmed by the full electromagnetic and hydrodynamic theory (see Chap. 9). Consider for simplicity a one-dimensional geometry and assume that sea surface elevations are described by a simple sinusoidal profile along one horizontal direction, say $\eta(x, t) = a\cos(kx - \omega t)$. The electric field produced by a far monochromatic source impinging at grazing incidence on the sea surface can be locally assimilated to an incident plane wave of the form $E_i \sim e^{i(K_0 x - \omega_0 t)}$, so that two backscattered waves emanating from successive crests on the sinusoidal surface will be out of phase by an amount $2K_0\lambda$, where λ is the wavelength of the ocean wave (Fig. 2.3). Thus, the backscattered electromagnetic waves from the surface will interfere constructively if this phase shift is an integer multiple of 2π, that is if the condition (2.3.17) is fulfilled. As we will see in Chap. 9, this also holds for the two-dimensional geometry where the incident electric field and the sinusoidal surfaces can propagate into a variety of directions. Among the many waves that compose the sea surface, there is only one that allows constructive interference in the backscattered electric field, namely the wave which is aligned with the electromagnetic wavevector $\mathbf{K_0}$ and has half the radar wavelength. The specific wavevector:

$$\mathbf{k_B} = -2\mathbf{K_0} \tag{2.3.19}$$

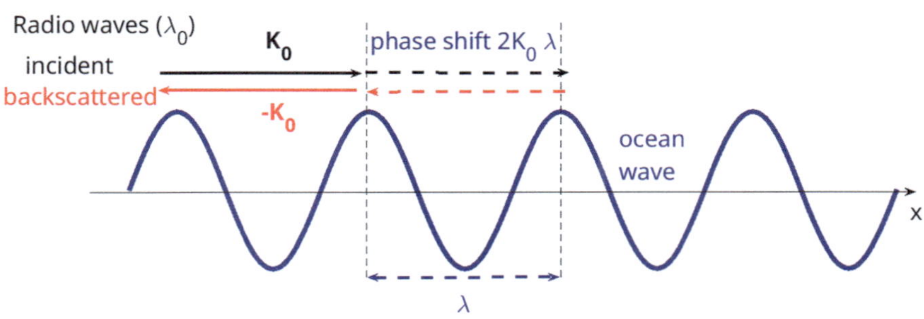

Fig. 2.3 The phase shift between the rays reflected from 2 successive crests is $2K_0\lambda$, where λ is the ocean wave period and K_0 the electromagnetic wavenumber. Constructive interferences with a 2π phase shift are obtained for $\lambda = \lambda_0/2$

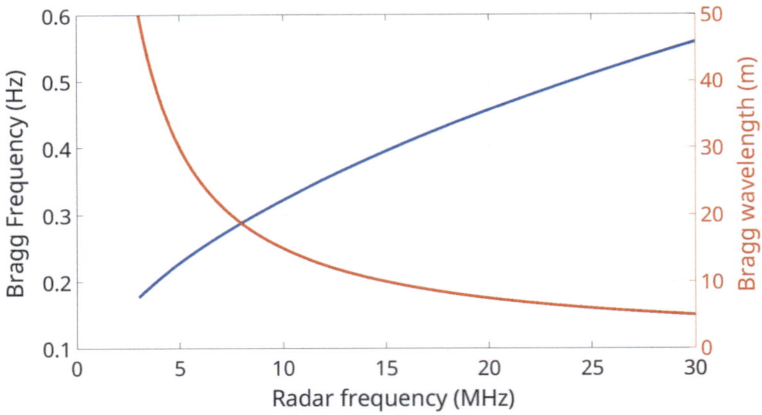

Fig. 2.4 Variation of the Bragg frequency f_B and the Bragg wavelength λ_B with the radar frequency in the HF radio band (3−30 MHz)

is called the *Bragg wavevector* and plays an important role in scattering theory. Note that we applied the Doppler shift formula (2.2.10) to a gravity wave as if it were a moving target. It is in fact not a moving object because there is no actual transport of matter, just a synchronized change in sea surface elevations. However, from the radar receiver prospective, the phase shifts in the electromagnetic field caused by the deformation of the sea surface are equivalent to those of a rigid target moving at the phase velocity of the Bragg wave.

In the HF range, the Bragg frequency spans from about 0.18 to 0.55 Hz and the Bragg wavelength from 50 to 5 m (Fig. 2.4). Note that in invoking the Bragg resonance mechanism to explain the sea echo, an implicit important assumption has been made. As we will see in Chap. 3, an appropriate processing of the radar signal allows to associate the measured Doppler spectra to given sea patches placed at given distances from the radar. The grating effect can only occur if there is a sufficient number of Bragg waves in the interior of the sea patch. This is always verified because the sea patches for HF radio waves are in general of kilometric extension while the Bragg waves are decametric. In his 1955 experiment, Crombie only measured one-sided Doppler spectra, that is he could not distinguish between the positive and negative frequency shifts.[4] In a later experiment about a decade later [57], Crombie deduced the simultaneous presence of both positive an negative Doppler shifts through the

[4] Although the 1955 paper does not give any information on the radio frequency equipment, it is very likely that the received signal was compared to the local oscillator with a single mixer, thus providing only the absolute value of the frequency shift. A second mixer fed in quadrature with the transmitted signal would have been required to resolve the ambiguity, see Chap. 3. It is also interesting to note that the spectral analysis of time records was performed mechanically by rotating a photographic paper into an optical-electrical device [9]. At that time, computing even a simple Fourier transform was a challenge !.

interference patterns of the sea echo in a range-time representation. Positive Doppler shifts around the Bragg frequency f_B were interpreted as Bragg scattering from wave approaching the radar while negative Doppler shift are caused by receding Bragg waves. A few years later, the technological advances made it possible to measure two-sided frequency spectra and it became apparent that the spectral sea echo generally has two main Bragg peaks of opposite sign and unequal amplitude [1,23,53,55,58,181], the magnitude of the peaks depending on the direction in which the radar is looking with respect to the wind direction. These experiments confirmed the obvious fact that waves mostly propagate downwind (or at least in the downwind half-space) but also established the far less intuitive result that a small fraction of them can propagate against the wind, as evidenced by the presence of a second Bragg peak even when the radar view is aligned with the wind direction [56].

Thus, the general pattern observed for ocean Doppler spectra is a set of prominent peaks around the symmetric frequencies $\pm f_B$, whose relative magnitude depends on whether the radar is looking upwind or downwind (these configurations are illustrated in Fig. 2.5). The positive (resp. negative) peak is stronger (weaker) when the radar is looking upwind (resp. downwind) and the two peaks are of similar magnitude if the radar is looking cross-wind (Fig. 2.6). This qualitative result emanating from experimental observations has been given a solid theoretical foundations a few years later by Barrick [15], who showed that the magnitude of the positive/negative Bragg peak in the Doppler spectrum is proportional to the directional ocean wave spectrum at plus/minus the Bragg wavevector.

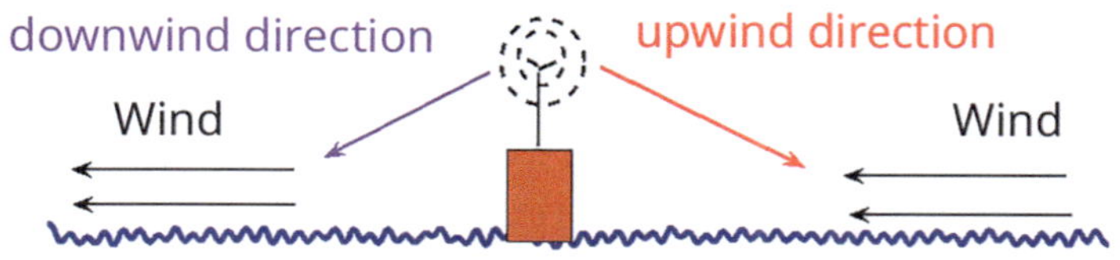

Fig. 2.5 Definition of the upwind and downwind directions for the radar

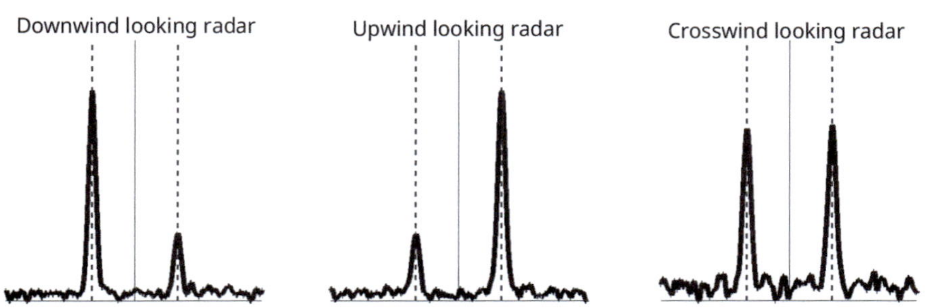

Fig. 2.6 The magnitude of the positive and negative Bragg peaks depends on the relative orientation of the directional ocean wave spectrum and the radar look direction

2.4 The Effective Surface Current

A uniform surface current **U** can be viewed as a rolling pad beneath the waves, changing their phase velocity (Fig. 2.8). Therefore, the apparent velocity $c_B = \sqrt{g/k_B}$ of the Bragg wave seen from the radar is increased by the radial velocity of the current, which is the projection of the current vector on the radar look direction (Fig. 2.7).

Hence, the Bragg wave crests propagate with the modified velocity $c_B + U_r$, causing an additional Doppler shift:

$$f_B \rightarrow f_B - \frac{U_r}{\lambda_B} \tag{2.4.20}$$

This additional shift is positive if the surface current is traveling to the radar ($U_r < 0$) and negative otherwise. This is the key idea in measuring surface current with HFR: it is sufficient to measure the displacement of the main Doppler peak relative to the Bragg frequency. This holds for the positive ($+f_B$) and negative ($-f_B$) Bragg peaks, which are translated by the same amount. Again, Crombie pioneered this observation and first put it into practice in an experiment in the Gulf of Mexico in 1969 [53,55]. He observed that the Doppler shifts of the approaching and receding Bragg waves were not exactly the same and found that their values were consistent with the assumed current velocity. This was confirmed a few years later by subsequent

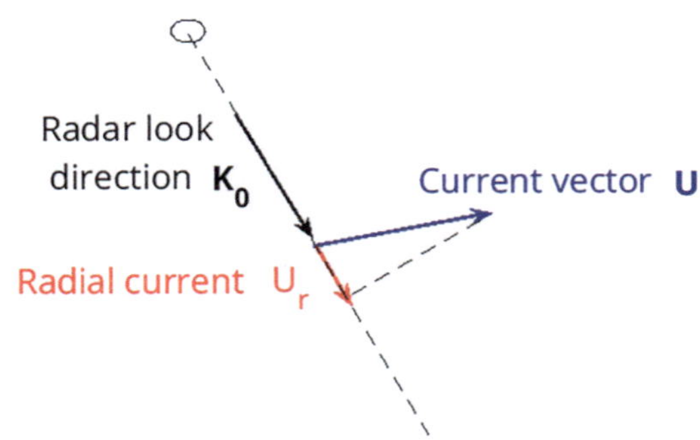

Fig. 2.7 The radial current U_r is defined as the projection of the current vector on the radar look direction. It is positive if it is flowing away from the radar and negative otherwise

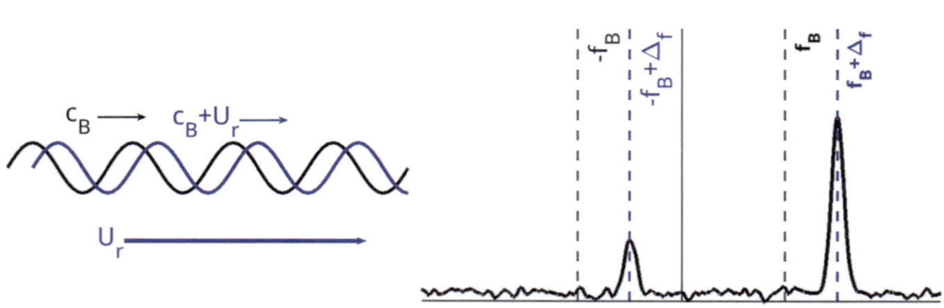

Fig. 2.8 In the presence of an additional surface current the apparent velocity of the Bragg wave seen from the radar is increased by the radial velocity of the current (U_r)

experiments at San Clemente Island, where currents were measured in two different directions [23,251]. Shortly afterwards, Barrick developed at the NOAA the first HFR specifically designed for measuring ocean currents. He generalized the measurement of surface current to an arbitrary number of directions and by 1977 the first surface current maps were available [22]. From the beginning, it was clear that the phase velocity of the Bragg wave was modified by the presence of the current. However, since the latter is in general not uniform in depth, the question rapidly arose as to which surface layer and which current component was being measured when calculating the additional Doppler shift. In their famous 1974 paper, Stewart and Joy [251] provided a solid hydrodynamic foundation to this problem. For a vertical profile of radial current $U_r(z)$, they showed using a perturbation approach (assuming that the current velocity is much smaller than the wave celerity) that the Bragg wave celerity c_B is augmented by an effective current U_e:

$$U_e = 2k_B \int_{-\infty}^{0} e^{2k_B z} U_r(z) dz, \qquad (2.4.21)$$

where the z axis is directed upward and the mean surface level is assumed to be at $z = 0$. In the special but important case of a linearly decreasing current the above formula is analytically solvable leading to:

$$U_e = U_r \left(\frac{\lambda_B}{4\pi} \right) \qquad (2.4.22)$$

This amounts to replace the current $\mathbf{U}$ in the dispersion relation (2.1.4) with a uniform effective current $\mathbf{U_e} = \mathbf{U}(\lambda_B/(4\pi))$. For an arbitrary profile of current this is only an approximation that holds as long as U_r is monotonically decreasing over a layer larger than $\lambda_B/(4\pi)$. Therefore, the effective depth of the surface current measured by a HFR is a fraction of the radar wavelength $(\lambda_0/(8\pi))$ (Fig. 2.9). It is 75 cm at the typical HFR frequency of 16 MHz and ranges from 44 cm to 2.4 m from the highest (27 MHz) to the lowest (5 MHz) frequencies that are assigned to oceanographic radars.

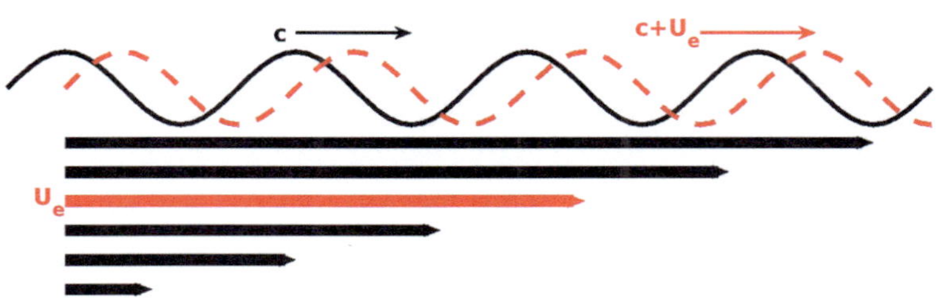

Fig. 2.9 The depth of the effective current sensed by the radar is a fraction of the radar wavelength $(\lambda_0/(8\pi))$

2.5 The Ocean Wave Spectrum

As seen above with simple heuristic arguments, the positive and negative main Doppler peaks are caused by the Bragg wave traveling towards and away from the radar, respectively, and experiments show that these peaks exist simultaneously, albeit with different amplitudes. How can waves travel in two opposite directions simultaneously? This is because there are always a variety of wave directions (which can even be opposite) and amplitudes in a sea state. These are described by what is known as the ocean wave spectrum. Even though this is a very classical concept (e.g. [61, 134, 207]), the definitions and conventions regarding wave spectra can vary depending on the context, which makes it an ambiguous notion. Therefore, we provide here a synthesis of what is necessary to know in the framework of HFR.

In short, the ocean wave spectrum gives the distribution of the ocean waves squared amplitudes as a function of the spatial or temporal frequencies. Water waves in the open sea are generally described by stochastic fields. Under the assumption that the random fluctuations at the sea surface can be described by a linear superposition of independent waves propagating along a continuum of directions and frequencies, one can represent the surface elevation with respect to the mean water plane by a stationary Gaussian random process in time and space. It can be written as a stochastic Fourier integral:

$$\eta(\mathbf{r}, t) = \int a(\mathbf{k}, \omega) e^{i(\mathbf{k}\cdot\mathbf{r}-\omega t)} \mathbf{dk} d\omega, \tag{2.5.23}$$

where $a(\mathbf{k}, \omega)$ are independent random variables. Note that the involved Fourier coefficients are complex; however, since the surface is real, they satisfy the hermitian symmetry $a(-\mathbf{k}, -\omega) = a^*(\mathbf{k}, \omega)$. The surface elevations can be equivalently represented in terms of real sinusoidal components with random phases, a description which is less frequently employed [182, 183]. Due to the random nature of the wave field, the complex amplitude coefficient is not a "true" function but a stationary random process which can be manipulated using the following symbolic rules[5]:

$$\begin{aligned} \langle a(\mathbf{k}, \omega)\rangle &= 0 \\ \langle a(\mathbf{k}, \omega) a^*(\mathbf{k}', \omega')\rangle &= S(\mathbf{k}, \omega)\, \delta(\mathbf{k}-\mathbf{k}', \omega-\omega') \\ \langle a(\mathbf{k}, \omega) a(\mathbf{k}', \omega')\rangle &= 0 \end{aligned} \tag{2.5.24}$$

where $\delta(\mathbf{k}, \omega)$ is the three-dimensional Dirac function, the brackets stand for the ensemble average and a^* is the complex conjugate. The three-dimensional positive function $S(\mathbf{k}, \omega)$ is called the *wavenumber-frequency spectrum* or simply the (k, ω) spectrum. The total energy of waves is quantified by their mean square wave elevation:

$$\langle \eta^2 \rangle = \int S(\mathbf{k}, \omega) \mathbf{dk} d\omega \tag{2.5.25}$$

[5] A rigorous mathematical basis can be given in the framework of the so-called Fourier-Stieltjes integrals.

Hence the wave spectrum is the energy density of the wave components in the $(\mathbf{k}, \omega)$ space. From (2.5.25) we see that it has the physical dimension of $[M]^4[S]$. By construction, the (k, ω) spectrum is a positive even function of its three-dimensional variable, i.e. it satisfies the central symmetry:

$$S(-\mathbf{k}, -\omega) = S(\mathbf{k}, \omega). \tag{2.5.26}$$

Note, however, that this property does not apply to the "slices" of the spectrum at fixed $\mathbf{k}$ or ω, i.e. $S(\mathbf{k}, \omega) \neq S(-\mathbf{k}, \omega)$ in general, indicating that waves do not have the same amount of energy propagating in opposite directions. The *wave frequency spectrum* is defined as the integration of the wavenumber spectrum over the $\mathbf{k}$ space:

$$S(\omega) = 2 \int S(\mathbf{k}, \omega)\mathbf{dk}, \ \omega \geq 0, \tag{2.5.27}$$

with a physical dimension in $[M^2][S]$. Likewise, the *directional ocean wavenumber spectrum* is obtained by integrating the wavenumber spectrum over the frequencies:

$$S(\mathbf{k}) = 2 \int_0^{+\infty} S(\mathbf{k}, \omega) d\omega, \tag{2.5.28}$$

which has the physical dimension of $[M^4]$. In both case, the factor 2 is due to the restriction to positive frequencies and the need to conserve the total energy:

$$\langle \eta^2 \rangle = \int_0^{+\infty} S(\omega) d\omega = \int S(\mathbf{k})\mathbf{dk} \tag{2.5.29}$$

Note that the directional ocean wavenumber spectrum is in general not symmetric ($S(-\mathbf{k}) \neq S(\mathbf{k})$). More details can be found in the Appendix A regarding the definitions and properties of ocean wave spectra and their most common models. Here we will only point out some universal features of wave spectra that are relevant for HFR. The dominant parameter shaping the ocean wave spectrum in the open ocean is the wind vector. Ocean waves obviously grow under the wind input and the largest wavelength that can be generated at a given constant wind speed is called the *dominant wave* or the *peak wave*; it scales approximately as the square of wind speed,[6] $\lambda_p \sim U_{10}^2$. Frequency wave spectra have a strong maximum around the peak frequency $\omega_p = \sqrt{g/(2\pi\lambda_p)}$ and a rapid cut-off below this frequency. Above the peak frequency, they have a weak dependence on wind speed and are found to follow a quasi power-law decay $\omega^{-\alpha}$ over several decades with an exponent transitioning from $\alpha = 4$ to $\alpha = 5$. This frequency range is called the *equilibrium range* or the *saturation range*, because there is a balance between the wind energy input to the waves and the dissipation by breaking and nonlinear mechanisms. A classical Pierson-Moskowitz spectrum (see Appendix A) illustrates these features in Fig. 2.10.

[6] Wind speed over the ocean is traditionally measured at 10 meter above the mean sea surface (U_{10}).

2.6 First-Order Doppler Spectra

In the framework of linear wave theory, deep-water gravity waves satisfy the dispersion relation (2.1.2). It follows that the wavenumber-frequency spectrum can be reduced essentially to a spatial frequency variable, namely:

$$S(\mathbf{k},\omega) = \frac{1}{2}S(\mathbf{k})\delta(\omega - \sqrt{gk}) + \frac{1}{2}S(-\mathbf{k})\delta(\omega + \sqrt{gk}) \tag{2.6.30}$$

As first derived in [10,15], the normalized Doppler spectrum in its leading approximation, say $\sigma_1(\omega)$, is proportional to the wavenumber-frequency spectrum at the Bragg wavenumber. Precisely,

$$\sigma_1(\omega) = \mathcal{N}S(\mathbf{k_B})\delta(\omega - \omega_B) + \mathcal{N}S(-\mathbf{k_B})\delta(\omega + \omega_B), \tag{2.6.31}$$

where $\mathcal{N}$ is a normalization factor that will be discussed in Chap. 10. This formula explains quantitatively the first observations of the ocean HFR Doppler spectrum by Crombie and confirms the heuristic derivation of the Bragg resonance mechanism. The delta-like spectrum is a theoretical model corresponding to the ideal case of an infinite stationary sea surface observed over an infinite time. Practical observations always involve a finite radar sea patch and limited integration time, so that the Bragg peaks are finite in width and amplitude. In addition, the entraining cyclic motion of long waves modulates the velocity of the Bragg waves, further broadening the peaks. All these aspects are discussed in detail in Chap. 3. In presence of a uniform or effective current $\mathbf{U}$, the modified wave dispersion relation (2.1.4) implies that the peaks of the Doppler spectrum are shifted by an amount $\mathbf{k_B} \cdot \mathbf{U}$:

$$\sigma_1(\omega) = \mathcal{N}S(\mathbf{k_B})\delta(\omega - \omega_B - \mathbf{k_B} \cdot \mathbf{U}) + \mathcal{N}S(-\mathbf{k_B})\delta(\omega + \omega_B - \mathbf{k_B} \cdot \mathbf{U}), \tag{2.6.32}$$

Note that $\mathbf{k_B} \cdot \mathbf{U} = -k_B U_r$ where U_r is the radial velocity defined with the outwards radial vector from the radar. This is consistent with the well-known fact that currents flowing to the radar cause a positive Doppler shift with respect to the Bragg frequency and negative otherwise. As mentioned above, this principle is used to map surface currents (see Chap. 3). The interesting property of Bragg waves in the HF frequency range (3–30 MHz) is that they always lie within the saturation range of the ocean wave spectrum (see Fig. 2.10), except for very rare extreme cases. It follows that ocean waves at the Bragg wavelength are always present in the ocean wave field.

2.7 Second-Order Doppler Spectra

In many applications of oceanography and remote sensing, the linear wave theory is sufficient to describe the desired phenomena. One can thus assume a representation (2.5.23) of the field of elevations as a random superposition of mutually independent harmonics. However, in the context of HFR, which primarily measure the Doppler spectrum, nonlinear interaction between waves must also be taken into account to

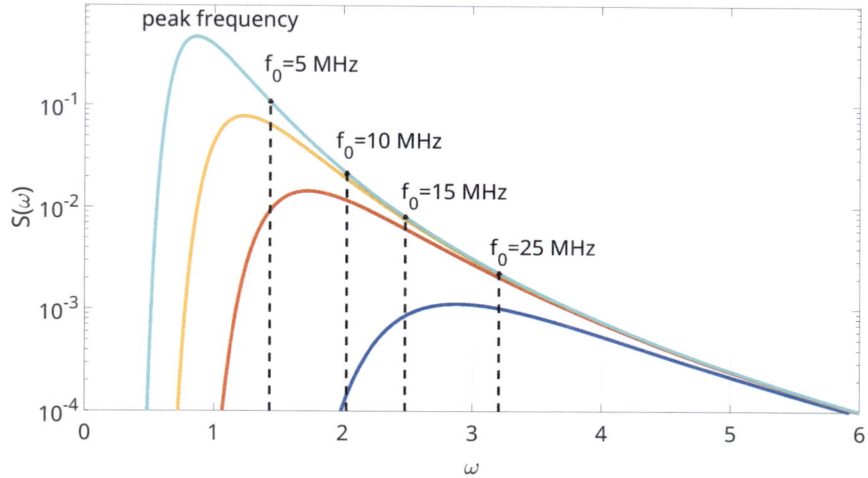

Fig. 2.10 A Pierson-Moskowitz spectrum at wind speed 3 m/s (blue line), 5 m/s (red line), 7 m/s (yellow line) and 10 m/s (cyan line). The Bragg frequency ω_B corresponding to different HFR operating frequencies is shown with black arrows. It most often falls inside the saturation range of the ocean wave spectrum

interpret the sea echo. As explained above, the Bragg lines are the prominent features of oceanic HFR Doppler spectra and are well described by the Bragg resonance mechanism and the linear gravity wave dispersion relation (2.1.2). However, in addition to the Bragg lines, experimental Doppler spectra also display a weaker but visible continuous component that depends on the sea state and which was already observed by Crombie in its pioneering experiments. Ward [281] first suggested (rather allusively, however) that the continuous component comes from nonlinear sea waves. A little later, Hasselmann[7] [121] gave a precise and broader interpretation of the continuous Doppler spectrum in terms of both hydrodynamic and electromagnetic quadratic interactions. He proposed to write the second-order Doppler spectrum in the form:

$$\sigma_2(\omega) = \sum_{n_1,n_2=\pm 1} \int \mathbf{dk_1} T\ S(n_1\mathbf{k_1})S(n_2\mathbf{k_2})\delta(\omega - n_1\sqrt{gk_1} - n_2\sqrt{gk_2}) \quad (2.7.33)$$

where the summation runs over all sign combinations $n_1 = \pm 1$ and $n_2 = \pm 1$ (4 possibilities) and T is some coupling coefficient that combines hydrodynamic and electromagnetic effects. In this integral the two wavevectors are related by the resonant Bragg condition:

$$\mathbf{k_1} + \mathbf{k_2} = \mathbf{k_B} \quad (2.7.34)$$

---

[7] Klaus Hasselmann (1931–) is a prominent German geophysicist. He has made major contributions to ocean and climate modeling, for which he received the Nobel prize in 2021. In oceanography he is well known for his pioneering mathematical description of nonlinear ocean wave interactions and for leading the JONSWAP project to measure ocean wave spectra.

Hasselmann also suggested that the second-order Doppler spectrum should reproduce the ocean frequency spectrum starting at the Bragg frequency, an idea that has since been used and refined for the sea state inversion from the Doppler spectrum [17,250]. This form of the continuous Doppler spectrum was confirmed and specified by Barrick [10], who obtained a rigorous and explicit analytical derivation of the coupling coefficient in (2.7.33) using a general analytical approach known as *perturbation theory* applied to both the wave elevations and the scattered field (Chap. 10). Perturbation expansion is a very old technique for finding analytical solutions of differential equations in the form of a power series with respect to a small parameter. Approximate solutions are found iteratively at each order of the expansion. In the context of ocean waves, this approach was used in the 1960s to describe the free surface at the air-sea boundary assuming irrotational gravity waves and implicitly taking the wave steepness as the small parameter [32,118–120,182,183,216,261,283]. The classical boundary value problem for water waves (e.g. [51,150]) is solved jointly up to second or even third order for the velocity potential and the surface elevation. Accordingly, the complex Fourier coefficients of the wave elevations are expanded as:

$$a(\mathbf{k}, \omega) = a_1(\mathbf{k}, \omega) + a_2(\mathbf{k}, \omega) + \cdots \tag{2.7.35}$$

where a_1 are the first-order "linear" coefficients, a_2 are the second-order corrections, etc. At first order, these Fourier coefficients are mutually independent and comply to the usual calculation rules (2.5.24) together with the linear gravity wave dispersion relation:

$$a_1(\mathbf{k}, \omega) = a_1(k, \sqrt{gk})\delta(\omega - \sqrt{gk}) + a_1(k, -\sqrt{gk})\delta(\omega + \sqrt{gk}). \tag{2.7.36}$$

The second-order coefficients appear as quadratic combinations of first-order coefficients that satisfy a frequency coupling condition ($\omega_1 + \omega_2 = \omega$ and $\mathbf{k_1} + \mathbf{k_2} = \mathbf{k}$):

$$\begin{aligned} a_2(\mathbf{k}, \omega) = \int \mathbf{dk_1 dk_2} d\omega_1 d\omega_2 \, \mathrm{i}\Gamma_H(\mathbf{k_1}, \mathbf{k_2}, \omega_1, \omega_2) a_1(\mathbf{k_1}, \omega_1) a_1(\mathbf{k_2}, \omega_2) \\ \times \, \delta(\omega - \omega_1 - \omega_2)\delta(\mathbf{k} - \mathbf{k_1} - \mathbf{k_2}). \end{aligned}, \tag{2.7.37}$$

The kernel Γ_H involved in this integral is called the "hydrodynamic" coupling coefficient and is given by[8]:

$$\Gamma_H(\mathbf{k_1}, \mathbf{k_2}, \omega_1, \omega_2) = -\frac{\mathrm{i}}{2}\left(k_1 + k_2 - \frac{\omega_1\omega_2}{g}(1 - \widehat{\mathbf{k_1}} \cdot \widehat{\mathbf{k_2}})\frac{g\,\|\mathbf{k_1} + \mathbf{k_2}\| + (\omega_1 + \omega_2)^2}{g\,\|\mathbf{k_1} + \mathbf{k_2}\| - (\omega_1 + \omega_2)^2}\right), \tag{2.7.38}$$

where the frequencies ω_1, ω_2 can be positive or negative. This formula first appears as such in the 1977 companion papers by Barrick and Weber [27,283] but was already derived in slightly different forms by Hasselmann [118] and Longuet-Higgins in 1963

[8] We use the notation $\widehat{\mathbf{k}} = \mathbf{k}/k$ for the unit vector along $\mathbf{k}$.

[183] (see the discussion in [206]). Note that due to the various Dirac constraints, the multiple integral (2.7.37) can be reduced to a single integral.

When forming the covariance of $a(\mathbf{k}, \omega)$ and using the calculation rules (2.5.24) and (9.3.23) we obtain the expression of the second-order ocean wave spectrum:

$$S_2(\mathbf{k}, \omega) = \frac{1}{2} \sum_{n_1, n_2 = \pm 1} \int \mathbf{dk_1} \left| \Gamma_H(\mathbf{k_1}, \mathbf{k_2}, n_1\sqrt{gk_1}, n_2\sqrt{gk_2}) \right|^2 \\ \times S(n_1\mathbf{k_1}) S(n_2\mathbf{k_2}) \delta(\omega - n_1\sqrt{gk_1} - n_2\sqrt{gk_2}), \tag{2.7.39}$$

with the conditions $\mathbf{k_1} + \mathbf{k_2} = \mathbf{k}$. This last expression was derived by Barrick and Weber in [27] but was implicitly used in earlier work from 1972 by Barrick, where the second-order Doppler spectrum was already given in its final form [10]. The expression (2.7.33) proposed by Hasselmann is recovered, up to a constant factor, by specifying the above wave spectrum to the Bragg wavenumber. In addition to the nonlinear hydrodynamic effects, there is also a contribution from the second-order scattering process, which is formally similar and can be absorbed by an additional term in the kernel (2.7.33). In the end, the second-order normalized Doppler spectrum can be written:

$$T = \mathcal{N} \left| \Gamma_H(\mathbf{k_1}, \mathbf{k_2}, n_1\sqrt{gk_1}, n_2\sqrt{gk_2}) + \Gamma_{EM}(\mathbf{k_1}, \mathbf{k_2}) \right|^2, \tag{2.7.40}$$

where

$$\Gamma_{EM}(\mathbf{k_1}, \mathbf{k_2}) = \frac{1}{2} \frac{(\mathbf{k_1} \cdot \widehat{\mathbf{k_B}})(\mathbf{k_2} \cdot \widehat{\mathbf{k_B}}) - 2\mathbf{k_1} \cdot \mathbf{k_2}}{\sqrt{\mathbf{k_1} \cdot \mathbf{k_2}} + Z_r} \tag{2.7.41}$$

is the so-called "electromagnetic" coupling coefficient [10] and $\mathcal{N}$ is the same normalization factor as that of first-order component (2.6.31). Here,[9] Z_r is a small complex number referred to as the "impedance term", which depends on the conductivity of sea water ($Z_r \simeq 0.011{-}0.012\text{i}$ is a good approximation across the HF band, see [173]). The demonstration requires the use of the Rayleigh-Rice theory for rough surface scattering and will be discussed in detail in Chap. 9. From the fonctional form (2.7.33) we can already derive some general qualitative properties of the second-order spectra:

- The wavenumber (or frequency) spectra have a dominant value around the peak wavenumber k_p (or frequency ω_p) that is generally much smaller than the Bragg wavenumber k_B (or frequency ω_B). Therefore, we can keep the values around $k_1 = k_p$ or $k_2 = k_p$ as the dominant contribution to the integral (2.7.33). With the rough approximation $\omega_1 \simeq \omega_p$ and $\omega_2 \simeq \omega_B$ (or *vice versa*) we see that $\sigma_2(\omega)$ has

[9] We corrected a sign error in the impendance term ($+Z_r$ instead of $-Z_r$), which is commented in Sect. 10.1. Many papers use the notation "Δ" for the impedance term. We prefer to use the notation "Z_r" and to reserve the Δ symbol for a small variation of a physical quantity (e.g. Δf, ΔR, etc.).

4 maxima at approximately the frequencies $\pm\omega_B \pm \omega_p$ surrounding the first-order Bragg peaks.

- Due to the low-frequency cut-off of wave spectra below the peak frequency, the small wavenumbers $k_1 << k_p$ or $k_2 << k_p$ do not contribute to the integral. Since $\omega_2 \to \omega_B$ when $\omega_1 \to 0$ (and vice versa), this implies that the second-order Doppler spectrum is quasi-zero in the vicinity of the positive and negative Bragg frequencies. Thus, the first-order Bragg peak are well separated from the second-order continuous components.
- The contour integration domain in the **k**-plane that satisfies the delta-like constraint in (2.7.33) has a degeneracy for $\omega = \pm 2^{1/2}\omega_B$ (see Chap. 10), resulting in a sharp maximum of $\sigma_2(\omega)$ at this value, originally called the "hydrodynamic singularity" [10], although it is not specifically caused by the hydrodynamic coupling coefficient. These additional peaks are not always visible as they are often close or even merged with the 2 peaks caused by the dominant frequency of the wave spectrum.
- In the case of a mixed sea with a long swell that adds up to wind waves, the wave spectrum has an additional low-frequency component $\omega_{sw} < \omega_p$ that is well separated from the wind wave spectrum. The swell spectrum is usually very narrow and can be represented by a delta-like component $\delta(\mathbf{k} - \mathbf{k}_{sw})$ [87]. With the same arguments as before, this implies 4 additional sharp peaks surrounding the first-order Bragg peaks at the approximate frequencies $\pm\omega_B \pm \omega_{sw}$ (see Sect. 12.2.3). The swell peaks can be often observed in the oceans (e.g. [278,279]) where the swell frequency may be well below the dominant wind wave frequency but are much rarer in closed seas.
- The electromagnetic coupling coefficient (2.7.41) has a sharp maximum (which would be a singular value without the impedance term) whenever the vector $\mathbf{k_1}$ and $\mathbf{k_2}$ are perpendicular to each other ($\mathbf{k_1} \cdot \mathbf{k_2} = 0$). This is interpreted as a "corner reflector" effect and causes another sharp peak in the Doppler spectrum at the frequency $2^{3/4}\omega_B$, referred to as the "electromagnetic singularity" (see Chap. 10).
- In the presence of a surface current **U**, which is uniform in space and depth, the gravity wave dispersion relation is modified according to (2.1.2) and the Dirac constraint in (2.7.33) is changed to $\delta(\omega - n_1\sqrt{gk_1} - n_2\sqrt{gk_2} - \mathbf{k_B} \cdot \mathbf{U})$. This has the effect of shifting the entire spectrum (first and second order) by the same frequency $\mathbf{k_B} \cdot \mathbf{U}$.

Figure 2.11 illustrates these features of the second-order Doppler spectrum (except for the swell) with a real example taken from the HFR station of Fort Peyras near Toulon, France, that operates at 16.15 MHz. The data shown where acquired and processed over an hour for a short range (15 km) through a day of heavy sea (March 1st, 2020). The different peaks are well visible and marked with colored lines. We remind the reader that this is not a "typical" Doppler spectrum, as the second-order peaks depend on sea state and are rarely so clearly visible. In particular, the $2^{3/4} f_B$ peaks are often blurred in the background. However, in the case of strong sea states, the second-order spectrum is sufficiently well-formed that most of the peaks can be

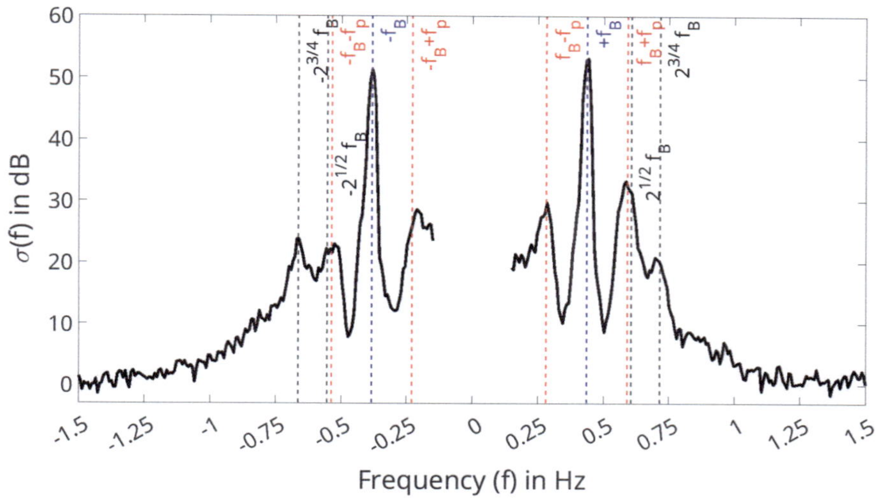

Fig. 2.11 An experimental hourly Doppler spectrum as measured at the station of Fort Peyras, Toulon. Shown with the dashed vertical color lines are the first-order Bragg peaks $\pm f_B$ (blue), the $2^{1/2} f_B$ "hydrodynamic" and $2^{3/4} f_B$ "electromagnetic" peaks (black) and the dominant wave frequency peaks at approximately $\pm f_B \pm f_p$. The central part around the zero frequency is corrupted by environmental noise and has been removed. All the peaks have been shifted by 0.026 Hz to account for the additional shift of a small radial current ($\simeq$ 24 cm/s)

identified, at least in the near range. The properties of second-order Doppler spectra are discussed in detail in Chap. 10.

The semi-analytic functional form of the second-order Doppler spectrum makes it applicable to the inversion of ocean wave spectra, as first proposed by Hasselmann [121], but the task is hampered by the somewhat complicated coupling coefficient T. Barrick provided further approximations of the second-order integral, which allows a fully analytical inversion of the main sea state parameters [17] and the non-directional ocean wave spectrum [18] from one single radar station. Directional ocean wave spectra can also be inverted numerically using parametric models [171] or sophisticated nonlinear inversion techniques (see e.g. [289] for a review). This will be further discussed in Chap. 12.

Range and Doppler Processing

3

When a radar illuminates the sea, it receives simultaneously the echoes arising from various patches at the surface. However, an appropriate processing of the received signal makes it possible to distinguish the patches at different distances, on operation which is called *range processing*. It is also possible to classify the echoes in terms of their induced Doppler shift, an operation which is called *Doppler processing*. In this chapter we describe these two fundamental operations, which are common to all types of HFR. We explain how the Doppler spectrum can be estimated from a range-resolved radar time series, using a Fourier transform of the windowed signal. We introduce the concepts of coherent summation over an integration time and incoherent summation over an observation time. We recall some classical artefacts of windowing and sampling in the calculation of Doppler spectra and discuss various factors that are responsible for the broadening of the Bragg peaks as well as the expected accuracy in measuring their position.

3.1 The Radar Signal

When the sea surface is illuminated by a radar beam, the simultaneous backscattered EM fields from different parts of the surface interfere to produce the received signal at the antenna. HFR are *coherent on receive*, i.e. they are capable of measuring both the amplitude and the phase of the backscattered field, where the phase is defined relative to the incident field. *Non-coherent radars*, on the other hand, can only measure the intensity of the signal and lose track of its phase. Such radars are generally of simpler and cheaper design. Only coherent radars can measure Doppler shifts, which are used for both range and Doppler processing.

In order to extract the geophysical parameters such as surface currents or waves, it is necessary to distinguish the contributions of the various sea patches in the

C.-A. Guérin, *High-Frequency Radars for the Oceanographer*, Geophysics and Environmental Physics, https://doi.org/10.1007/978-3-031-95642-3_3

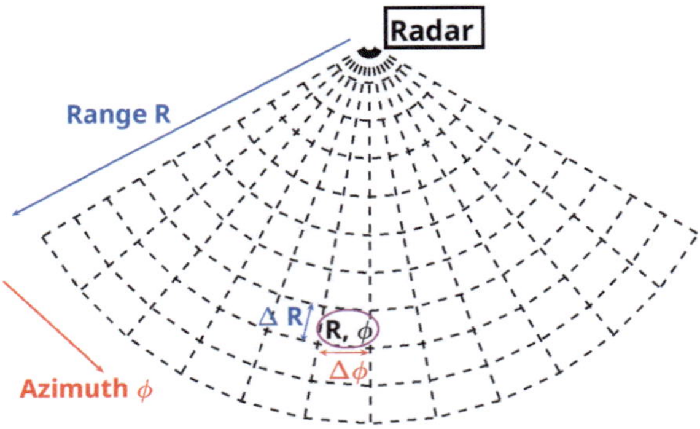

Fig. 3.1 The radar cells are defined on a radial grid where the range R is the distance to the radar and the azimuth ϕ is the angle with respect to a reference direction. A cell at range R has an area $R\Delta R\Delta\phi$

radar signal. This can be achieved by digital processing of the antenna signals (an operation that was carried out analogically before the advent of computers). The smallest surface area that can be distinguished in the radar echo is called the *radar cell*, also sometimes referred to as the *scattering patch*. The natural coordinate system used to identify the radar cells is a polar coordinate system (R, ϕ) centered at the radar position (Fig. 3.1). The distance R to the radar is called the *range* and the angle ϕ with respect to the main radar aim direction (which is the normal to the received array for phased array radars) is called the *azimuth* or the *bearing*. The radar cell is limited along one dimension by the *range resolution* ΔR and along the other dimension by the *azimuthal resolution* $\Delta\phi$. The range resolution depends only on the frequency bandwidth of the transmitted signal. For phased-array radars, the azimuthal resolution depends essentially on the array length in wavelength units. The *radar coverage* is limited in range by the decrease of the measurable signal. It is limited in azimuth by the antenna gain of the transmit and receive antennas.

The most important and universal operations in radar remote sensing are the modulation and demodulation of the transmitted and received signals by the carrier radio frequency. A so-called *waveform* with a certain prescribed functional form is produced by the generator within a low frequency *baseband* $[-B/2\ B/2]$ which is the allocated radar transmission bandwidth (typically $B = 50–100$). The waveform is multiplied by a high frequency monochromatic signal at the radar frequency f_0 (3–30 MHz), an operation known as *modulation*, before its power is amplified and radiated by the antennas. At the receiver, the backscattered echo is processed back to the baseband frequencies, an operation called *demodulation*. Demodulation allows for the measurement of the physical phase and Doppler shifts, which provide information on the distance and speed of the target; it also allows the signal to be recorded and processed at a manageable sampling rate (a few tens of kHz instead of a few tens of MHz) Demodulation is achieved by means of a device called a *mixer*, which multiplies the received signal by a copy of the transmitted signal. To get the main idea, consider the case of a continuous wave (CW) signal with a constant waveform, in which case the transmitted signal is of the form $e_1(t) = e_0 \cos(2\pi f_0 t)$ for some amplitude e_0. If we consider an echo from a single distant and moving target, the backscattered signal is attenuated, delayed and Doppler shifted in frequency, so that it can be written as $s(t) = s_0 \cos(2\pi(f_0 + f_D)t - \varphi)$ for some other amplitude s_0, frequency shift f_D and phase φ. Using the trigonometric formula $\cos a \cos b = \frac{1}{2}(\cos(a+b) + \cos(a-b))$ we have:

$$e_1(t) \times s(t) = \frac{e_0 s_0}{2} \left[\cos(2\pi f_D t - \varphi) + \cos(2\pi(2f_0 + f_D)t - \varphi)\right] \qquad (3.1.1)$$

where the term on the right hand side oscillates at about twice the carrier frequency and can therefore be eliminated with a low pass filter. The first term is a low frequency signal in the base band. The Doppler frequency shift can be identified with a simple Fast Fourier Transform (FFT). Note, however, that there is an ambiguity in the sign $\pm f_D$ when identifying the frequency from a simple sine or cosine. To specify the sign of the Doppler shift, a second mixer is required with a $\pi/2$ phase-shifted copy of the transmitted signal, $e_2(t) = e_0 \sin(2\pi f_0 t)$. Using the other trigonometric formula $\sin a \cos b = \frac{1}{2}(\sin(a+b) + \sin(a-b))$ we get this time:

$$e_2(t) \times s(t) = \frac{e_0 s_0}{2} \left[\sin(2\pi f_D t - \varphi) + \sin(2\pi(2f_0 + f_D)t - \varphi)\right] \qquad (3.1.2)$$

where the second-term on the right-hand side can again be low-pass filtered (Fig. 3.2). The complex resulting signal $e_1(t) + i\ e_2(t) \sim e^{2i\pi f_D t}$ thus allows the complete unambiguous determination of the Doppler frequency shift f_D by spectral analysis.

This method for determining a small frequency shift (compared to the carrier frequency) has been illustrated in the case of a moving target illuminated with a purely monochromatic source. It can be used in the same way for determining the small frequency shift induced by a fixed target illuminated with a frequency-modulated transmitted waveform (this will be described in Sect. 3.2.2).

The output e_1 of the first mixer in the receiver is called the *I signal* (I for "in Phase") and the output e_2 of the second mixer is called the *Q signal* (Q for "in Quadrature"). The major advantage of the mixer operation is that it can be performed in a purely analogical way so that no high frequency sampling is required. In the following, the radar signal will refer to this demodulated signal.

3.2 Range Processing

3.2.1 Pulse Radars

The universal principle for estimating the distance of a scatterer is to measure the time delay between the transmitted signal and the backscattered signal. The conceptually most simple way to achieve this is to send short pulses and record their echoes. The backscattered signal from a target at a distance R has a delay $t = 2R/c_0$ with respect

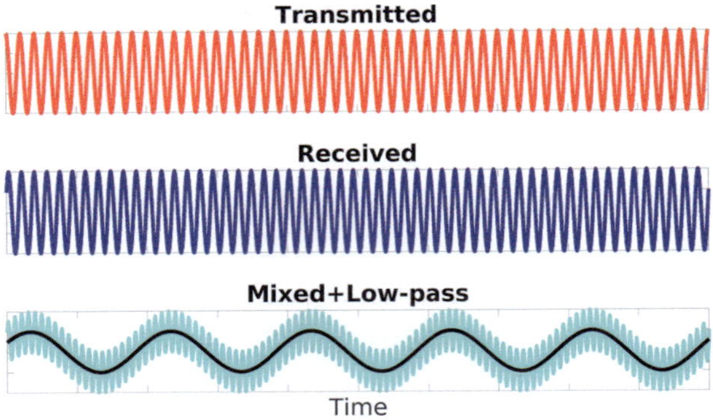

Fig. 3.2 The mixer multiplies a copy of the transmitted signal (red line) with the received signal (blue line). It results in a slow modulation at the Doppler frequency (black line) together with a fast signal (cyan line) that can be eliminated with a low-pass filter

to the transmit time, where c_0 is the speed of light in vacuum. Now, the transmitted pulse has a short, but finite duration τ, which limits the accuracy with which this time delay can be evaluated. This results in a range resolution $\Delta R = \frac{1}{2}c_0\tau$. Since the bandwidth B required to generate a pulse of length τ is $B \simeq 1/\tau$, the range resolution is inversely proportional to the available bandwidth around the operating frequency, independent of the latter:

$$\Delta R = \frac{c_0}{2B} \tag{3.2.3}$$

After the transmission of a short pulse, a longer time T is required to "listen" to the echo, and binning in time achieves range sorting, with a theoretical maximum range that is related to the listening time. The ratio τ/T is called the *duty cycle* of the radar system and is usually very small (less than 1%). This operation is repeated periodically over a large number of cycles at a rate called the *Pulse Repetition Frequency* (PRF). The level of noise in the received signal can thus be reduced by averaging a large number of pulse returns. There are therefore 2 time scales in the processing of a radar signal. There is the "fast" time, which corresponds to the sampling rate of the Analog-to-Digital-Converter (ADC) that converts the received demodulated electrical signal at the antenna output to a digital signal; it is typically of the order of a few microseconds. There is the "slow" time which corresponds to the number of cycles required to process the radar signal in range and noise. It is of the order of a few tenths of a second, a time which is still short enough to assume that the sea surface is "frozen" during the measurement. In the simplest and cheapest design, the pulse waveform is made of an unmodulated continuous waves (CW), that is a quasi-monochromatic signal with a stable frequency, that is interrupted periodically (Fig. 3.3). A very high peak power is necessary to deliver the required energy within a very short transmission time.

The choice of the various parameters of this radar system is the result of several compromises. Reducing the operating frequency bandwidth makes it possible to increase the pulse duration and therefore the transmitted power, thus increasing the

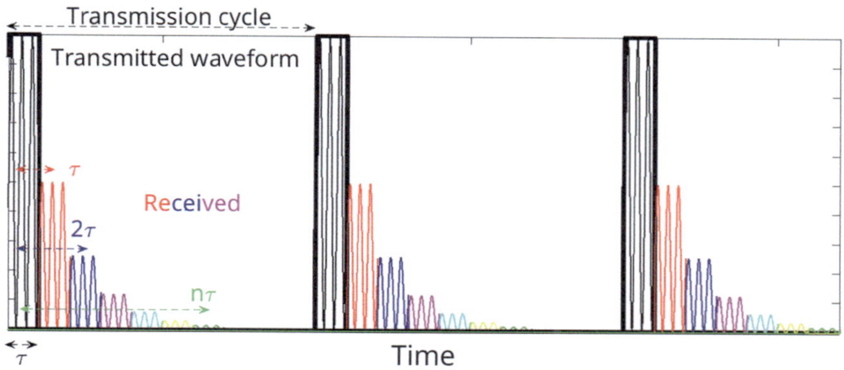

Fig. 3.3 Short and intense pulses are transmitted periodically (black lines). Echoes (color lines) are recorded during the listen time and binned in time delays (hence in range) in each transmission cycle

range and reducing the noise in the receiver (which is proportional to the bandwidth). However, this reduces the range resolution. Increasing the number of cycles that are used for the summation increases the magnitude of the signal relative to the noise level; the downside is that it increases the sampling period of signal on the slow time, which reduces the maximum available frequency in the Doppler spectrum, and hence the maximum current or wave velocity that can be measured.

3.2.2 FMCW Radars

Another solution for range processing, which is more efficient from an energetic point of view, is obtained by modulating the intra-pulse continuous waveform. Doing this, it is possible to proceed with a continuous wave emission while keeping a time marking to measure the time delay, and the transmission time can be increased to a 100% duty cycle while maintaining the range resolution. It requires a much lower peak power to maintain the same range and the same signal level. This technique, which is described in detail in many textbooks, is called *Frequency Modulated Continuous Waves* (FMCW). The main idea is to transmit a *chirp* waveform [157], where the instantaneous operating frequency sweeps a band around the central frequency at a linear rate (Fig. 3.4). The chirps are repeated periodically at the *sweep rate*, without interruption. After demodulation the backscattered signal from the sea patches at range R therefore takes the form of a delayed identical chirp (except for an amplitude factor). The key property of chirps is that a time delay t_D can be converted into a Doppler frequency shift, $f_D = \alpha t_D$, where α is the slope of the chirp. Therefore, sorting the echoes in the fast time amounts to sorting the Doppler shifts at the output of the mixers, an operation which can be done efficiently using the FFT. The range resolution is therefore related to the frequency resolution of the FFT, $\Delta R = c\Delta f/(2\alpha)$. Since $\alpha = B/\tau$, where B is the swept frequency band and τ the duration of the chirp, and $\Delta f = 1/\tau$, we recover the same formula (3.2.3). The number of available

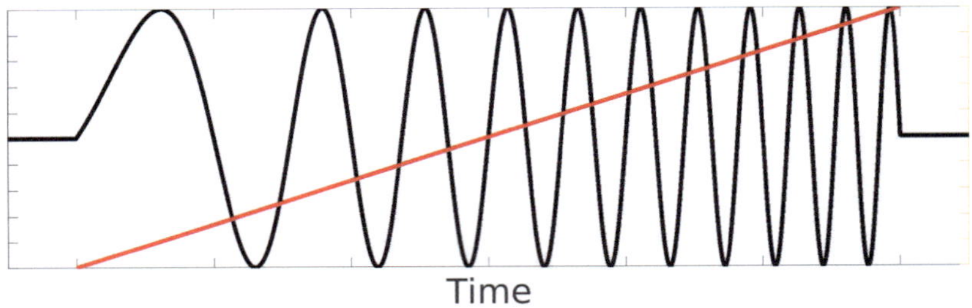

Fig. 3.4 The "chirp" waveform is frequency modulated signal (black lines) with linearly increasing instantaneous frequency over a time interval (red solid line)

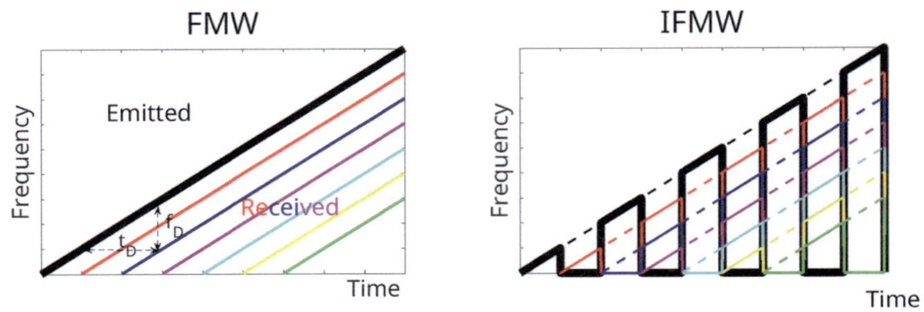

Fig. 3.5 In the FMCW technique, the transmitted waveform is a long continuous chirp and the received echoes are recorded simultaneously, which implies separate transmit and receive antennas. The instantaneous frequency of the received signal is shifted by an amount f_D proportional to the time delay, which is itself proportional to the range. In the IFMCW technique, the emitted waveform is periodically interrupted in such a way that the same antenna can be used alternatively for transmitting and receiving. The measurement of the time delay remains the same

range gates is related to the fast time sampling rate of the demodulated signal on the receiver. It is of the order of a few thousand Hz. In general the signal in the most distant range gates is well below the noise level and is not relevant for a geophysical analysis. It can therefore be discarded (and the volume of data reduced). Since the FMCW waveform is transmitted continuously, a separate receiving antenna is required to process the received signal. However, the technique can be adapted by periodic interruptions of the chirps in such a way that the same antenna can be used for both transmitting and receiving. This variant is called the *Interrupted Frequency Modulated Continuous Waves* technique (IFMCW, see Fig. 3.5).

3.3 Doppler Processing

Once the radar signal has been sorted in range using a FFT on the fast time, range-resolved time series are obtained at the slow time, which is the chirp repetition period (typically on the order of a quarter or half of a second), and a Doppler spectrum can therefore be calculated for every range bin. As seen in Chap. 2, the Doppler spectrum is defined as the Fourier transform of the correlation function of the radar signal time series, with the implicit assumption that the latter is a stationary random process. In principle, the correlation function is based on an ensemble average, whose estimation requires disposing of a very large number of independent and statistically identical samples, a condition which is impossible to achieve in a practical experiment. It is therefore usually assumed that an ensemble average (henceforth denoted by the brakets $\langle \cdot \rangle$) can be obtained by averaging the data over time instead, a property which is known as *ergodicity* in the theory of Probability. In that case the correlation function can be estimated with ergodic time averages:

$$\gamma(t) = \langle s(t'+t)s^*(t') \rangle = \lim_{T \to \infty} \frac{1}{T} \int_{-T/2}^{T/2} s(t'+t)s^*(t')dt', \tag{3.3.4}$$

and the Doppler spectrum is given by its Fourier transform:

$$\sigma(\omega) = \int_{-\infty}^{+\infty} e^{-i\omega t} \gamma(t) \tag{3.3.5}$$

Evaluating such a quantity implies a *coherent summation* (i.e. with amplitude and phase) of the complex radar signal over an infinite duration. In practice, this can only be done over a finite time T, called the *integration time*. When doing this, it is assumed that the random process describing the signal is stationary during the time interval under consideration, which is typically on the order of 20 min to 1 h for oceanographic applications. The power spectrum can be estimated with (3.3.4) but it is simpler to evaluate it by using a windowed Fourier transform of the signal:

$$\widehat{s_T}(\omega) = \int_{-\infty}^{+\infty} e^{-i\omega t} W_T(t)s(t)dt, \tag{3.3.6}$$

where W_T is some positive function that limits the integration domain to the interval $[-T/2\ T/2]$. The resulting function is often referred to as the *complex Doppler spectrum* or the *amplitude spectrum*. Its squared modulus:

$$\sigma_T(\omega) = |\widehat{s_T}(\omega)|^2 \tag{3.3.7}$$

is referred to as the *instantaneous Doppler spectrum* to distinguish it from the power spectrum $\sigma(\omega)$. Let $\widehat{W_T}$ be the Fourier transform of the window and let:

$$g_T(\omega) = \frac{1}{2\pi} \left|\widehat{W_T}(\omega)\right|^2 \tag{3.3.8}$$

The ensemble average of the instantaneous Doppler spectrum can be calculated, leading to:

$$\langle \sigma_T(\omega) \rangle = (\sigma * g_T)(\omega) \tag{3.3.9}$$

where $*$ is the convolution operation.[1] With a normalization of the window such that:

$$g_T(\omega) \to \delta(\omega),\; T \to \infty \tag{3.3.10}$$

this show that the instantaneous Doppler spectrum converges to the power spectrum in the mean sense:

$$\langle \sigma_T(\omega) \rangle \to \sigma(\omega),\; T \to \infty, \tag{3.3.11}$$

In the idealized case where the Doppler spectrum is a delta-like function ($\sigma(\omega) = A\delta(\omega - \omega_B)$, as it follows from the diffraction model presented in Chap. 2), the mean instantaneous Doppler spectrum is a finite peak at the spectral resolution of the window:

$$\sigma_T(\omega) = |A|^2\, g_T(\omega - \omega_B) \tag{3.3.12}$$

The window[2] is usually taken as a dilated version of a reference shape $W(t)$:

$$W_T(t) = T^{-1/2} W(T^{-1}t), \tag{3.3.14}$$

where the scaling factor $T^{1/2}$ and the normalization $\int |W(t)|^2\, dt = 1$ are required to ensure the limit (3.3.10) When W is a positive function, its Fourier Transform reaches its maximum at $\omega = 0$. It follows that the mean instantaneous Doppler spectrum has a main peak about the Bragg frequency ω_B which becomes sharper and sharper as the integration time is increased. Precisely, the amplitude of the peak amplitude increases like T while its width decreases like $1/T$ (Fig. 3.6). In signal processing, the width of a peak is often defined by its value at the half maximum; it is called the *Full Width at Half Maximum* (FWHM).

The radar signal $s(t)$ is most often modeled as the independent superposition of a physical signal $s_0(t)$ representing the sea echo and a noise component $\zeta(t)$:

$$s(t) = s_0(t) + \zeta(t), \tag{3.3.15}$$

where $\zeta(t)$ is a stationary white noise, i.e. a purely decorrelated random process ($\langle \zeta(t'+t)\zeta^*(t') \rangle = \zeta_0^2 \delta(t)$). Since the signal and noise components are assumed to be uncorrelated, their respective Doppler spectra add up, so that the "physical" Doppler spectrum is increased by a small positive quantity:

$$\sigma(\omega) \to \sigma(\omega) + \zeta_0^2 \tag{3.3.16}$$

[1] We recall that $(f * h)(x) := \int f(x-y)h(y)dy$.

[2] For the particular case of the rectangular window ($W(t) = 1$ if $|t| < \frac{1}{2}$, $W(t) = 0$ otherwise), we recover the well-known result expressing the power spectrum as an ergodic time average (e.g. [214] p. 343), which is known as Wiener-Khinchine Theorem:

$$\sigma(\omega) = \lim_{T\to\infty} \frac{1}{T} \left\langle \left| \int_{-T/2}^{T/2} \mathrm{e}^{-i\omega t} s(t)dt \right|^2 \right\rangle \tag{3.3.13}$$

.

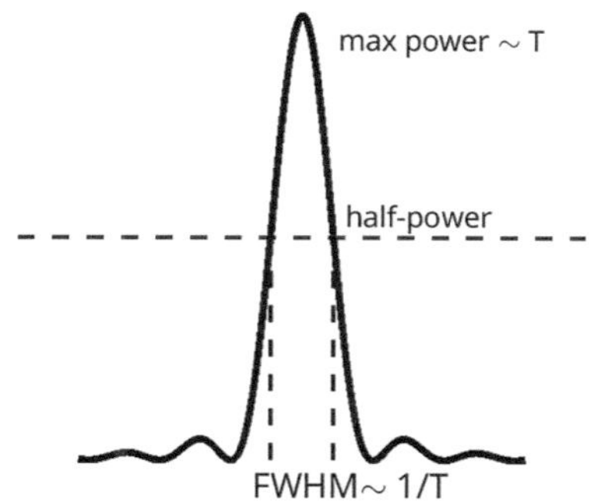

Fig. 3.6 The maximum height of the Doppler spectrum is proportional to the observation time (T) while its FWHM is inversely proportional

In the case of a delta-like Doppler spectrum $A\delta(\omega - \omega_B)$, this leads to:

$$\langle \sigma_T(\omega) \rangle = |A|^2 \, g_T(\omega - \omega_B) + \zeta_0^2 \tag{3.3.17}$$

Note that the noise power density ζ_0^2 does not depend on the integration time, whereas the signal power increases like T. Hence the *Signal to Noise Ratio* (SNR), which is defined as the ratio of the maximum signal power to the mean noise power, also increases in the same way. Therefore, a long integration time improves both the Bragg peak resolution ($\sim 1/T$) and the SNR ($\sim T$). It is thus tempting to increase the integration time as much as possible. However, there are some limitations that prevent making it arbitrarily large. The first is the need to consider a stationary sea state, i.e. to assume that the wave spectrum and the currents do not vary significantly over a radar observation cycle. The second reason is more profound: the instantaneous Doppler spectrum converges in mean to the Doppler spectrum as the integration time increases but with a possibly large variance, i.e. with large fluctuations from one sample to the next. This is problematic for the continuous components of the Doppler spectrum whose second-order peaks (which are finite amplitude peaks rather than delta-like peaks such as the Bragg peaks) are more difficult to distinguish from the random oscillations. To reduce the variance of the estimator, it is necessary to perform an additional averaging over several independent and identical sample spectra. This operation can only be achieved by dividing the available time window into a sequence of sub-intervals, at the cost of a reduced integration time (see Fig. 3.7). A trade-off must therefore be found between the frequency resolution $\sim 1/T$ and the variance reduction.

While the summation that is used to evaluate the Fourier transform in (3.3.6) is coherent because it uses the complex amplitude of the signal, the additional summation that is needed to average the sample spectra is said to be an *incoherent summation* because it only deals with their intensity. The total time required to process a Doppler spectrum, including coherent and incoherent summation, is called the *observation time*. It is, by definition, larger than the integration time unless there is only one sample spectrum. The Doppler spectrum is therefore estimated using an empirical mean:

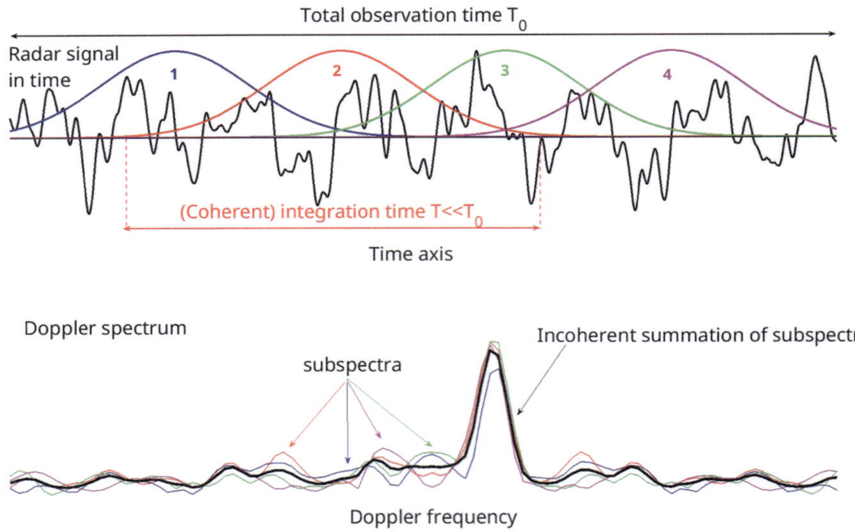

Fig. 3.7 The signal to analyze is split in shorter sub-samples using overlapping windows (top panel) and sub-spectra are obtained by coherent summation inside each window. An incoherent summation is performed over the sub-spectra to obtain an estimate of the Doppler spectrum during the total observation time (bottom panel)

$$\overline{\sigma}(\omega) = \frac{1}{N}\sum_{n=1}^{N}\sigma_T^{(n)}(\omega) \tag{3.3.18}$$

where the $\sigma_T^{(n)}$ are a series of N instantaneous "short" spectra calculated over sliding intervals of width T. Each interval is identified by an integration window and the windows overlap at the edges in order to maximize the useful signal (Fig. 3.7).

If one assumes the sample spectra to be independent (which is not exactly true if an overlap is allowed), the variance is reduced by a factor N as it well known in Probability theory (e.g. [214]):

$$\text{Var}[\overline{\sigma}(\omega)] = \frac{1}{N}\text{Var}[\sigma_T(\omega)], \tag{3.3.19}$$

where the right-hand side is the variance which is common to all sub-spectra. Since the complex Doppler spectrum, which is obtained by integrating a random process, can be regarded as a complex normal variable, its square modulus $\sigma_T(\omega)$ is an exponential random variable (see e.g. [214]). It follows that the variance of the instantaneous Doppler spectra is equal to the square of their mean and thus:

$$\text{Var}[\overline{\sigma}(\omega)] = \frac{1}{N}\left[(\sigma * g_T)(\omega)\right]^2 \simeq \frac{1}{N}\sigma^2(\omega) \tag{3.3.20}$$

The number of samples N and the coherent integration time T vary in the inverse ratio because the observation time is fixed. Therefore, the choice of the integration time for the estimation of Doppler spectra is a trade-off between the required frequency resolution and SNR and the variance reduction. A natural question arises when considering this dilemma. Why not simply estimate the instantaneous power spectrum with the maximal integration time in order to maintain a high SNR and then kill the oscillations by smoothing the resulting "long" spectrum over frequencies? It turns out that this is not as good in terms of variance reduction, for the same frequency resolution. Calling T_0 the observation time and T a much shorter integration time, the smoothed instantaneous power spectrum at the frequency resolution $1/T >> 1/T_0$ is obtained by convolution of the instantaneous power spectrum with a window of the same type g_T:

$$(\sigma_{T_0} * g_T)(\omega) = (|\widehat{s}|^2 * g_{T_0} * g_T)(\omega) \tag{3.3.21}$$

Since the first convolution window is much sharper than the second (for $T_0 >> T$), it acts as a delta function and therefore the smoothed version of the high-resolution instantaneous power spectrum is close to the instantaneous power spectrum at a coarser resolution:

$$(\sigma_{T_0} * g_T)(\omega) \simeq (|\widehat{s}|^2 * g_T)(\omega) = \sigma_T(\omega) \tag{3.3.22}$$

It follows that the $1/N$ reduction factor of the variance is lost when smoothing the high-resolution power spectrum, showing that it is preferable to split the observation time in a series of smaller integration times. The number N of samples spectra cannot be very large due to the simultaneous constraint on the integration time and the observation time. It is typically of the order $10 - 30$.

3.4 Windowing and Sampling

As explained above, Doppler processing of the radar signal requires the application of a tapering window $W(t)$ to limit the time interval. From (3.3.12), we see that this operation transforms an idealized delta-like Doppler spectrum $\delta(\omega - \omega_B)$ into a finite peak that has the shape of $\left|\widehat{W}_T(\omega - \omega_B)\right|^2$. The most elementary window is the rectangular one, which we define on a unit interval:

$$W(t) = \begin{cases} 1, & |t| < 1/2 \\ 0, & |t| > 1/2 \end{cases} \tag{3.4.23}$$

Its Fourier transform is given by:

$$\widehat{W}(\omega) = \text{sinc}\left(\frac{\omega}{2}\right) = \text{sinc}(\pi f) \tag{3.4.24}$$

where sinc is the *cardinal sine function*:

$$\text{sinc}(\omega) = \frac{\sin(\omega)}{\omega} \tag{3.4.25}$$

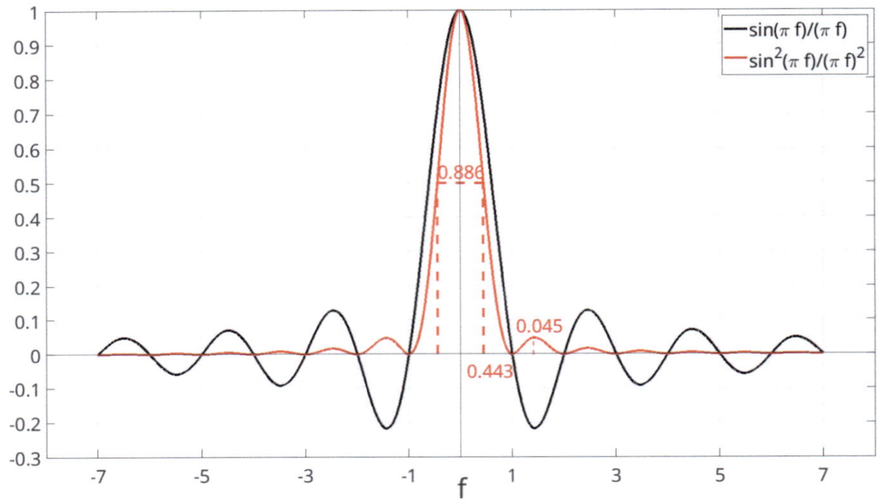

Fig. 3.8 The sinc function sinc(πf) (black line) is the Fourier transform of the unit rectangular window (3.4.23). It is represented as function of frequency, together with its square modulus (red line). The FWHM of the latter is ~ 0.89 Hz

This function, which is represented in Fig. 3.8, has strong oscillations and a slow decrease ($\sim 1/\omega$) which can introduce artificial high-frequency components in the signal. These oscillations can be removed or attenuated by flattening the discontinuity at the edges through a smooth tapering window. The counterpart is a slight broadening of the main peak, hence a small loss in frequency resolution. Popular tapering windows $W(t)$ are the triangular (Bartlett [29]), cosinus squared (Hann) and weighted cosinus (Hamming and Blackman) windows, whose definitions and properties of the different windows are recalled in Table 3.1. They are plotted together with the squared modulus of their Fourier Transform in Fig. 3.9. The spectral bandwidth of each window can be defined with a usual 3 dB criterion, which is the FWHM. The level of oscillations is characterized by the level of the first side-lobe as compared to the main peak. The rectangular window has the best spectral resolution but high side-lobes.

The Blackman window has the strongest side-lobe rejection (-58 dB) at the cost of a spectral resolution deteriorated by a factor about 2 with respect to the rectangular window. The Hann and Hamming windows are good compromises to maintain a high spectral resolution with an acceptable level of side-lobe rejection.

Side-lobes are in general undesirable when analyzing the Doppler spectrum for characterizing the sea surface. They are harmless as far as the surface current extraction is concerned because they do not alter the detection of the main peak but they "spoil" the components of the Doppler spectrum outside the Bragg frequency which carry information on the wave spectrum (Fig. 3.10).

After range processing, the radar signal on the slow time is obtained at a time sampling rate Δt (equivalently, at the *sampling frequency* $1/\Delta t$, which is equal to the sweep rate) corresponding to the duration of each transmitted pulse, typically a few tens of seconds for HFR. Hence the Doppler spectrum must be estimated from

Table 3.1 Classical tapering windows $W(t)$ one the interval $|t| < \frac{1}{2}$ corresponding to a unit integration time $T = 1$ and their Fourier Transforms $\widehat{W}(\omega) = \widehat{W}(2\pi f)$. Also given are the FWHM of $\left|\widehat{W}(2\pi f)\right|^2$ (expressed in Hz) and the side-lobe level (SL, expressed in dB) relative to the main peak. We use the notation $S_a = \text{sinc}(\omega/2) = \text{sinc}(\pi f)$

	Rect.	Bartlett	Hann	Hamming	Blackman
$W(t)$	1	$1 - 2\|t\|$	$\cos^2(\pi t)$	$0.5435 + 0.4565\cos(2\pi t)$	$0.42 + 0.5\cos(2\pi t) + 0.08\cos(4\pi t)$
$\widehat{W}(\omega)$	S_a	S_a^2	$\frac{2\pi^2}{4\pi^2-\omega^2} S_a$	$\frac{2.16\pi^2-0.08\omega^2}{4\pi^2-\omega^2} S_a$	$\left(\frac{1.68\pi^2+0.08\omega^2}{4\pi^2-\omega^2} - \frac{0.08\omega}{16\pi^2-\omega^2}\right) S_a$
FWHM	0.89	1.27	1.44	1.30	1.64
SL	–13.5	–26.5	–31.5	–43	–58

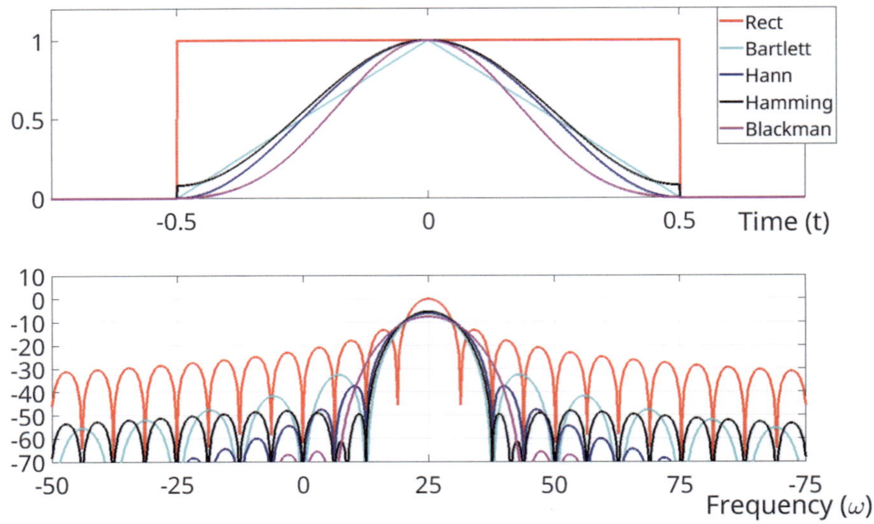

Fig. 3.9 Comparison of various windows on a unit time interval ($W(t)$, upper panel) together with the square modulus of their Fourier Transforms ($20\log_{10}(|\widehat{W}(\omega)|)$, lower panel). The sinc function associated to the rectangular window has a narrower peak but suffers from strong oscillations and a slow decay

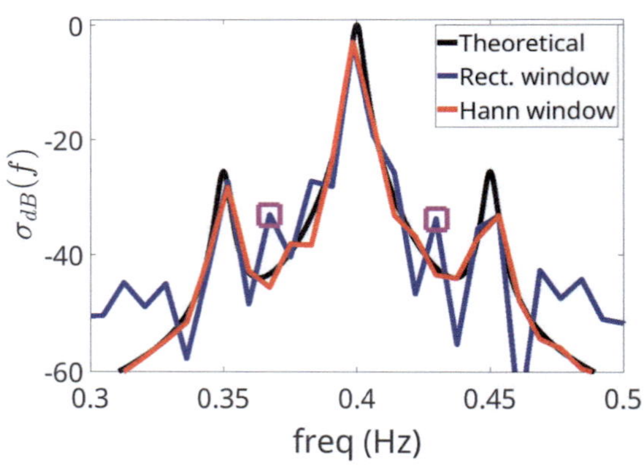

Fig. 3.10 Effect of the time window in evaluating a theoretical Doppler spectrum in presence of secondary peaks. The side-lobes (marked with magenta squares) associated to the rectangular window can be confused with the satellite peaks around the Bragg frequency. This ambiguity is removed by using a smooth window such as the Hann window

a time series $s_n = s(n\Delta t)$ with $N = T/\Delta t$ points over the integration time.

$$\sigma_T(\omega) \simeq (\Delta t)^2 \left| \sum_{m=0}^{N-1} e^{-\mathrm{i}\omega m \Delta t} s_m \right|^2 \tag{3.4.26}$$

This operation is done with the help of the *Discrete Fourier Transform*

$$\widehat{s}[n] = \sum_{m=0}^{N-1} e^{-\frac{2i\pi mn}{N}} s_m \tag{3.4.27}$$

Fig. 3.11 The peak frequency of the "true" Doppler spectrum (black solid lines) can only be determined from the discrete time series within the accuracy given by the frequency resolution $\Delta\omega$ of the FFT (in red solid lines)

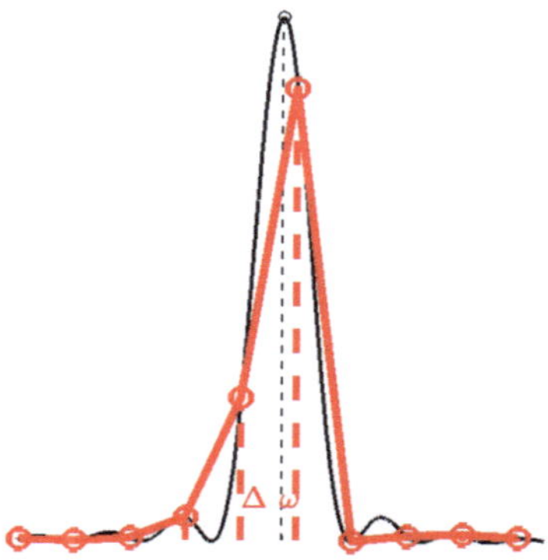

whose efficient calculation is obtained by FFT. The Doppler spectrum is therefore evaluated on a set of discrete frequencies $\omega_n = 2\pi n/T$, which are multiple integers of the frequency resolution $\Delta\omega = 2\pi/T$. This last quantity, which we have already encountered as the typical width of a spectral peak for a continuous signal of duration T, is for a discrete time series the smallest distance between two discrete frequencies or, to put it differently, the maximal error with which a given frequency can be approached (Fig. 3.11). In the context of surface current extraction which is based on measuring the shift of the Bragg frequency, the frequency resolution has an important impact on the accuracy of the estimation (this will be discussed in more details in Sect. 3.5). Note that the frequency resolution does not depend on the sampling rate Δt but only of the integration time.

The main effect of time sampling is that it limits the available Doppler bandwidth. As it is well known in signal theory, the N-point Discrete Fourier Transform of a sampled signal at the time step Δt is the periodized and discretized version of its continuous Fourier Transform at the period $1/\Delta t$ and the frequency step $\Delta f = 1/(N\Delta t)$ (Fig. 3.12). It thus coincides with the sampled Fourier Transform of the continuous signal in the first period if the signal is band-limited to half the sampling frequency, $|f| < 1/(2\Delta t)$. This is the well-known sampling theorem and the maximal available (regular or circular) frequency f_{max} or ω_{max} is the so-called *Nyquist frequency*:

$$f_{max} = \frac{1}{2\Delta t}, \quad \omega_{max} = \frac{\pi}{\Delta t} \tag{3.4.28}$$

If the Doppler spectrum contains frequencies higher than the Nyquist frequency, the tails of the translated versions will add up and make the value of the estimated Doppler spectrum quite different from what it should be, a phenomenon which is known as *aliasing*; in particular some "ghost peaks" can appear at low frequencies due to the folding of some high-frequency peak beyond the Nyquist frequency (Fig. 3.13). This means that a minimal sampling frequency of $2f_B$ is required for the correct estimation of the Doppler spectrum around the Bragg frequency. As we will see in Chap. 10, the physical signal contained in sea Doppler spectra extends well beyond the Bragg frequency, up to, say, $2f_B$. Hence, a minimal sampling frequency of about

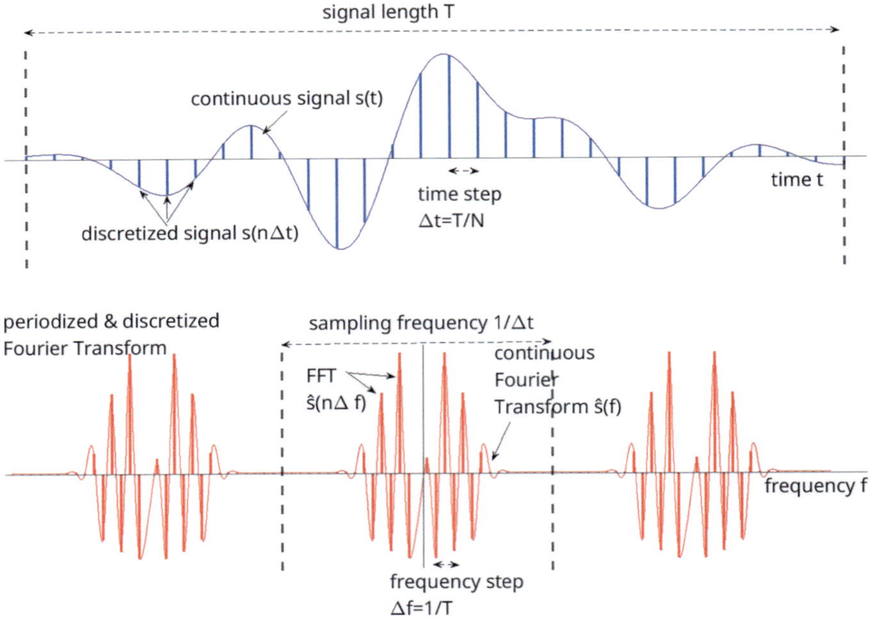

Fig. 3.12 When a continuous signal $s(t)$ is sampled at a step Δt over an interval of length T, its FFT is the periodized and discretized version of the continuous Fourier transform $\widehat{s}(f)$ with a period $1/\Delta t$ and a frequency step $\Delta = 1/T$

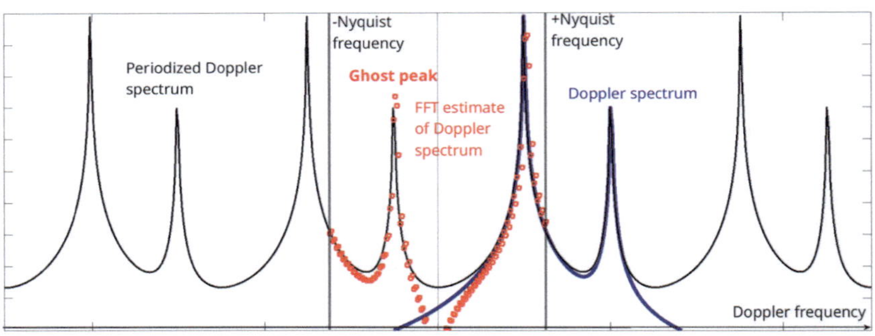

Fig. 3.13 The Doppler spectrum of the sampled signal (black lines) is made of periodic replica of the actual spectrum (blue lines) with a period equal to the sampling frequency. The Doppler spectrum calculated from a FFT of the sampled signal (red squares) is relevant only in a bandwidth limited by the Nyquist frequency (black vertical lines). Any frequency content of the Doppler spectrum that is beyond value is folded onto the Nyquist interval and can create "ghost" peaks

$4f_B$ is required. The sweep rates for oceanographic HFR systems are typically 2 or 4 Hz. Since the Bragg frequency is on the order $f_B \sim 0.4 - 0.5$ Hz, the sampling criterion is met in both cases.

3.5 Resolving the Bragg Peaks

The experimental Bragg peaks can be very sharp but they have always a finite bandwidth that depends on several physical and instrumental factors. Broadening of the Bragg peaks can be due to:

- Insufficient integration time.
- Orbital velocity of the Bragg wave induced by the long waves.
- Spatial variations of the surface current inside the radar cell.
- Fast temporal variations of the surface current during the observation time.

Except at very low HF frequency and for very calm seas, there are always longer waves than the Bragg waves present in the ocean wave field. Assuming the classical picture of a linear superposition of free gravity waves with different wavelengths, Bragg waves are thus riding on the facets of long waves; their instantaneous celerity is therefore modified by the orbital motion of the latter, that acts like a local cyclic surface current, say $U_0 \cos(\omega_p t)$, where ω_p is the period of the long wave (Fig. 3.14). It follows that the local Bragg frequency undergoes cyclic fluctuations $2U_0 \cos(\omega_p t)/\lambda_0$ around the theoretical Bragg frequency f_B, which has the effect of broadening the main Bragg peak to a FWHM of about $2U_0/\lambda_0$. The order of magnitude of the long wave orbital speed U_0 can be estimated easily by considering the extreme scenarios of a very calm and a strong sea state. A particle circling at the dominant wave period (4–10 s) with a crest-to-trough diameter of about 0.25–2 m has a mean tangential speed of about 0.2–0.6 m/s. The FWHM of the Bragg peak will therefore range from $0.4/\lambda_0$ to $1.2/\lambda_0$. At the typical HFR frequency $f_0 = 16$ MHz ($\lambda_0 = 18.75$ m), this is between 0.02 and 0.06 Hz, which would be obtained with a non-modulated Bragg wave and a short integration time between 17 and 50 s. It follows that even in the most favorable case of a calm sea state, the limiting factor for the sharpness of the Bragg peak is not the integration time but the long wave modulation. This is another reason why there is no point increasing the integration time beyond a few minutes.

In principle, the accuracy with which one can estimate the current is bound to the accuracy with which one can measure the position of the shifted Bragg peaks. This depends on:

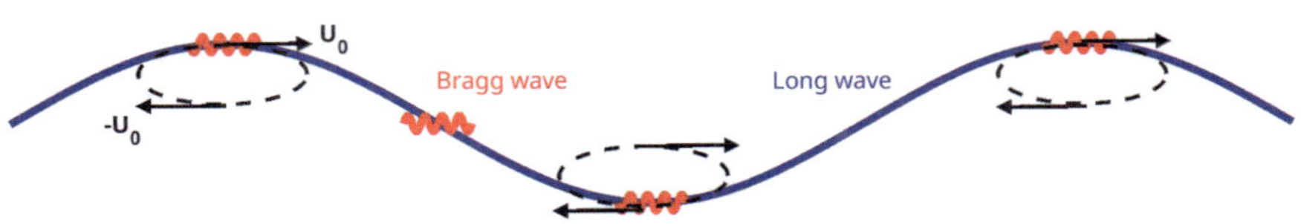

Fig. 3.14 The amplitude and speed of the Bragg wave (red ripples) is modulated by the orbital motion of longer waves (black dashed lines). This causes a broadening of the main Bragg peak and a variation of its amplitude

- the integration time T, which controls the frequency discretization step $\Delta f = 1/T$.
- the number N of sub-spectra that are incoherently summed to form a Doppler spectrum. Since the variance of this sum decreases like $1/N$, this parameter controls the convergence of the estimation.
- the FWHM of the first-order Doppler spectrum. Depending on whether it is larger or smaller than the frequency step Δf, the Bragg peaks will be resolved or not.

The most simple way to locate the Bragg peak is to search for the ray of maximal amplitude after FFT processing of the range- and azimuthally-resolved complex voltage time series. However, this can be quite inaccurate if the frequency resolution is too coarse, because the available discrete frequency $f_n = n\Delta f$ are integer multiples of the frequency resolution. Indeed, in determining the shifted Bragg frequency f_B from the maximum of the Doppler spectrum on the positive or negative frequency side, the exact peak location will be missed by an amount of the order $\delta f_B \sim \Delta f/2$. The integration time can be increased to reduce the sampling error. However, for a fixed observation time, this implies reducing the number of independent sample spectra, which increases the variance of the estimated Doppler spectrum. If the latter is insufficiently converged, peaks can occur at the wrong frequency bins, an effect that is more pronounced if the Doppler spectrum is broad (FWHM> Δf). Hence, there is no obvious choice of the parameters T and N when trying to optimize the estimation of the peak location.

The estimation can be greatly improved by using the "barycenter method" (Fig. 3.15). Instead of taking the maximum of the Doppler spectrum on the available frequency grid, one can better evaluate the actual peak by forming a barycenter mean over a few points around the maximum (say, p values on each side of the maximum obtained at $n = n_0$):

$$\overline{f_B} = \frac{\sum_{n=n_0-p}^{n_0+p} f_n S(f_n)}{\sum_{n=n_0-p}^{n_0+p} S(f_n)} \tag{3.5.29}$$

This has the double advantage of allowing off-grid values to estimate the Bragg frequency, which is useful if the frequency resolution is too coarse, and reducing the variance of the estimate, which is useful if the number of sample sub-spectra is too low. In practice, a few units of p are enough to improve significantly the accuracy of the estimation. One should be careful, however, not to exceed the actual bandwidth of the Bragg peak otherwise one could "catch" part of the Doppler spectrum that is not related to the main Bragg peak (typically, second-order spectrum). Precise statistical performances of the barycenter method have been quantified by Barrick [21] who found the following root-mean-square error δf_B for the estimate $\overline{f_B}$ of the Bragg frequency, assuming a Gaussian shape for the Bragg peak:

$$\delta f_B = \langle (\overline{f_B} - f_B)^2 \rangle^{1/2} = \frac{1}{2} \Delta f \sqrt{N_0/N_s}, \tag{3.5.30}$$

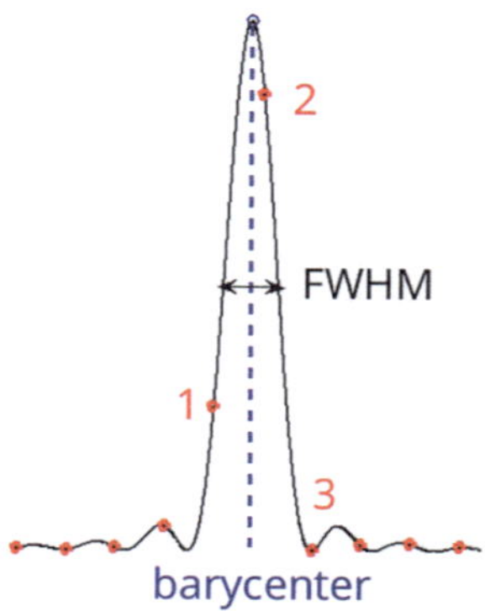

Fig. 3.15 The location of the peak frequency of the "true" Doppler spectrum (black solid lines) can be accurately determined from the discrete Doppler spectrum by taking a 3 point barycenter around the maximum. In this example, there is only one sampling point ($N_0 = 1$) to describe the peak

where N_0 is the number of sampling points inside the FWHM of the first-order Doppler spectrum and N_s is the number of independent sample sub-spectra used to estimate the latter. Because these quantities evolve in inverse ratio when the number of sample spectra is increased, the result depends only on the total observation time and not in the way it is split in shorter intervals with coherent integration. If the first-order peak is well resolved (Δf <<FWHM) one has $N_0 \simeq \text{FWHM}/\Delta f$ and the root-mean-square error δf_B on the peak location is thus:

$$\delta f_B = \frac{1}{2}\sqrt{\frac{\text{FWHM}}{T}} \tag{3.5.31}$$

where N_s has been taken to 1 and T is therefore the total observation time. This shows that the actual error in the estimation of the Bragg peak is not simply proportional to the frequency resolution Δf (this would only be true for a perfectly Dirac-shaped Doppler spectrum) but, due to the finite width of the Doppler spectrum, decreases only like its square root. In terms of radial velocity this yields the following root-mean-square error:

$$\delta U_r = \lambda_B \delta f_B = \frac{1}{2}\lambda_B\sqrt{\frac{\text{FWHM}}{T}} \tag{3.5.32}$$

To give an order of magnitude of this quantity, consider a HFR operating at $f_0 = 16$ MHz. At this radar frequency, the Doppler peak broadening due to the wave orbital motion was estimated to FWHM = 0.02–0.06 Hz depending on sea state. With 20 min observation time, which is a classical duration to assume stationary sea state in oceanography, this yields $\delta U_r \sim 2$–3 cm/s, which is quite accurate (see Fig. 3.16). However, this result has been obtained under idealized assumptions discarding the effect of noise, the sub-cell variations of the surface current and the non-stationarity of the latter, so that it is rather a limiting bound to what can be obtained. It must be kept in mind that being able to determine the peak position accurately does not mean that one is able to reach the same accuracy for the surface current, because it is only a mean quantity that is measured. As we have seen, the instantaneous current undergoes variations in time, which are mainly due to the orbital motion of long waves, as well as spatial variations inside the radar cell, not to mention the variations with depth

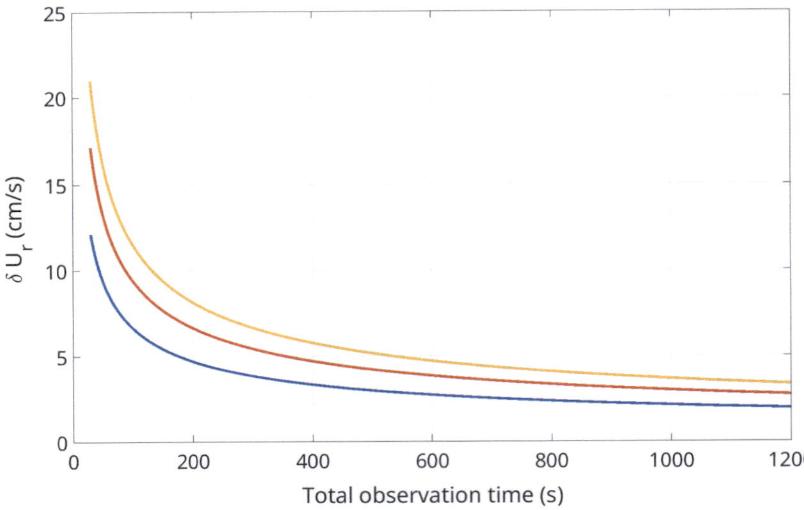

Fig. 3.16 Root-mean-square error of the radial surface current estimated from the barycenter of the first-order Bragg peak as a function of the total observation time for a radar frequency of 16 MHz. The FWHM of the Doppler spectrum due to the wave orbital motion is taken to be 0.02 Hz (blue line), 0.04 Hz (red line) or 0.06 (yellow line) corresponding to increasing sea state

inside the effective top layer. These spatio-temporal fluctuations are responsible for the broadening of the Bragg peaks in a given radar cell. Hence, a radar measurement is only able to provide a mean value of the current, which is identified with the maximum or the barycenter of the Bragg peak. It follows that the actual surface current that would be obtained with a "pointwise"[3] measurement in space and time (for example using a drifter) always differ from the HFR measurement by an amount which is related to the bandwidth of the actual Doppler spectrum. This will be further discussed in Chap. 12.

[3] There is always a finite scale and a finite duration of measurement attached to any instrument. However, with a drifter these are several orders of magnitude below the HFR scales of measurement, so that they can be considered point-like in space and time.

Azimuthal Processing with Arrays

4

While the range and Doppler processing steps can be performed individually for each antenna signal, azimuthal processing (i.e., the operation of sorting the radar signal in various directions of arrival) requires a collective use of various antenna signals. In this chapter, we describe the classical technique employed with phased-array radars, for which the receive system consists most often of regular linear antenna arrays.

4.1 The Array Factor

We consider a receive array made of N identical elements with periodic spacing d and we suppose that it is illuminated by a source at "infinity", that is at a very large distance from the receiver (this will be made precise in Sect. 4.3). The geometry of the problem is bidimensional in the horizontal plane. The EM field reaching the receive antennas can therefore be assimilated to a plane wave with an horizontal wave vector $\mathbf{K_0}$ incoming from the direction ϕ_s of the source, measured from the normal to the array (Fig. 4.2). It differs by a constant geometrical phase shift $K_0 d \sin\phi_s$ from one antenna to another, where $K_0 = 2\pi/\lambda_0$ is the EM wavenumber in vacuum (Fig. 4.1). Assuming that all antennas have identical gains, the complex signal X_n on the nth antenna therefore depends only on its position in the array and the direction of the sources:

$$X_n = e^{\mathrm{i}(n-1)K_0 d \sin\phi_s} X_1, \tag{4.1.1}$$

where the signal X_1 from the first antenna is taken as reference. It follows that all the antenna signals remain identical if $\phi_s = 0$ so that the coherent summation:

$$Y = X_1 + X_2 + \cdots X_N \tag{4.1.2}$$

C.-A. Guérin, *High-Frequency Radars for the Oceanographer*, Geophysics and Environmental Physics, https://doi.org/10.1007/978-3-031-95642-3_4

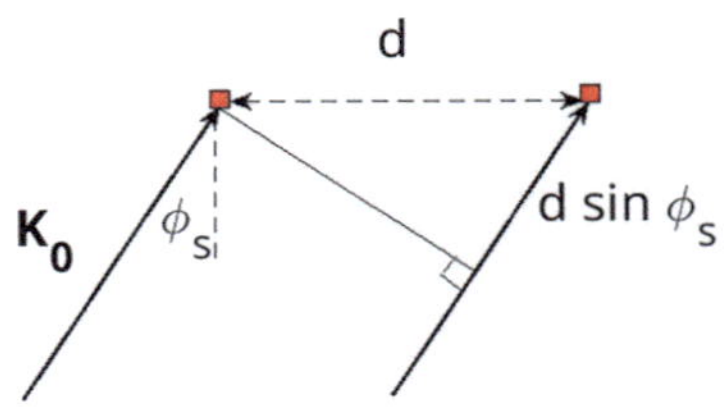

Fig. 4.1 The phase shift of the incident field on 2 antennas at a distance d is $K_0 d \sin\phi_s$, where ϕ_s is the incident direction of the plane wave measured from the normal to the array

is maximum in amplitude if the direction of the source is normal to the array axis. The related function:

$$\mathrm{AF}(\phi_s) = 1 + e^{\mathrm{i}K_0 d \sin\phi_s} + \cdots + e^{\mathrm{i}(N-1)K_0 d \sin\phi_s} \tag{4.1.3}$$

is called the *array factor* [8]; it depends only on the geometry of the array and not on the type of the individual antennas. A simple calculation using geometrical series shows that:

$$|\mathrm{AF}(\phi_s)|^2 = \frac{\sin^2(N\Psi/2)}{\sin^2(\Psi/2)} \tag{4.1.4}$$

where $\Psi = K_0 d \sin\phi_s$. At small incidence angle ϕ_s, it can be well approximated by a squared cardinal sine function, $|\mathrm{AF}(\phi_s)|^2 \simeq N^2 \mathrm{sinc}^2(N\Psi/2)$. Such a power distribution reaches its maximum value N^2 at $\phi_s = 0$ and displays some decreasing oscillations with secondary lobes besides the main peak, as depicted schematically in Fig. 4.2. The array factor for a classical value $K_0 d = \pi$ and various length N of the array is shown in Fig. 4.3.

As seen in Sect. 3.4, the FWHM of $\mathrm{sinc}^2(N\Psi/2)$ is obtained for $N\Psi/2 = 0.443\pi$ corresponding to an angular width:

$$\Delta\phi \simeq 0.89 \frac{\lambda_0}{Nd} \tag{4.1.5}$$

Hence, averaging the signal on the array elements (4.1.2) enhances the contribution of any source in the far field that is found in the main lobe $|\phi_s| < \Delta\phi$ and eliminates

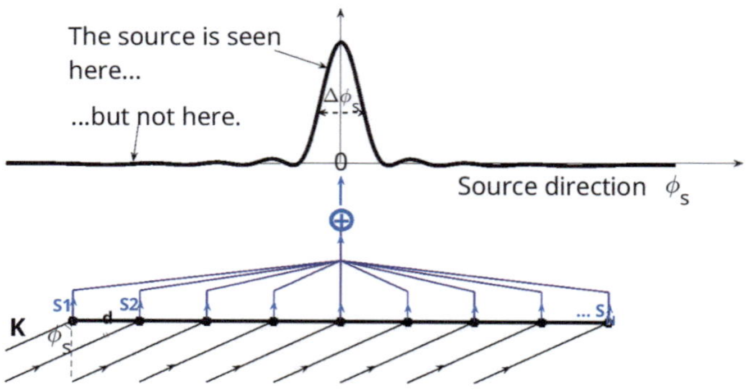

Fig. 4.2 The array factor is formed by summing the complex signals received on the various antenna channels. These add constructively when the source direction is normal to the array ($\phi_s = 0$)

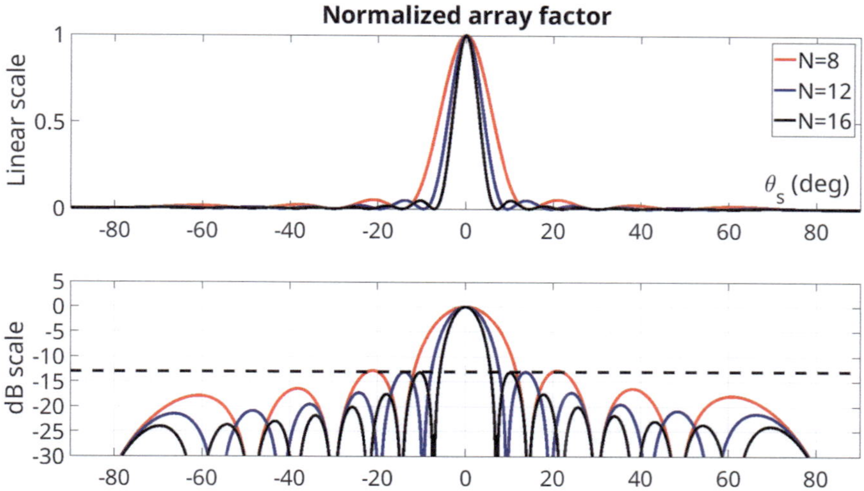

Fig. 4.3 Normalized array factor (4.1.3) shown in linear (top panel) and dB scale (bottom panel) as a function of the source direction ϕ_s for $K_0 d = \pi$ and for various lengths N of the antenna array. The dashed black line shows the peak level of the secondary lobes at -13 dB

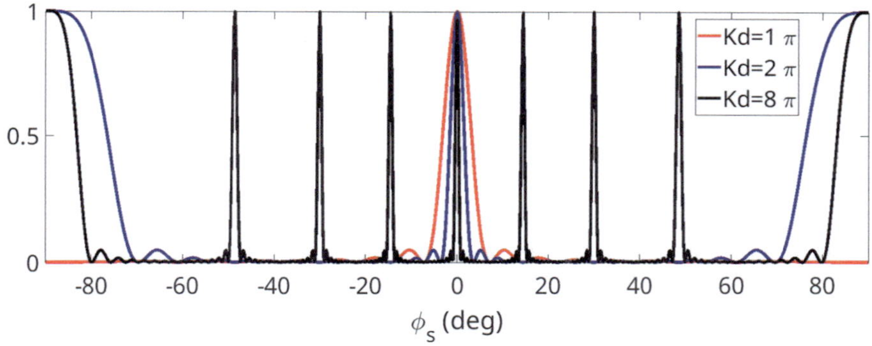

Fig. 4.4 Normalized array factor (4.1.3) for $N = 16$ and various values of the array spacing ($K_0 d = \pi$, $K_0 d = 2\pi$ or $K_0 d = 8\pi$). Increasing $K_0 d$ decreases the width of the central lobe but creates grating lobes

the other directions. Note that the sharpness of this main peak increases with both the array length N and the array period d. However, because the array factor is a 2π-periodic function of Ψ, the array factor can reach its maximum value at non-zero angles as soon as $K_0 d \geq 2\pi$ (or $d \geq \lambda_0$). This introduces additional peaks, referred to as the *grating lobes*, in addition to the central lobe at $\phi_s = 0$, resulting in an ambiguity of the angular position of the source. For this reason the array spacing cannot be increased beyond $d = \lambda_0$ (in fact, $d = \lambda_0/2$, see further). Figure 4.4 shows the effect of increasing $K_0 d$. It diminishes the width of the central peak but at the same time introduces grating lobes for $K_0 d \geq \pi$. The array factor characterizes the antenna array in transmission as well as in receiving mode. This is because the expression 4.1.3 is also proportional to the far-field that would be produced in direction ϕ_s by

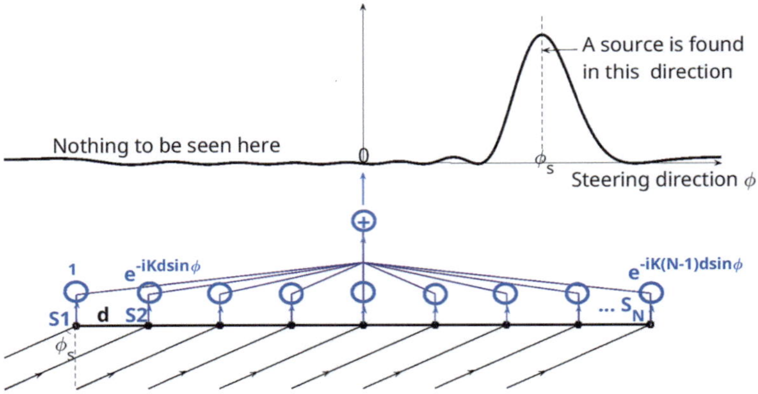

Fig. 4.5 After appropriate phase shifting and summing of the antenna signals, the array factor (4.2.7) expressed as a function of the steering angle ϕ is maximal in the direction ϕ_s of the source

N identical sources at the antenna locations.[1] The array factor is maximal when the source direction is normal to the array ($\phi_s = 0$) and, with the additional condition $K_0 d < 2\pi$, there is no other direction where this can happen so that it allows to filter out this single direction.

4.2 Beam Forming at Infinity

Now what about an oblique direction? It suffices to compensate the phase shifts of each antenna in such a way to maximize the array factor at a given direction $\phi_s \neq 0$. This can be done by multiplying the antenna signal X_n by an extra factor $e^{-i(n-1)K_0 d \sin\phi}$ before summation:

$$Y(\phi) = X_1 + X_2 e^{-iK_0 d \sin\phi} + \cdots X_N e^{-i(N-1)K_0 d \sin\phi} \tag{4.2.6}$$

in which case the array factor is changed to:

$$\mathrm{AF}(\phi) = \sum_{n=1}^{N} e^{i(n-1)K_0 d(\sin\phi_s - \sin\phi)}, \tag{4.2.7}$$

With the same calculation as previously we have (4.1.4) with $\Psi = K_0 d(\sin\phi - \sin\phi_s)$. The array factor is now maximal in the oblique direction $\phi = \phi_s$ (Fig. 4.5), which corresponds to the actual source direction. However, since $\Delta\Psi = K_0 d \cos\phi_s \Delta\phi$, the width of the main lobe increases with the side angle:

$$\Delta\phi = \frac{\lambda_0}{Nd\cos\phi} \tag{4.2.8}$$

[1] This is a general consequence of a principle known as the reciprocity of light, which tells that the received signal is unchanged after swapping the location of the source and receiver.

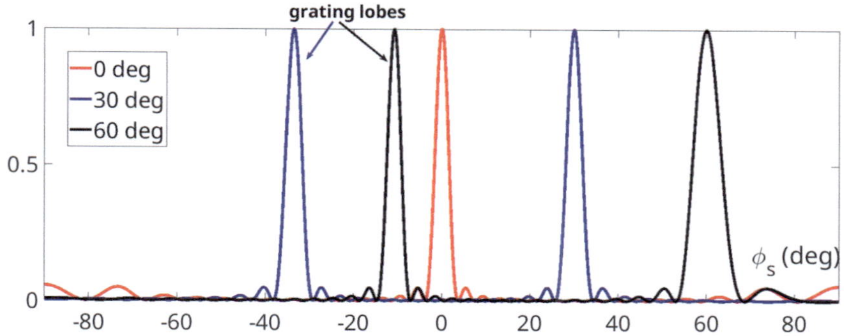

Fig. 4.6 Normalized array factor after BF for $N = 16$ and $K_0d = 1.9\pi$ and various incidence angles of the source ($\phi_i = 0$, 30 and 60°). For sources at an oblique angle, the width of the central lobe is increased. For $\pi < K_0d < 2\pi$, grating lobes can appear on the opposite side of the source direction

so that the angular resolution deteriorates by a factor of 2 at 60° and a factor of 6 beyond 80°. Figure 4.6 shows the effect of the source angle on the array factor. Also, it shows that another grating lobes can appear on the opposite side of the source whenever $\pi < K_0d < 2\pi$. For this reason, the limitation on the array spacing is more stringent than previously found for one source in normal direction ($\phi_s = 0$) and one usually takes:

$$d = \lambda_0/2 \tag{4.2.9}$$

for the optimal array period. The summation that is performed in (4.2.6) is referred to as *Beam Forming*, henceforth abbreviated to BF. This is actually the simplest version of a family of processing techniques referred to with the same terminology (see e.g. [198]). Here we will only consider what is known as *Conventional Beam Forming* or *Delay-and-Sum Beam Forming* (which amounts to *Phase-Shift-and-Sum* for a monochromatic field), where the complex weights in (4.2.6) are purely geometrical and do not depend on the dataset unlike more advanced methods such as adaptive beam forming.

This sum-and-delay operation (4.2.6) amounts to virtually re-radiate the received complex signals $(X_1, \ldots, X_N)$ in specific azimuthal directions at infinity while continuously steering the aim angle ϕ. In the case of a perfect linear array with identical antenna in the absence of noise, the focus spot is given by the array factor (4.2.7).

Like Doppler processing, BF proceeds through a coherent summation of the antenna signals and therefore increases the SNR. One generally assumes that the actual antenna signal X_n on each array channel is perturbed by an additional complex noise b_n:

$$X_n \rightarrow X_n + b_n, \tag{4.2.10}$$

where X_n is the signal produced by a remote source (and eventually multiplied by the antenna gain), hence of the form $X_n = \mathrm{e}^{\mathrm{i}(n-1)K_0d\sin\phi_s} X_1$. It follows that the delay-and-sum beam-formed signal can be written as:

$$Y(\phi) = X_1 \sum_{n=1}^{N} \mathrm{e}^{\mathrm{i}(n-1)K_0 d(\sin\phi_s - \sin\phi)} + \sum_{n=1}^{N} b_n, \tag{4.2.11}$$

If we assume that the complex noises from the various antenna channels have zero mean and are mutually independent processes with identical variance σ_b^2:

$$\langle b_m b_n^* \rangle = \begin{cases} \sigma_b^2 \text{ if } n = m, \\ 0 \text{ else} \end{cases} \tag{4.2.12}$$

then we have:

$$\mathrm{Var}[Y(\phi)] = |X_1|^2 \, |\mathrm{AF}(\phi)|^2 + N\sigma_b^2 \tag{4.2.13}$$

Since $|\mathrm{AF}(\phi)|^2 \sim N^2$ it follows that the SNR$\sim N \, |X_1|^2 \, / \sigma_b^2$ is increased by a factor of N with respect to the SNR measured on a single antenna. Hence, performing a BF processing with a large number of antenna increases the SNR together with the azimuthal resolution.

Note that the array factor can also be defined for irregular geometries. This is important as the site topography often prevents from installing the complete array along a straight line. Writting $\mathbf{d_n}$ the vector position of the nth antenna relative to the first one, the delay-and-sum beam-formed signal for an incoming source in direction ϕ_s can be written as:

$$Y(\phi) = X_1 \sum_{n=1}^{N} e^{+i K_0 (\mathbf{u}(\phi) - \mathbf{u}(\phi_s)) \cdot \mathbf{d_n}} \tag{4.2.14}$$

where $\mathbf{u}(\phi)$ is the outgoing unit vector in direction ϕ. At this point it is convenient to introduce the so-called *steering vector* $\mathbf{a}(\phi)$:

$$\mathbf{a}(\phi) = \left(1, e^{\mathrm{i}K_0 \mathbf{u}(\phi) \cdot \mathbf{d_1}}, \ldots, e^{\mathrm{i}K_0 \mathbf{u}(\phi) \cdot \mathbf{d_N}}\right) \tag{4.2.15}$$

If we introduce the vector $\mathbf{X} = (X_1, \ldots, X_N)$ to describe the complex antenna signals, we can simply express the sum- and delay operation (4.2.6) as a scalar product with the steering vector:

$$Y(\phi) = \mathbf{X} \cdot \mathbf{a}^*(\phi) \tag{4.2.16}$$

The same holds for the array factor, which can be rewritten as:

$$\mathrm{AF}(\phi) = \mathbf{a}(\phi) \cdot \mathbf{a}(\phi_s)^* \tag{4.2.17}$$

Irregular arrays perform generally less well than regular arrays with the same total extension because the corresponding array factors have stronger side lobes. Figure 4.7 shows the example of an irregular array obtained by a small random displacement of the antenna spacing along the array axis. As seen, the width of the main lobe is unchanged but side lobes with less than 10 dB attenuation are visible even at large angles.

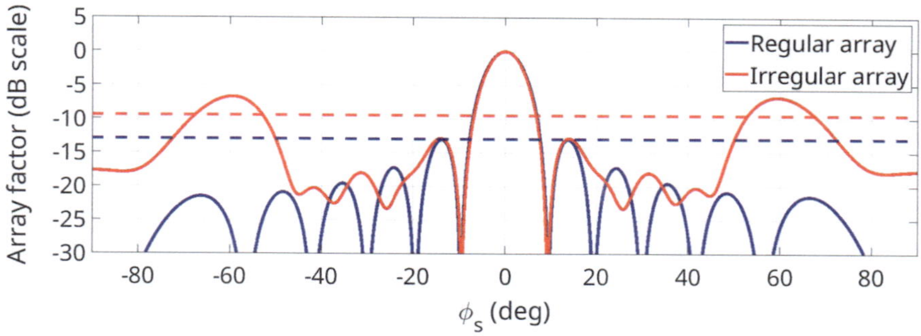

Fig. 4.7 Normalized irregular array factor obtained by perturbation of a regular array (here, $K_0 d = \pi$ and $N = 16$), assuming a $\pm 10\%$ random variation of the spacing between consecutive antennas. While the central lobe is unchanged, strong side-lobes appear. The horizontal dashed lines show the peak level of the secondary lobes

4.3 Beam Forming at Finite Distance

At this point we have used the simplifying assumption that the sources of illumination, which in fact are the various scattering patches on the sea surface, are sufficiently far so that the plane wave approximation can be used in the delay-and-sum operation. We will now evaluate the actual distance beyond which one can consider this true. For this, consider a source M at a distance R from the center O of the array, seen with an angle ϕ_s from the array and radiating a spherical wave in the far field (Fig. 4.8). Let A be one edge of the array, at a distance $L/2$ from O. The phase difference Ψ_{A0} between the center and the edge of the array is given by:

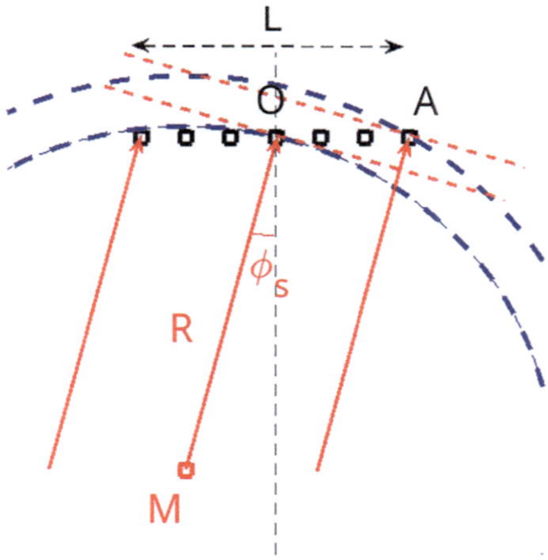

Fig. 4.8 At finite distance from the receiver, the spherical shape of the wave front must be taken into account and there is a quadratic correction $\sim L^2/R^2$ to the antenna phase shift predicted with the plane wave approximation

$$\Psi_{AO} = K_0(AM - OM) = K_0\left[(R^2 + LR\sin\phi_s + L^2/4)^{1/2} - R\right] \quad (4.3.18)$$

Since the length of the array is small compared to the observation distance ($L/R <<$ 1), one can perform a Taylor expansion of the square root, leading to:

$$\Psi_{AO} = \frac{1}{2}K_0 L\sin\phi_s + \frac{K_0 L^2}{8R}\cos^2\phi_s + O(L^3/R^3) \quad (4.3.19)$$

The first term on the right-hand side is the phase shift that would be obtained with a plane wave in direction ϕ_s. The second term is the first finite distance correction. Following the standard criterion, we will neglect its contribution to the phase shift as soon as it is smaller than $\pi/8$. This imposes the condition:

$$R \geq \frac{2L^2}{\lambda_0} \quad (4.3.20)$$

This lower bound is called the *Fraunhofer distance*. It defines the *Fraunhofer zone* in which one can neglect the quadratic terms in the phase shift and assimilate the spherical wave front to a plane wave front, as opposed to the *Fresnel zone* in which the latter must be taken into account. Now considering a typical phase-array with 16 antennas at $\lambda_0/2$ spacing, we have $L \simeq 8\lambda_0$ and thus the Fraunhofer distance is obtained at $R = 128\lambda_0$. For a 15 Mhz HFR ($\lambda_0 = 20$ m), this is about 2.5 km, which is of the order the first range cell which is in general discarded. It follows that BF "at infinity" is in general sufficient for HFR applications. Calculating the phase shift in the Fresnel zone might only be necessary in some extreme case with very large array at very low radar frequency (e.g. [106]).

4.4 Side-Lobes Reduction

In order to separate properly the signals arising from the various radar cells, it is important to reduce the side-lobes of the array factor. The array factor (4.1.3) can be thought as a discrete Fourier transform of the antenna signal with respect to the angular variable and displays the same sinus cardinal shape (4.1.4). In a way very similar to the Fourier Transform, these oscillations can be removed or at least attenuated if the different antenna signals are weighted according to their position on the array, with a damping on the two ends. Hence, the BF delay-and-sum operation is usually performed with a set of weights $W(n)$:

$$Y(\phi) = W(1)X_1 + W(2)X_2 e^{-\mathrm{i}K_0 d\sin\phi} + \cdots W(N)X_N e^{-\mathrm{i}(N-1)K_0 d\sin\phi}, \quad (4.4.21)$$

which modifies the array factor (for a given source in direction ϕ_s):

$$\mathrm{AF}(\phi) = \sum_{n=1}^{N} W(n)e^{\mathrm{i}(n-1)K_0 d(\sin\phi_s - \sin\phi)}, \quad (4.4.22)$$

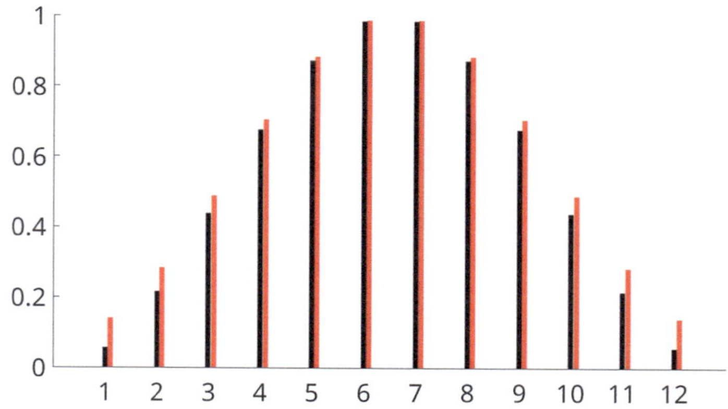

Fig. 4.9 Weights of a 12-antenna array according to the Hann (black lines) and Hamming (red lines) window

Customary antenna weights are given by the discrete version of the Hann window:

$$W(n) = \sin^2\left(\frac{\pi n}{N+1}\right) \tag{4.4.23}$$

or the Hamming window:

$$W(n) = 0.5435 - 0.4565\cos\left(\frac{2\pi n}{N+1}\right) \tag{4.4.24}$$

These weights are shown in Fig. 4.9 for a 12-antenna array. The case of unweighted arrays is recovered with the uniform (or rectangular) window ($W(n) = 1$).

With application of these weights, the relative magnitude of the main side-lobes can be decreased from –13 dB (uniform window) to –31.5 dB (Hann window) or –35.4 dB (Hamming window), at the expense of an increase of the width of the central peak (Fig. 4.10). The respective values of the side-lobe level and central width are summarized in Table 4.1 for various array sizes and window types.

4.5 Azimuthally Resolved Doppler Spectra

Performing the BF operation (4.2.16) for each range cell and each slow time step, we end up with a complex time series $Y(t, R, \phi)$ for every pair of range (R) and bearing (ϕ). The complex Doppler spectrum associated to a given radar cell can therefore be obtained by a Fourier Transform along the time dimension, leading to the so-called *Directional Doppler spectrum* or *azimuthally resolved Doppler spectrum*, as opposed to the *omnidirectional Doppler spectrum* which is obtained from the range-resolved radar time series of each receive antenna without azimuthal processing and which therefore integrates the backscattered echos from all directions. Note that Doppler and azimuthal processing are commutative operations, that is they can be

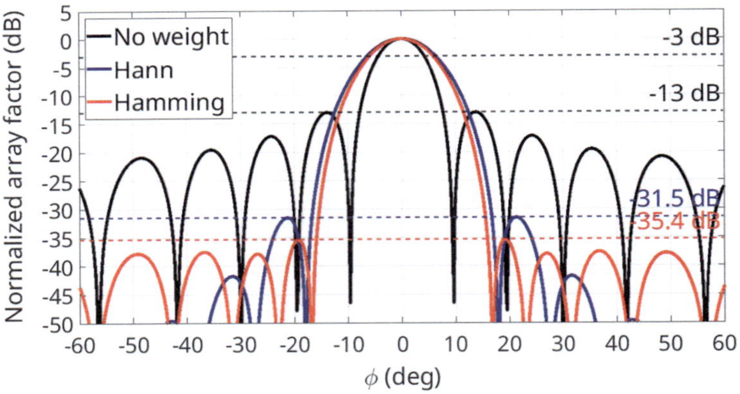

Fig. 4.10 Normalized array factor (4.1.3) for a 12-antenna array with various weighting windows. The peak levels of the secondary side-lobes are marked with horizontal dashed lines

Table 4.1 Characteristics of the weighted array factor for various number N of antennas and various windows: FWHM $\Delta\phi$ of the main lobe around the central direction and maximum level SL of the secondary lobe, in dB

N	6	8	12	16
Uniform window				
$\Delta\phi$ (deg)	17.1	12.8	8.5	6.3
SL (dB)	−12.4	−12.8	−13	−13
Hann window				
$\Delta\phi$ (deg)	23.6	18.6	12.8	9.5
SL (dB)	−32.4	−31.4	−31.5	−31.5
Hamming window				
$\Delta\phi$ (deg)	22.5	17.1	11.7	8.8
SL (dB)	−30.9	−32.8	−35.4	−37.3

applied in either order to the radar time series. Since the omnidirectional first-order Doppler spectrum cumulates Bragg peaks with various frequency shifts due to the variation of the radial current inside the radar coverage, it is much broader than the directional first-order Doppler spectrum at one given bearing, which is associated to a single shifted Bragg peaks. The frequency region where the omnidirectional first-order Doppler spectrum extends is often referred to as the *First-order Bragg Region* (FOBR) (Fig. 4.11). As we will see in the next chapter, it plays an important role in high-resolution direction finding techniques.

BF is the most straightforward technique to resolve the radar signal in azimuth but has a major practical limitation since it requires sufficient physical space to install a linear antenna array. As seen from (4.2.8), the theoretical angular resolution $\Delta\phi$ of the radar signal after BF is given by the width of the main lobe of the array factor, which is of the order of λ_0/L, where L the array size. It is optimal in the direction normal to the array axis and deteriorates by a factor $1/\cos\phi$ as the steering angle

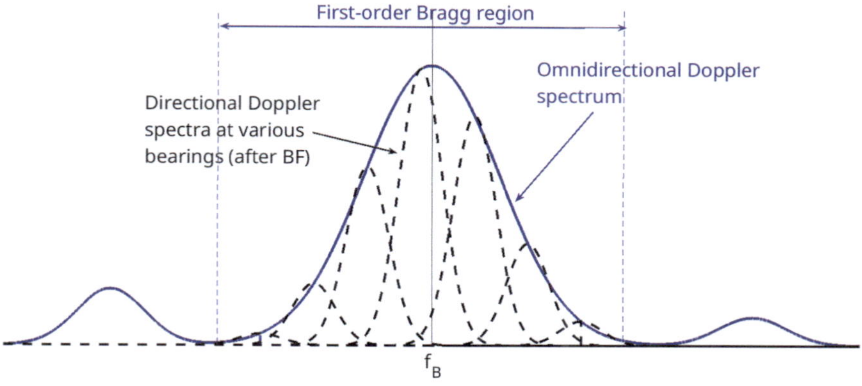

Fig. 4.11 The omnidirectional Doppler spectrum is unresolved in azimuth and cumulates all possible shifted Bragg peaks. It is thus broader than the directional Doppler spectra that correspond to single shifted Bragg peaks at various bearings

ϕ departs from the normal direction. For this reason, the angular sector swept by BF is usually limited to $\pm 60°$ from the normal, so that the azimuthal resolution at the edge of the sector is only twice coarser than in the central direction. With an optimal typical spacing $\sim \lambda_0/2$ to prevent from the grating lobes, the width of the main angular lobe is at best $\Delta\phi = 1.76/N$. This allows an angular resolution of the order of 9° (in fact, rather 12° if a tapering window is applied) in the central direction for 12 antennas. Note that the step size at which the focusing direction is swept should not be confused with the actual angular resolution. This angular step, which is used to define the radial grid of surface current and can be made arbitrarily small, is usually taken of the order of 1 or 2°, which is much smaller than the actual angular resolution.

Since both the directional ocean wave spectrum and the radial surface current (which is the projection of the current along the direction of observation) vary with the azimuth, the directional Doppler spectrum is quite sensitive to the bearing ϕ. In particular, the Doppler shift and the amplitude of the main Bragg lines vary with the azimuth. The angular width of the array factor determines the accuracy and the quality of the azimuthal processing and impacts the width of the Bragg lines in the directional Doppler spectra. Secondary lobes that go along with BF can be an important source of error if they are insufficiently rejected since they may "capture" strong sources (through the Bragg lines) away from the focusing direction, with the consequence that the signals arising from different sea patches can mix up. As seen, side-lobes can be efficiently rejected by using tapering windows in the array but the azimuthal resolution remains bound to the array extension.

5 Direction Finding

As opposed to phased-array radars, compact HFR systems do not have receive antenna arrays and therefore cannot make use of standard BF techniques to process the radar signal in azimuth. Instead, they use a method that belongs to the family of *Direction Finding* techniques. This is a statistical approach, based on the covariance properties of the set of antenna signals, to identify the different bearings associated with the Doppler rays in the First-Order Bragg Region, which in the end allows the surface currents to be resolved in azimuth. It can be applied to both compact and phased-array radar systems and can achieve high azimuthal resolution in the optimal cases. We describe the method in a general algebraic framework that also includes BF techniques and discuss its performance and limitations in the context of HFR surface current mapping.

5.1 High Resolution Techniques

As we have seen in the previous chapter, azimuthal processing of the radar signal can be performed using the beamforming technique. However, this involves the deployment of an extended linear receive arrays, a practical requirement that it is often difficult to achieve on the shore side. In the HF regime, the radar wavelength is typically on the order of 10–20 m while a sufficient azimuthal resolution for surface current mapping requires at least 12 antennas spaced at half-wave length. As a result, the length of the array is in typically greater than 100 m, making it challenging to install in densely urbanized coastal areas. Nevertheless, azimuthal processing can also be achieved with compact antenna systems by resorting to alternative techniques of azimuthal processing, namely a family of methods usually referred to as *high-resolution techniques* or Direction Finding (DF) methods. The main idea is that the determination of the azimuth for the measured surface radial current can be

C.-A. Guérin, *High-Frequency Radars for the Oceanographer*, Geophysics and Environmental Physics, https://doi.org/10.1007/978-3-031-95642-3_5

reformulated as a problem of *goniometry*, that is finding the *direction of arrival* (DoA) of different *sources*. Goniometry, which etymologically means "measure of angles", is a very old science for measuring the direction of all types of incoming waves (optical, acoustic, electromagnetic, ...). The first instruments date back to at least the 18th century. In the context of surface current extraction, the goniometric problem can be formulated as follows. Due to the variation of the surface current over the radar coverage and due to the different angles of view, which change its radial component, the first-order Bragg rays undergo many different frequency shifts depending on the sea patch under consideration. This results in a broad omnidirectional Doppler spectrum. Now, for a given range, what are the azimuths of the sea patches which are responsible for a given shifted Bragg ray observed in the omnidirectional complex Doppler spectrum (see Fig. 5.1)? In this context, the source is the associated radar cell, the DoA is its azimuth and the measured signal is the complex value of the selected Doppler ray for each antenna.

How does DF work? It is simply based on the fact that the different receiving antennas "see" the same incoming signal from each direction ϕ with their own directive gains, a set of complex values that form the steering vector:

$$\mathbf{a}(\phi) = (g_1(\phi), \ldots, g_N(\phi)) \tag{5.1.1}$$

In the case of a linear array with N identical antennas, we have seen in the previous chapter that the complex directive gains $g_1, \ldots, g_N$, differ only by a phase shift ($g_n = e^{iK_0 d(n-1)\sin\phi}$). For a compact antenna system the receive antennas also differ by their amplitude. The application of DF processing techniques is a key element of the success of the SeaSonde CODAR system for surface current mapping [26]. This

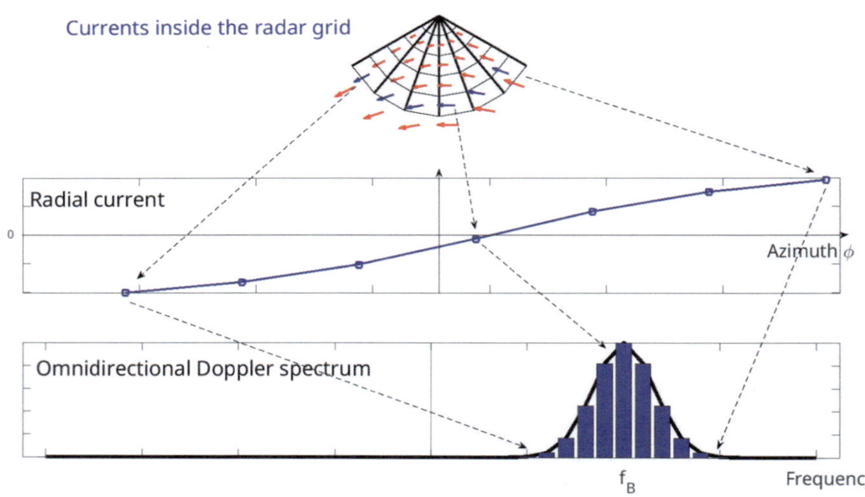

Fig. 5.1 The surface current vector inside the radar coverage (top panel) has a varying radial projection with the azimuth (middle panel). For every given range (blue arrows) this leads to a broad omnidirectional first-order Doppler spectrum (bottom panel) whose different rays mirror the various values of the radial current

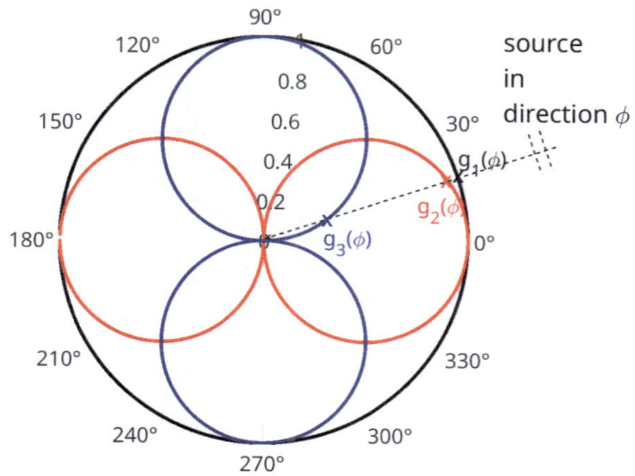

Fig. 5.2 The CODAR antenna system is made of one monopole antenna with uniform ideal gain (black circle) and 2 crossed-loop antennas with sine and cosine ideal gains (read and blue circles). The received field on the 3 antennas, emanating from a source in direction ϕ, is proportional to the steering vector $(g_1(\phi), g_2(\phi), g_3(\phi))$

compact HFR system, which is one of the most widely used instruments, has only 3 antennas with the following theoretical gains:

$$\begin{aligned} g_1 &= 1 \\ g_2 &= \cos(\phi - \phi_b) \\ g_3 &= \sin(\phi - \phi_b) \end{aligned} \tag{5.1.2}$$

where ϕ_b is a reference bearing. The first antenna is an isotropic monopole antenna and the other 2 are magnetic cross-loop antennas (Fig. 5.2). An angle ϕ is uniquely determined by its sine and cosine so that there is unique steering vector associated to every specific incoming direction over 360° azimuth. Note that in the case of a linear array, there is a $\phi/\pi - \phi$ ambiguity in the steering vector which is uniquely determined by the sinus of bearing. However, the HFR receiving arrays are typically deployed along a coastline that divides the bearings in one half-space of land and one half-space of sea, so in practice there is no ambiguity (except for very specific configurations involving islands or narrow peninsulas). The idea of DF is to invert the DoA from the measurement of the steering vector (Fig. 5.3), provided there is a one-to-one correspondence between the N-dimensional values of this vector and the different source bearings (in which case the array is said to be *unambiguous*, a property that is generally satisfied). Due to the random nature of the antenna signal, this can only be done in a statistical way.

5.2 Bartlett Estimator

We start with the simplest case of a single source. This means that for a given range, there is a single sea patch bearing, say ϕ_1, which is responsible for a given value of the radial current, which corresponds to a single ray of the first-order omnidirectional complex Doppler spectrum (see Fig. 5.1). Let X_n denote the complex value of this ray for the nth antenna, which we recall is obtained by a simple FFT of the complex

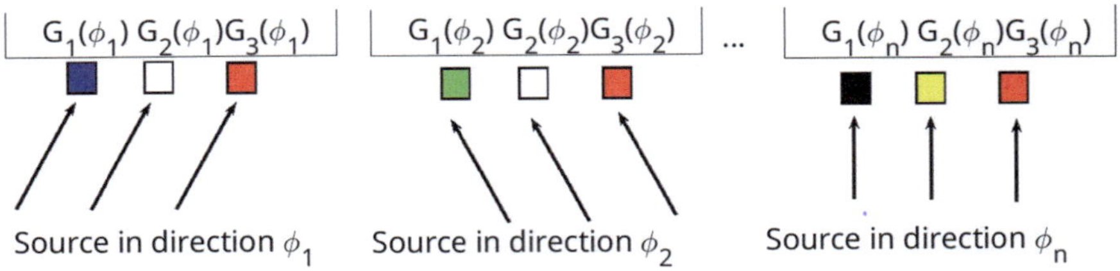

Fig. 5.3 Each illumination from a different source direction produces a specific set of complex antenna gains, called the steering vector (here, represented with colors). When the antenna array is unambiguous, there is a one-to-one correspondence between a source direction and a steering vector

range-sorted radar signal over the integration time T. Ignoring for the moment the noise component on the antennas, we can write this complex signal in the form:

$$X_n = g_n(\phi_1)Z_1, \tag{5.2.3}$$

where $g_n(\phi_i)$ is the complex gain of the nth receiving antenna in direction ϕ_i and Z_1 is a random variable that is common to all antennas and describes the incoming signal after Doppler processing. To extract the angle ϕ_1 from the received signals X_n, we form the *antenna covariance matrix* Σ, which is a $N \times N$ matrix with elements:

$$\Sigma_{mn} = \langle X_m X_n^* \rangle - \langle X_m \rangle \langle X_n^* \rangle \tag{5.2.4}$$

which involves the ensemble average $\langle \cdot \rangle$ of the random variable X_n. As usual, the ensemble average is an idealized quantity that can only be approached by taking successive samples in time, assuming that they are sufficiently independent and stationary over the duration of observation time. Under these assumptions, the antenna covariance matrix Σ takes the form:

$$\Sigma_{mn} = \alpha_1 g_m(\phi_1) g_n^*(\phi_1) \tag{5.2.5}$$

where $\alpha_1 = \text{Var}[Z_1]$. This is a rank-one matrix as can be seen by rewriting it in tensor notations[1]:

$$\Sigma = \alpha_1 \mathbf{a}(\phi_1)\mathbf{a}^*(\phi_1) \tag{5.2.6}$$

where $\mathbf{a}^*$ is the conjugate of the steering vector. It follows that the steering vector in the direction ϕ_1 is an eigenvector, i.e.:

$$\Sigma \mathbf{a}(\phi_1) = \lambda_1 \mathbf{a}(\phi_1), \tag{5.2.7}$$

[1] We recall that for vectors $\mathbf{u}$, $\mathbf{v}$, $\mathbf{w}$, the tensor $\mathbf{uv}$ is an operator such that $(\mathbf{uv})\mathbf{w} = (\mathbf{v} \cdot \mathbf{w})\mathbf{u}$.

It is associated with the unique nonzero eigenvalue $\alpha_1 \|\mathbf{a}(\phi_1)\|^2$, which is proportional to the magnitude α_1 of the source. We recall the property that the largest eigenvalue of an hermitian matrix is given by:

$$\lambda_1 = \max_{\mathbf{u}} \frac{\mathbf{u}^* \cdot \Sigma \mathbf{u}}{\mathbf{u} \cdot \mathbf{u}^*}, \tag{5.2.8}$$

where $\mathbf{u}$ is any vector. If the bearing ϕ_1 of the source is unknown, this simple property can be applied to find the DoA from the steering vector:

$$\phi_{\mathrm{DoA}} = \mathrm{Argmax}_{\phi} \frac{\mathbf{a}^*(\phi) \cdot \Sigma \mathbf{a}(\phi)}{\mathbf{a}(\phi) \cdot \mathbf{a}^*(\phi)}. \tag{5.2.9}$$

Noting that

$$\mathbf{a}^*(\phi) \cdot \Sigma \mathbf{a}(\phi) = \langle |Y(\phi)|^2 \rangle, \tag{5.2.10}$$

where $Y(\phi) = \mathbf{X} \cdot \mathbf{a}^*(\phi)$ is the sum-and-delay operation resulting from BF (see Eq. 4.2.16), we can understand (5.2.9) as maximizing the average power of the beam-formed signal to find the direction of the source. This DoA estimator is referred to as the *conventional beamformer* or *Bartlett estimator*.[2] It has the same azimuthal resolution as the BF.

5.3 Subspace Methods

We now consider the case of multiple sources, which occurs when sea patches at different bearings can generate the same radial current. This situation can occur in the presence of meandering or eddy surface currents. In this case, the contributions from different radar cells merge in the same spectral bin of the omnidirectional Doppler spectrum and one has:

$$X_n = g_n(\phi_1) Z_1 + \cdots + g_n(\phi_M) Z_M, \tag{5.3.11}$$

where M is the number of sources and the Z_m are independent random variables that represent the incoming signal from the different sources. Then the antenna covariance matrix can be written:

$$\Sigma = \sum_{m=1}^{M} \alpha_m \mathbf{a}(\phi_m) \mathbf{a}^*(\phi_m) \tag{5.3.12}$$

with $\alpha_m = \mathrm{Var}[Z_m]$. It is clear that the Bartlett estimator can only provide one DoA, namely that associated with the strongest eigenvalue. To identify multiple sources, one must refine the analysis of the covariance matrix and resort to a family of methods

[2] Bartlett's theory [28,30] actually deals with spectral analysis and periodogram estimates rather than beamforming but its algorithm has been adapted to array theory, see e.g. [144,158].

known as *subspace methods*. If the number of sources is less than the number of antennas ($M \leq N$) and if the steering vectors $a(\phi_m)$ ($m = 1 \cdots M$) are linearly independent (which holds for unambiguous arrays), it is a matrix of rank M. The steering vectors in the source directions are no longer eigenvectors but since they are independent they span an eigenspace of dimension M, which is called the *signal subspace*. Any vector $\mathbf{u}$ that is orthogonal to all the steering vectors ($\mathbf{a}(\phi_m) \cdot \mathbf{u} = 0$) satisfies $\Sigma\mathbf{u} = 0$ and thus belongs to the *null subspace* of Σ, which is the eigenspace of dimension N-M associated with the zero eigenvalue. We have so far ignored the influence of noise but the conclusion is essentially the same as long as it is not too strong. If we assume random noise components on the antennas, which for simplicity we assume to be independent and identical, then the antenna covariance matrix has an additional term:

$$\Sigma = \sum_{m=1}^{M} \alpha_m \mathbf{a}(\phi_m)\mathbf{a}^*(\phi_m) + \beta \mathrm{I}_N, \tag{5.3.13}$$

where β is the variance of noise and I_N is the identity matrix in dimension N. Because of these additional diagonal elements, the antenna covariance matrix is now full rank, i.e. it has the maximum dimension N of the space. Any vector $\mathbf{u}$ that is orthogonal to the steering vectors is now an eigenvector associated with the nonzero eigenvalue β. However, if the variance of noise is much smaller than the variance of the signal ($\beta << \min \alpha_m$), this additional eigenvalue is much smaller than the eigenvalues of the signal subspace. The corresponding subspace of dimension N-M is referred to as the *noise subspace* instead of the null subspace (Fig. 5.4).

In summary, the DF can be expressed in a purely algebraic way from the antenna covariance matrix, which is obtained by cross-averaging the complex omnidirectional Doppler spectra computed at a given range and frequency: the DoA are those for which the corresponding steering vector belongs to the signal subspace of the antenna covariance matrix. The signal subspace can be identified as the eigenspace associated to the largest eigenvalues as long as there is a clear separation between the "large" signal and "small" noise eigenvalues. However, this is not always obvious and the practical identification of the signal and noise subspaces is one difficulty of the method.

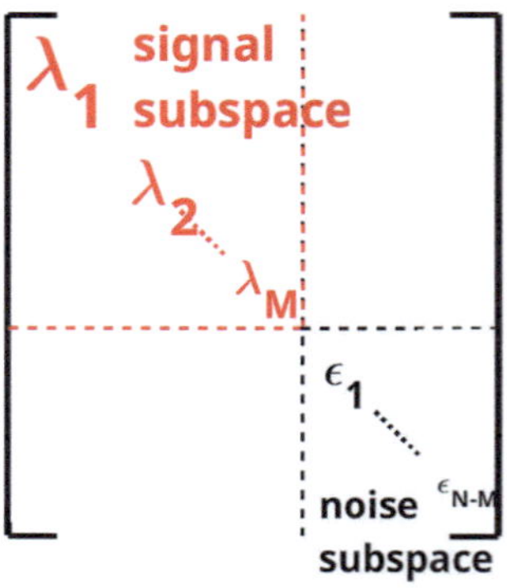

Fig. 5.4 After Singular Value Decomposition, the antenna covariance matrix is reduced to a diagonal matrix. The "large" eigenvalues, sorted in descending order, are those of the signal subspace while the "small" eigenvalues are associated to the noise subspace

5.4 The MUSIC Algorithm

The algebraic definition of the DoA paves the way for their identification in the general case where they are unknown: it is sufficient to estimate the antenna covariance matrix and to identify the eigenvectors corresponding to the strongest eigenvalues. For this, a *Singular Value Decomposition* (SVD) of the covariance is sought. The SVD is a matrix factorization that generalizes the matrix diagonalization to arbitrary complex rectangular matrices. In the case of a covariance matrix (which by construction is hermitian and positive semi-definite), the SVD writes:

$$\Sigma = \mathrm{U}\Lambda\mathrm{U}^*, \tag{5.4.14}$$

where $\Lambda = \mathrm{Diag}(\lambda_1, \ldots, \lambda_N)$ is a diagonal matrix whose elements $\lambda_1 \geq \cdots \geq \lambda_N \geq 0$ are non-negative and sorted in descending order; they are referred to as the *singular values*. The matrix $\mathrm{U} = (\boldsymbol{u}_1, \ldots, \boldsymbol{u}_N)$ is *unitary*[3] and contains the associated normalized eigenvectors $\boldsymbol{u}_j$. In the absence of noise, the M positive singular values $\lambda_1 \geq \cdots \geq \lambda_M > 0$ identify the number and strength of the different sources and the associated eigenvectors $(\boldsymbol{u}_1, \ldots, \boldsymbol{u}_M)$ are a basis of the signal subspace, which is the eigenspace spanned by the steering vectors in the directions of the source. The remaining N-M eigenvalues $\lambda_{M+1} = \cdots = \lambda_N = 0$ and associated eigenvectors $(\boldsymbol{u}_{M+1} \cdots \boldsymbol{u}_N)$ define the null subspace. In the presence of noise, these last eigenvalues are nonzero but if their values are significantly lower than the signal eigenvalues $(\lambda_N \leq \cdots \leq \lambda_{M+1} << \lambda_M)$ they can be identified without ambiguity; the corresponding subspace is the noise subspace. If the parametric form of the steering vector $\mathbf{a}(\phi)$ as a function of the bearing ϕ, then the source directions can be determined by searching the steering vectors $\mathbf{a}(\phi_m)$ that fall inside the signal subspace while continuously sweeping over all possible bearings. The most natural way to do this is to search for the steering vectors that have a maximum projection onto the signal subspace:

$$\phi_{\mathrm{DoA}} = \mathrm{ArgMax}\, P(\phi), \tag{5.4.15}$$

where P is a skill function:

$$P(\phi) = \frac{|| \sum_{m=1}^{M} [\boldsymbol{a}(\phi) \cdot \boldsymbol{u}_m^*] \boldsymbol{u}_m ||^2}{||\boldsymbol{a}(\phi)||^2}. \tag{5.4.16}$$

We will refer to this DoA estimator as the *signal subspace projection* (SSP) estimator. The function P reaches its theoretical maximum of 1 for each DoA if the steering vector fully belongs to the signal subspace. However, it turns out that the peak of the function P around a maximum can be very broad, so that the position of the maximum is not accurately defined. A much more efficient and accurate method to estimate the DoA is given by the-so called *MUSIC* (MUltiple SIgnal Classification)

[3] We recall that a matrix U is unitary if $\mathrm{UU}^* = \mathrm{U}^*\mathrm{U} = \mathrm{I_N}$, where U^* is the adjoint matrix, i.e. the transpose conjugate of U).

algorithm, first introduced by Bienvenu and Klopp [33] and popularized under this name by Schmidt [232]. The idea of the MUSIC algorithm is to identify the DoA by minimizing the projection of the steering vector onto the noise subspace instead of maximizing the projection on the signal subspace. This amounts to maximizing the inverse projection with the following skill function:

$$Q(\phi) = \frac{||\boldsymbol{a}(\phi)||^2}{|| \sum_{m=M+1}^{N} [\boldsymbol{a}(\phi) \cdot \boldsymbol{u}_m^*] \boldsymbol{u}_m ||^2} \tag{5.4.17}$$

This quantity is referred to as the *MUSIC DoA function*; its first M maxima above some threshold identify the sources. Projection onto the signal subspace or projection onto the complementary of the noise subspace appear to be equivalent characterizations. However, they are spectacularly different in terms of performance as the latter method allows a much finer azimuthal resolution.

Consider the school case example of DF using the simple model (5.3.12) with a set of 3 CODAR-type antennas. The antenna signal comes from 1 or 2 sources and the variance of the smallest source is at least 4 times larger than the noise variance ($\alpha_m \geq 2\beta$). The covariance matrix is estimated with 30 identical samples. Figure 5.5 shows the resulting skill functions $P(\phi)$ and $Q(\phi)$ together with the Bartlett estimator (5.2.10). For a unique source that is consistently assumed (case a), the Bartlett and SSP estimators are close and show a very broad skill function around the source bearing while the MUSIC estimator gives a very sharp peak at this very value. The assumption of an additional source (case b) does not change the MUSIC nor SSP pattern. If we assume a single source where they are actually 2 (case c), the 3 skill functions peak around the strongest source (with a possible bias, however) with the same pattern as in case a). If a second source is assumed (case d), the MUSIC estimate is able to accurately detect the two bearings while the Bartlett skill function is still broadly centered about the strongest source and the SSP skill function is unable to resolve the two peaks. If the number of sources is greater than or equal to the maximum dimension of the signal subspace (i.e., the number of antennas), then the SSP estimator takes a constant value while both the Bartlett and MUSIC skill function peak about the strongest source, but with very different accuracy.

5.5 The Limits of Direction Finding

In view of the enormously improved theoretical performances of the MUSIC estimator over BF estimators, one might think that this method "kills the competition" and that there is no need to install linear arrays, since excellent azimuthal accuracy can be achieved with a simple compact system. However, reality makes things a little more complicated than that and there are several practical limitations that makes the MUSIC method far less magical than it appears at first glance. In reality, the MUSIC DoA function shows multiple peaks and one has to decide which one are relevant and set thresholds to accept them as proxies for the source bearing or not. Why is the MUSIC function so different from what is expected ?

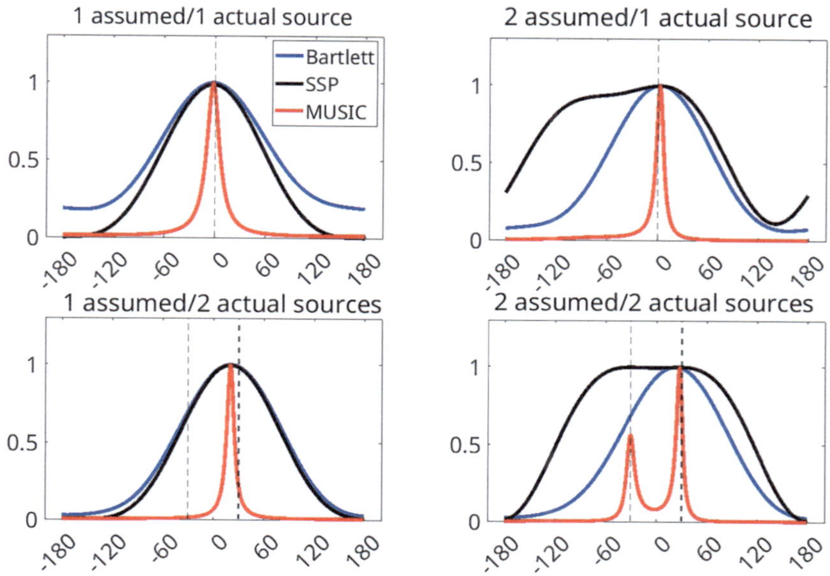

Fig. 5.5 Simulation of the skill functions of the Bartlett (blue solid line), SSP (black solid lines) and MUSIC (red solid lines) estimators as functions of the bearing (in degrees) for different sources number and directions (shown with the dashed vertical lines) in the case of an ideal CODAR-type 3-antenna system. The antenna covariance matrix is estimated with 30 independent samples and the noise variance is assumed at least 4 times smaller than the signal variance. The MUSIC DoA function has been normalized by its maximum to harmonize the vertical scales

The first reason is that the estimation of the covariance matrix requires a sufficient number of independent samples. In the context of HFR, these samples are the so-called *cross-spectra* $X_m(f_j)X_n^*(f_j)$, which are computed for each discrete frequency f_j belonging to the FOBR, by combining the complex omnidirectional Doppler spectra X_m and X_n arising from short time series recorded on any 2 antennas m and n. As the observation time is limited by the stationarity of the observed phenomena (surface current patterns, Bragg waves), the number of samples N_s can only be increased at the expense of the integration time, thereby deteriorating the frequency resolution and SNR. Typical artifact of a small number of samples is a deterioration of the MUSIC DoA function, whose peaks become smaller and broader and can pass below the established detection threshold.

Another major limitation is antenna distortion. Actual antenna patterns are generally quite different from their idealized, theoretical shape due to imperfect laboratory tuning, drift of the electronics over time or because of the surrounding electromagnetic environment. Figure 5.6 illustrates this problem schematically for a CODAR-type 3-antenna pattern. In contrast to the ideal pattern, the experimental pattern often has an asymmetric shape with secondary lobes and unbalanced loops. Performing DF on the basis of improper antenna patterns introduces errors in the detected bearings and reduces the performance. As shown in the numerical simulation of Fig. 5.7, using the ideal antenna pattern instead of the actual one results in biased estimates of

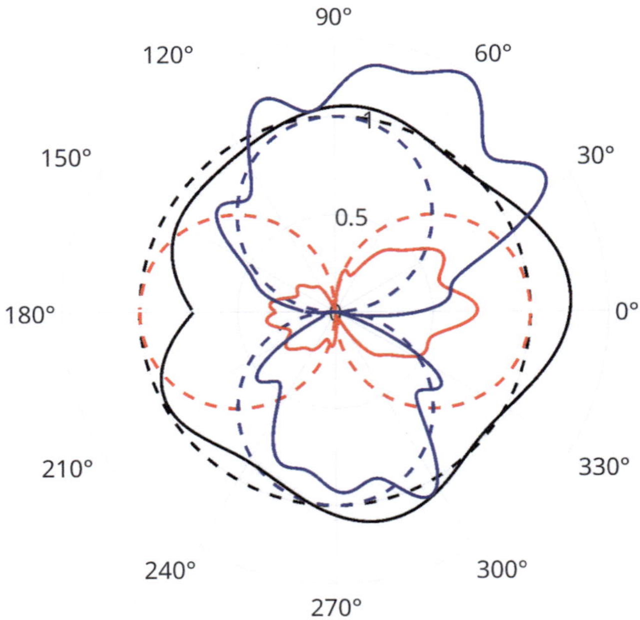

Fig. 5.6 Schematical distorted lobes (solid lines) as experimentally measured and comparison with the theoretical antenna pattern (dashed lines)

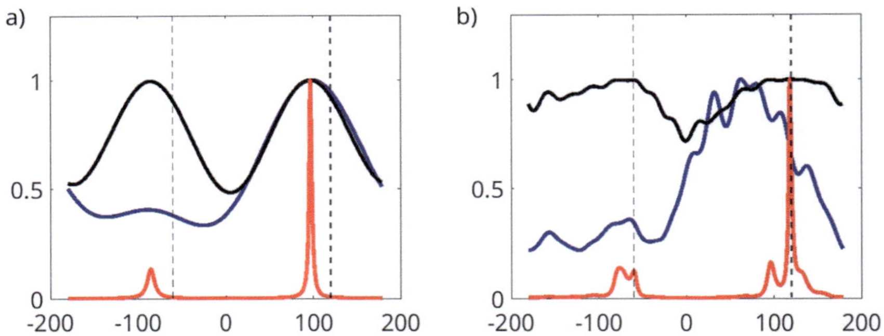

Fig. 5.7 Same as Fig. 5.5 in the 2-source case for the distorted antenna lobes shown in Fig. 5.6. The skill functions are calculated (1) using the ideal antenna pattern (left panel); (2) using the actual antenna pattern, assuming it is exactly known

the source bearings with all DF methods (Bartlett, SSP and MUSIC), although the MUSIC factor remains extremely peaked around its maxima. The use of the actual antenna pattern, under the optimistic assumption that it is perfectly known, corrects, at least partially, for the bias. For this reason, it is common practice to carry out a preliminary experimental measurement of the "true" antenna pattern, an operation which is generally done by navigating around the radar with a ship-borne transponder. However, even with a perfect knowledge of the experimental antenna pattern, the MUSIC performance is generally not as good as in the ideal case if there are too many secondary lobes or if the loops differ too much in magnitude. Depending on the antenna patterns, it may also be the case that the DF process is efficient in covering some angular sectors and almost "blind" to some other sectors, so that the radial current maps end up with some systematically blank regions.

A further issue of DF is determining the number of sources, which controls the dimension of the signal and noise subspaces. Ideally, the noise eigenvalue are much smaller than the signal eigenvalues, so that there is a clear identification of the two subspaces. However, as the noise level increases (mainly with range), the separation is less and less obvious and one must set some arbitrary thresholds to distinguish what is a source and what is noise. Overestimating the actual number of sources may result in incorrect DoA; an underestimation may result in missing structures in the surface current pattern (e.g. in the case of eddies or meanders). The number of sources is usually estimated by counting the local maxima of the MUSIC DoA function that exceed some predefined threshold. This requires some fine tuning because there may be many local maxima of comparable size.

Another limitation concerns the identification of the FOBR. The Bragg rays, that are combined for the DF analysis, are identified from the omnidirectional Doppler spectrum. Due to the variation of the radial current with azimuth and the resulting Doppler shift, the latter can be very broad in case of strong currents and overlap the second-order Doppler spectrum, leading to a false identification of first-order spectral rays. Several techniques have been proposed to improve the identification of the FOBR, based on image processing techniques [154], empirically defined thresholds [160], deep-learning methods [45] or pre-conditionning with a BF processing [62] (this last method only works for arrays). Nevertheless, it is sometimes difficult to disentangle the first and second-order components of the Doppler spectrum when the radial current undergoes strong azimuthal variations.

Finally, the DF analysis is not able to provide azimuthally resolved Doppler spectra as does the BF processing. This is an additional difficulty when it comes to estimating the directional ocean wave spectra, which are related to the second-order directional Doppler spectra. This is discussed in detail in Sect. 12.2.

5.6 DF for Arrays

Although phased-array systems are primarily designed for BF processing, it is also interesting to use them with DF processing techniques to increase the azimuthal resolution, especially when the array is too short (e.g., 8 antennas). The technique is identical to that for compact antenna systems, with the difference that the complex elements of the steering vector differ only by the antenna complex gains $g_n(\phi) = e^{i(n-1)K_0 d \sin\phi}$ and that the antenna covariance matrix has larger dimension. Figure 5.8 shows a numerical simulation of the skill functions for the 3 DoA methods (Bartlett, SSP and MUSIC) in the case of an 8-antenna array illuminated by 3 sources of various strengths and calculated on the basis of only 10 sample cross-spectra. As seen, the MUSIC DoA function is very sharply peaked around the source bearings and can detect them all once a sufficient number of sources is assumed; the

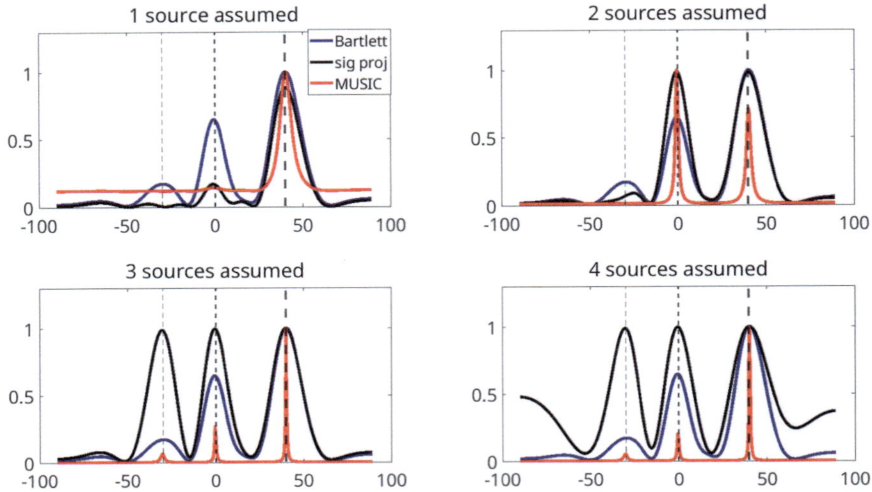

Fig. 5.8 Skill functions of the Bartlett (blue solid line), SSP (black solid lines) and MUSIC (red solid lines) for a perfect linear array of 8 antennas with $0.45\lambda_0$ spacing. There are 3 sources at the bearing $-30, 0, 40$ degrees with respective amplitude $1, 2, 3$ (arbitrary units) and the covariance matrix is calculated with 10 samples. The root-mean-square noise is 0.5 for each antenna. The skill functions are calculated using from 1 to 4 sources

Bartlett and SSP skill functions have a much broader angular width, corresponding to the width of the array factor. This last example was calculated with a moderate noise root-mean-square, equal to half the amplitude of the weakest source signal. However, more systematic studies with both numerical and field data show that, unlike the Bartlett estimator, the accuracy of the DF method is highly dependent on the SNR and only better than BF for large SNR [280].

The advantage of arrays is that they can accommodate many more sources (depending on the size of the array) that the 3-antenna compact systems (which can deal with at most 2 sources), hence improving the detection. They are also very robust to antenna phase distortion and noise (see [64] for more details). The use of DF with regular or irregular arrays has been found to be effective in increasing the azimuthal resolution of BF in the case of 8-[63, 196] and 12-antenna [64] systems. For 16-antenna systems, the two methods have been found to be very close [62].

Part II

From Sea Surface to Antenna Voltage

6 Antenna Radiation in Free Space

Radar remote sensing of the sea surface involves the radiation and propagation of electromagnetic (EM) waves in the atmosphere and their backscattering from the sea surface. These processes are governed by Maxwell's equations in the air together with the associated boundary conditions at the air/water interface. In this chapter we recall the elementary notions of electromagnetism, which are necessary to understand the physical description of the sea echo, starting with the propagation in the air.

6.1 The Electromagnetic Field in Free Space

6.1.1 Maxwell Equations

The EM field in any medium is described by four vector fields **D**, **E**, **B**, **H**, which are related to each other by the *macroscopic Maxwell's equations*:

$$\begin{aligned}(\nabla \cdot \mathbf{D}) &= \rho \\ (\nabla \cdot \mathbf{B}) &= 0 \\ (\nabla \times \mathbf{E}) &= -\frac{\partial \mathbf{B}}{\partial t} \\ (\nabla \times \mathbf{H}) &= \mathbf{j} + \frac{\partial \mathbf{D}}{\partial t},\end{aligned} \tag{6.1.1}$$

where ρ is the excess charge density and $\mathbf{j}$ is the current vector density. These last quantities are related by the continuity equation:

$$\frac{\partial \rho}{\partial t} + (\nabla \cdot \mathbf{j}) = 0, \tag{6.1.2}$$

C.-A. Guérin, *High-Frequency Radars for the Oceanographer*, Geophysics and Environmental Physics, https://doi.org/10.1007/978-3-031-95642-3_6

which is the local formulation of the conservation of charge when an electrical current is flowing. We refer to any classical textbook in electromagnetism (e.g. [79,143,145]) for the derivation and interpretation of these equations. **E** is the *electric field*, which is expressed in volt per meter (or $kg \cdot m \cdot s^{-3} \cdot A^{-1}$ in SI base units); **B** is the *magnetic flux density*, most often simply referred to as the *magnetic field*, which is measured in Tesla (or $kg \cdot s^{-2} \cdot A$ in SI units). The vector **D** is called the *electric displacement field* or *electric flux density* and has the unit of Coulomb per square meter ($A \cdot s \cdot m^{-2}$ in SI base units); it is an auxiliary field that takes into account the electric polarization of charges inside a dielectric material, that is the local field change induced by the formation of microscopic dipoles under an excitation field. The vector **H** is called the *magnetic field strength* and is measured in Ampere per meter ($A \cdot m^{-1}$); it is another auxiliary field that expresses the magnetic polarization of matter, i.e. magnetic dipole moments induced by microscopic currents. The auxiliary fields are related to the electric and magnetic fields by the so-called *constitutive relations*, which depend specifically on the matter under consideration:

$$\mathbf{D} = \varepsilon \mathbf{E}, \ \mathbf{B} = \mu \mathbf{H} \tag{6.1.3}$$

The parameter ε is called the *absolute permittivity* or simply *permittivity* of the medium and is measured in farads per meter (F/m) or $kg^{-1} \cdot m^{-3} \cdot s^4 \cdot A^2$ in SI base units; it is most commonly expressed through the dimensionless *relative permittivity*:

$$\varepsilon_r = \frac{\varepsilon}{\varepsilon_0}, \tag{6.1.4}$$

where $\epsilon_0 = 8.85 \ 10^{-12}$ F/m is the vacuum permittivity. Similarly, μ is called the *permeability* and is measured in Henry per meter (H/m) or $kg \cdot m \cdot s^{-2} \cdot A^{-2}$ in SI units; it is usually expressed in terms of its dimensionless value relative to vacuum, namely the *relative permeability*:

$$\mu_r = \frac{\mu}{\mu_0}, \tag{6.1.5}$$

where $\mu_0 = 4\pi \ 10^{-7}$ H/m. Let us recall that these two fundamental constants of physics are related to the speed of light in vacuum c_0 by:

$$\epsilon_0 \mu_0 c_0^2 = 1 \tag{6.1.6}$$

While the first set of Eq. (6.1.1) is universal, the constitutive relations depends specifically on the propagating medium through its electric permittivity and magnetic permeability. Maxwell's equations are the local forms of the integral laws of electromagnetism. The first equation in (6.1.1) is known as *Maxwell-Gauss* equations, which states that the flux of the electric field through a closed surface is proportional to the internal charge; the second divergence equation states that there is no magnetic charge. The third equation is the *Maxwell-Faraday* law of induction, which states that an electromotive force in a wire is induced by a varying magnetic flux; the fourth one is referred to as the *Maxwell-Ampere* equation and is the correction of Ampere's law with the continuity Eq. (6.1.2). It is often argued (see e.g. the discussion in [47,79]) that **H** should be considered as the "genuine" magnetic field because, unlike **B**, it has a direct relation to the macroscopic current density via Maxwell-Ampere's law, a quantity that can be controlled and measured experimentally. For this reason, the pair of fundamental fields that are retained to describe EM propagation in a medium are **E** and **H**, rather than **E** and **B**, and we will stick to this convention even though we will not be dealing with magnetic media.

6.1.2 Harmonic Solutions

In a stationary (i.e. non-transcient) regime, the EM field can be equivalently represented in the time or the frequency domain. Because of the linearity of Maxwell's equations, the EM field can therefore be solved for a given frequency, assuming a harmonic time dependence. As it is well known, this involves finding a mathematical solution in the form of a complex-valued EM field, where only the real part corresponds to a physical field:

$$\mathbf{E}(\mathbf{R},t) = \mathbf{E}(\mathbf{R})e^{-\mathrm{i}\omega_0 t}, \tag{6.1.7}$$

The harmonic Maxwell's equations write:

$$\begin{aligned} (\nabla \cdot \mathbf{D}) &= \rho \\ (\nabla \cdot \mathbf{B}) &= 0 \\ (\nabla \times \mathbf{E}) &= i\omega_0 \mathbf{B} \\ (\nabla \times \mathbf{H}) &= \mathbf{j} - i\omega_0 \mathbf{D} \end{aligned} \tag{6.1.8}$$

where the quantities involved no longer depends on time but only on the spatial variable **R**. We will start by describing the radiation and propagation of EM waves in the air before they reach the sea surface. In the context of HFR, the lower atmosphere can be identified to a very good approximation with the vacuum ($\epsilon_r = \mu_r = 1$) and the perturbations induced by refraction and turbulence are neglected. The charges and currents are those produced by a transmitting antenna. They are confined to a region called the *source region*, i.e. the radiating elements of the antenna. Outside the source region, the EM field satisfies Maxwell's equations in vacuum:

$$\begin{aligned} (\nabla \cdot \mathbf{E}) &= 0 \\ (\nabla \cdot \mathbf{H}) &= 0 \\ (\nabla \times \mathbf{E}) &= i\omega_0 \mu_0 \mathbf{H} \\ (\nabla \times \mathbf{H}) &= -i\omega_0 \epsilon_0 \mathbf{E} \end{aligned} \tag{6.1.9}$$

where we have eliminated the auxiliary fields. By combining the two curls in the above equations and using the well-known operator identity:

$$\nabla \times \nabla \times E = \nabla((\nabla \cdot \mathbf{E})) - \nabla^2 \mathbf{E} \tag{6.1.10}$$

we arrive at the *Helmholtz equation* for both the electric and magnetic field inside the medium:

$$\nabla^2 \mathbf{E} + K_0^2 \mathbf{E} = 0 \tag{6.1.11}$$

where $K_0 = \omega_0 / c_0$ is the vacuum wavenumber and $\nabla^2 \mathbf{E}$ is the *Laplacian* of the vector $\mathbf{E}$[1]. The same equation holds for the magnetic field. Canonical solutions of the Helmholtz equation can be sought in the form of plane waves:

$$\mathbf{E}(\mathbf{R}) = \mathrm{e}^{\mathrm{i}\mathbf{K}\cdot\mathbf{R}} \mathbf{E_0} \tag{6.1.12}$$

for some complex *wave vector* $\mathbf{K}$ with an arbitrary direction $\mathbf{E_0}$. Taking the Laplacian of this expression, we see that (6.1.11) holds provided that the wave vector satisfies the following relation:

$$\mathbf{K} \cdot \mathbf{K} = K_0^2. \tag{6.1.13}$$

When the wave vector is real, this implies that it has a fixed norm equal to the wavenumber, $K = K_0$ and $\mathbf{E}$ is a *propagating plane waves.*[2] However, solutions with complex wave vectors can also exist and will be discussed later. By re-injection of the plane wave (6.1.12) in Maxwell's equations we find that:

$$\begin{aligned} \mathbf{K} \times \mathbf{E}_0 &= \omega_0 \mu_0 \mathbf{H}_0 \\ \mathbf{K} \times \mathbf{H}_0 &= -\omega_0 \epsilon_0 \mathbf{E}_0 \end{aligned} \tag{6.1.14}$$

It follows that $(\mathbf{K}, \mathbf{E}_0, \mathbf{H}_0)$ is a direct triedre and that the electric and magnetic field magnitude have a fixed ratio:

$$\frac{E_0}{H_0} = Z_0 = \sqrt{\mu_0 / \epsilon_0} \tag{6.1.15}$$

The constant $Z_0 = 377$ Ohms is called the *vacuum impedance*. An important property of EM plane waves is their *transverse* nature , meaning that the directions of the electric and magnetic fields are perpendicular to the direction of propagation.

[1] We recall that the Laplacian of a scale function ψ is given by $\nabla^2 \psi = \frac{\partial^2 \psi}{\partial x^2} + \frac{\partial^2 \psi}{\partial y^2} + \frac{\partial^2 \psi}{\partial z^2}$; the Laplacian of a vector function is the vector formed with the Laplacian of each component.

[2] When adding the implicit harmonic time dependence $e^{-i\omega_0 t}$, it is seen that the wave fronts, that is, the surfaces with constant phases, are planes perpendicular to the $\mathbf{K}$ vector propagating at the speed of light $c_0 = \omega_0 / K_0$, therefore justifying the terminology.

6.1.3 The Field Polarization

The complex electric and magnetic field that compose the EM field are 3-dimensional functions of the space coordinate. The instantaneous vector direction of the physical electric field $\mathrm{Re}[\mathbf{E}(\mathbf{R},t)]$ is called the *polarization* . It turns out that the reflection of EM waves from a dielectric or conducting surface depends on the incident polarization and it is therefore important to go into more details. When considering a harmonic time-dependence, one can separate out the spatial and temporal dependence and write $\mathbf{E}(\mathbf{R},t) = \mathbf{E}(\mathbf{R})e^{-\mathrm{i}\omega_0 t}$, for some time-independent complex vector $\mathbf{E}(\mathbf{R})$ which can be decomposed into a real and imaginary parts:

$$\mathbf{E}(\mathbf{R}) = \mathbf{E}'(\mathbf{R}) + \mathrm{i}\mathbf{E}''(\mathbf{R}) \tag{6.1.16}$$

The real part of the time-varying electric field is now given by:

$$\mathrm{Re}[\mathbf{E}(\mathbf{R},t)] = \mathbf{E}'(\mathbf{R})\cos(\omega_0 t) - \mathbf{E}''(\mathbf{R})\sin(\omega_0 t) \tag{6.1.17}$$

and it is thus plain that the direction of this last vector varies with time but remains in a plane formed by the vectors $\mathbf{E}'(\mathbf{R})$ and $\mathbf{E}''(\mathbf{R})$, called the *polarization plane.* When $\mathbf{E}'(\mathbf{R})$ and $\mathbf{E}''(\mathbf{R})$ are arbitrary vectors with different directions and amplitudes, the tip of the real electric field describes an ellipse in time and the field is said to be *elliptically polarized*; When $\mathbf{E}'(\mathbf{R})$ and $\mathbf{E}''(\mathbf{R})$ are mutually perpendicular and have the same amplitudes, the tip of the real electric field describes a circle in time and the field is said to be *circularly polarized*, which is a particular case of elliptic polarization; when $\mathbf{E}'(\mathbf{R})$ and $\mathbf{E}''(\mathbf{R})$ are aligned, the direction of the real electric field remains constant and the field is said to be *linearly polarized.* It is plain from (6.1.17) that an elliptical polarized field can be written as a superposition of 2 linearly polarized fields. Hence solving the mathematical problem for any linearly polarized field solves it for an arbitrary polarization and we will only consider that case.

6.1.4 The Field Outside the Source Region

The most general way of writing the complex harmonic electric field associated to a fixed linear polarization is:

$$\mathbf{E}(\mathbf{R}) = f(\mathbf{R})\mathbf{E}^0 \tag{6.1.18}$$

for some constant real vector $\mathbf{E}^0$ and scalar complex function of the position, $f(\mathbf{R})$. As we have seen before, a consequence of Helmholtz equation (6.1.11) outside the source region is that the field can be written as a superposition of plane waves with a wave vector constrained by relation (6.1.13), thus running on an integration domain of reduced dimension 2. A simple way to integrate such vector is to consider its projection onto some reference plane, say $\mathbf{k} = (\mathbf{k}_x, \mathbf{k}_y)$ (which we refer to as the horizontal plane) and to define the remaining component (which we refer to as the vertical component) through the relation (6.1.13). We therefore make the decomposition:

$$\mathbf{K} = \mathbf{k} \pm q\mathbf{z}, \tag{6.1.19}$$

with $k^2 + q^2 = K_0^2$. Note that they are two sign possibilities for the vertical completion, so that there is a pair of upwards and downwards wave vector having the same horizontal projection. By integrating over the horizontal projection and accounting for the two possibilities, we can rewrite the scalar component of the field in the form:

$$f(\mathbf{R}) = f(\mathbf{r}, z) = \int_{\mathbb{R}^2} f^+(\mathbf{k})e^{\mathrm{i}(\mathbf{k}\cdot\mathbf{r}+qz)}d\mathbf{k} + \int_{\mathbb{R}^2} f^-(\mathbf{k})e^{\mathrm{i}(\mathbf{k}\cdot\mathbf{r}-qz)}d\mathbf{k}, \qquad (6.1.20)$$

where we have also decomposed the position vector into a horizontal and vertical components, $\mathbf{R} = \mathbf{r} + \mathbf{z}$ and we require the condition $k^2 + q^2 = K_0^2$ so that $f(\mathbf{R})$ satisfies the Helmholtz equation (6.1.11). Such representation, which will turn to be very useful for the half-space problem, is referred to as the *Rayleigh expansion* of the field. Note that the integrals (6.1.20) run over $\mathbb{R}^2$. For $k \leq K_0$, the elements $e^{\mathrm{i}\mathbf{k}\cdot\mathbf{r}}e^{\mathrm{i}qz}$ are propagating plane waves. Whenever $k > K_0$ we may nonetheless define an imaginary square root $q = \mathrm{i}\sqrt{k^2 - K_0^2}$. The corresponding elements $e^{\mathrm{i}\mathbf{k}\cdot\mathbf{r}}e^{-|q|z}$ are exponentially decreasing or increasing along the vertical axis. Because the field must remain bounded, it is clear that such solutions cannot exist in free space. However, exponentially decreasing solutions can be obtained in the case where the space is limited on one side (the so called half-space problem, which will be the object of the next section). Such solutions are referred to as the *evanescent waves*.

6.2 Antenna Radiation

6.2.1 The Far-Field Produced by Source of Current

Plane waves are the building bricks to represent the propagating EM waves outside the source region. We will now provide the explicit form of the field radiated by a source of current in vacuum. For simplicity we will only give the far-field expression, which is the only relevant in the context of sea surface scattering. We follow the standard procedure which involves the vector and scalar potentials. Since the magnetic induction is divergence free there exists some vector field $\mathbf{A}$ such that

$$(\nabla \times \mathbf{B}) = \mathbf{A}. \qquad (6.2.21)$$

Since $\mathbf{E} - i\omega_0\mathbf{A}$ is curl free there exists a scalar field V such that

$$\mathbf{E} - i\omega_0\mathbf{A} = -\nabla V. \qquad (6.2.22)$$

Because $\mathbf{A}$ is defined up to a divergence and V is defined up to a constant, there is one degree of freedom in the choice of these potentials, which can be relaxed by imposing a *gauge condition*. The choice of the *Lorenz gauge*:

$$(\nabla \cdot \mathbf{A}) - i\frac{\omega_0}{c_0^2}V = 0 \qquad (6.2.23)$$

leads to the standard equation for the potential vector after substitution of the latter in Maxwell's harmonic equations and use of the vector identity (6.1.10):

$$\nabla^2\mathbf{A} + K_0^2\mathbf{A} = -\mu_0\mathbf{j}, \tag{6.2.24}$$

A point-like source of current

$$\mathbf{j}(\mathbf{R}) = \delta(\mathbf{R})\mathbf{j_0}, \tag{6.2.25}$$

where $\mathbf{j_0}$ is some constant vector, is called a *dipole source* (because such an infinitesimal current can be obtained with an oscillating electric dipole). The associated potential vector is given by:

$$\mathbf{A}(\mathbf{R}) = -\mu_0 G(R)\mathbf{j_0}, \tag{6.2.26}$$

where

$$G(R) = -\frac{e^{\mathrm{i}K_0 R}}{4\pi R} \tag{6.2.27}$$

is the free space scalar Green's function (see Appendix B). In the general case of an extended distribution of current, the solution of (6.2.24) can be obtained by convolution with the source $\mathbf{j}$:

$$\mathbf{A}(\mathbf{R}) = -\mu_0 \int_{source} G(\mathbf{R} - \mathbf{R}')\mathbf{j}(\mathbf{R}')d\mathbf{R}', \tag{6.2.28}$$

Note that the function $G(R)$ is isotropic and thus only depends on the norm R of the vector $\mathbf{R}$. In the far field where the distance to the transmitter is large compared to both the radar wavelength and the size of the antenna we may approximate:

$$G(\mathbf{R} - \mathbf{R}') \simeq -\mu_0 G(R) e^{-iK_0\widehat{\mathbf{R}}\cdot\mathbf{R}'} \tag{6.2.29}$$

where $\widehat{\mathbf{R}} = \mathbf{R}/R$ is the direction of observation. It follows that the potential vector can be written as the product of a spherical wave and an angular factor:

$$\mathbf{A}(\mathbf{R}) \simeq -\mu_0 G(R)\mathbf{J}(\mathbf{K}), \tag{6.2.30}$$

where

$$\mathbf{J}(\mathbf{K}) = \int_{source} e^{-i\mathbf{K}\cdot\mathbf{R}'}\mathbf{j}(\mathbf{R}')d\mathbf{R}' \tag{6.2.31}$$

is the Fourier Transform of the current distribution along the wave vector $\mathbf{K} = K_0\widehat{\mathbf{R}}$. The magnetic field in the far field can thus be obtained by:

$$\mathbf{H} \simeq -\nabla \times (G\mathbf{J}) = -G(\nabla \times \mathbf{J}) - \nabla G \times \mathbf{J} \tag{6.2.32}$$

When calculating these gradients and curls several terms arise with a fall off like $1/R$ or $1/R^2$. Retaining only the terms that have the slower decay in $1/R$ we obtain easily:

$$\mathbf{H}(\mathbf{R}) \simeq -iK_0 G(R)\, \widehat{\mathbf{R}} \times \mathbf{J} \tag{6.2.33}$$

The electric field is given by the double curl of the vector potential:

$$\mathbf{E}(\mathbf{R}) = \frac{i}{\mu_0 \omega_0 \epsilon_0} \nabla \times (\nabla \times \mathbf{A}). \tag{6.2.34}$$

Taking this expression in the far field and retaining only the dominant $1/R$ terms, we obtain after some calculations:

$$\mathbf{E}(\mathbf{R}) \simeq -iK_0 Z_0 G(R) \widehat{\mathbf{R}} \times (\widehat{\mathbf{R}} \times \mathbf{J}). \tag{6.2.35}$$

This expression states that the electric field is a spherical wave (hence with a $1/R$ decay) multiplied by a direction dependent vector. The orientations of both the electric and magnetic vectors in the far-field are *transverse* to the direction of observation $\widehat{\mathbf{R}}$. This also tells us that the far-field radiated by any source can be locally assimilated to a plane wave with a wave vector along the direction of observation and an amplitude decreasing like $1/R$ (Fig. 6.1).

The EM power density is given by the complex Poynting vector:

$$\mathbf{P} = \frac{1}{2}\mathbf{E} \times \mathbf{H}^* = 2^{-5} Z_0 K_0^2 \frac{1}{R^2} \left\|\widehat{\mathbf{R}} \times \mathbf{J}\right\|^2 \widehat{\mathbf{R}} \tag{6.2.36}$$

and the radiated power per solid angle in direction (θ, ϕ), referred to as the *radiation intensity*, is given by

$$I(\theta, \phi) = R^2 P = 2^{-5} Z_0 K_0^2 \left\|\widehat{\mathbf{R}} \times \mathbf{J}\right\|^2 \tag{6.2.37}$$

To measure the ability of an antenna to radiate in a given direction, one uses the notion of *directive gain*, which is defined as the ratio of the radiated intensity in one direction (θ, ϕ) to the average power over 4π steradians:

$$g(\theta, \phi) = 4\pi \frac{I(\theta, \phi)}{\int I(\theta, \phi) d\Omega} \tag{6.2.38}$$

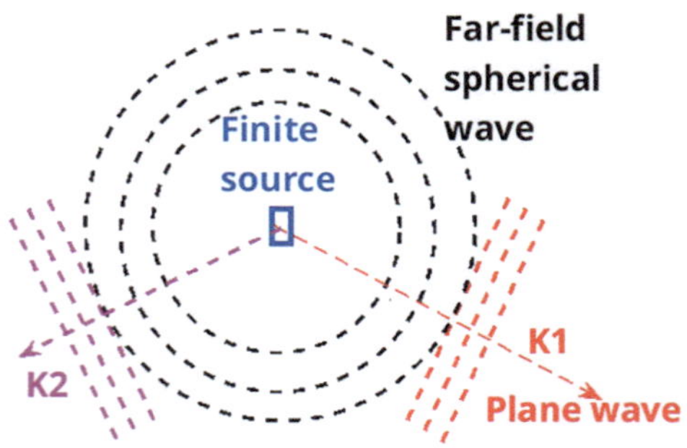

Fig. 6.1 The far-field produced by an arbitrary source of finite extent is proportional in any direction to a spherical wave and can be locally assimilated to a plane wave

where $d\Omega = \sin\theta d\theta d\phi$ is the element of solid angle. This amounts to compare the power of a given antenna to a fictitious isotropic antenna (that is, for which $I(\theta, \phi)$ =constant) that would radiate the same total power. The *directivity* of the antenna is defined as the maximum value of the gain. Of particular importance is the case of a very small antenna size with respect to the EM wavelength, which can be represented by a point-like source of current (6.2.25). In that case, **J** no longer depends on the direction **K**. Such a source is referred to as the *dipole antenna*. Its total radiated power increases like ω_0^4 and its directive gain is:

$$g(\theta, \phi) = \frac{3}{2}\sin^2\theta, \tag{6.2.39}$$

where θ is the angle with respect to the direction of **J**.

6.2.2 Linear Antennas

Most often the HFR antennas have a linear shape along one dimension, say z, and a negligible extension along the other 2 dimensions x, y. In that case the radiated intensity is isotropic in the (x, y) azimuthal plane and depends only on the polar angle θ with respect to the z-axis. The length L of the linear antenna needs not be small with respect to the radar wavelength so that it is no longer a dipole in general. By choosing the origin of coordinates at the middle of the antenna, we may therefore assume an idealized current density:

$$\mathbf{j}(\mathbf{R}) = j(z)\delta(x)\delta(y)\mathbf{z}, \quad -L/2 \leq z \leq L/2, \tag{6.2.40}$$

with the condition that the current must vanish at both ends ($j(-L/2) = j(L/2) = 0$). It follows that $J(\theta, \phi) = J(\theta)\mathbf{z}$ with:

$$J(\theta) = \int_{-L/2}^{L/2} e^{-iK_0 z\sin\theta} I(z)dz. \tag{6.2.41}$$

When $K_0 L$ is small, the exponential term can be well approximated with a few terms of its series expansion, leading to the so-called *multipole expansion*:

$$J(\theta) = J_0 - i\sin\theta J_1 - \frac{1}{2}J_2\sin^2\theta + \cdots \tag{6.2.42}$$

where the J_n are the moments of the current distribution inside the rod:

$$J_n = \int_{-L/2}^{L/2} (K_0 z)^n j(z)dz \tag{6.2.43}$$

If we assume that the current along the wire is symmetric on the two arms of the antenna, the even moments cancel and it follows that:

$$I(\theta, \phi) = I(\theta) = 2^{-5} Z_0 K_0^2 J_0^2 \left(\sin^2\theta - \frac{J_2}{J_0} \sin^4\theta + \cdots \right) \tag{6.2.44}$$

In the limit of an infinitely small antenna ($K_0 L \to 0$), only the first term is retained in the multi-pole expansion and the dipole radiation pattern (6.2.39) is recovered. It follows that a small linear antenna has a maximum of radiation perpendicular to its main axis and a null of emission along this axis. As the size parameter $K_0 L$ is increased, multi-pole corrections inter in play. The $\sin^4\theta$ term adds up a quadrupole correction with a $\sin^2\theta\cos^2\theta$ pattern, which is composed of four secondary lobes centered about $\pm\pi/4$ and $\pm 3\pi/4$ and so forth. Exact expressions of the radiation intensity can be obtained in the simple case of a sinusoidal current:

$$j(z) = I_0 \sin(kL/2 - k\,|z|) \tag{6.2.45}$$

for which [143]:

$$I(\theta) = \frac{Z_0 I_0}{8\pi^2} \left| \frac{\cos\left(\frac{kL}{2}\cos\theta\right) - \cos\left(\frac{kL}{2}\right)}{\sin\theta} \right|^2 \tag{6.2.46}$$

Figure 6.2 shows the radiation intensity for a few sizes of antennas. In the limit $K_0 L \to 0$, (6.2.46) is consistent with the dipole radiation intensity:

$$I(\theta) = 2^{-5} Z_0 K_0^2 J_0^2 \sin^2\theta \simeq 2^{-9} Z_0 (K_0 L)^4 I_0^2 \sin^2\theta \tag{6.2.47}$$

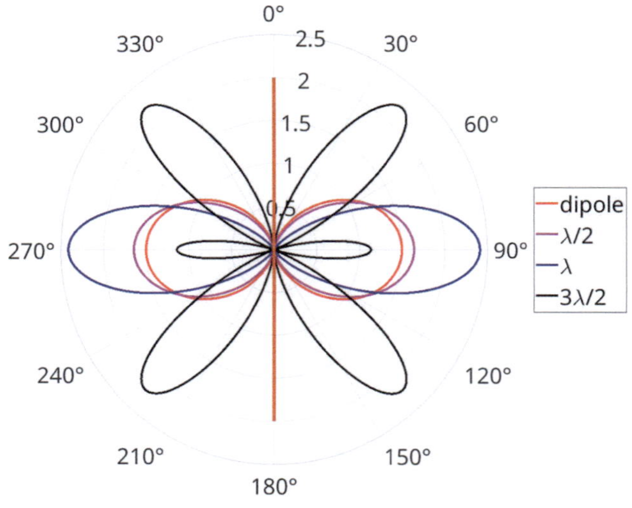

Fig. 6.2 Radiation intensity of a dipole antenna oriented along the vertical axis. The half-wavelength and full-wavelength antenna are more directive than the dipole antenna. For other antenna lengths (here, $3/2\lambda_0$), quadrupole lobes appear

The $\sin^2\theta$ dipole shape is not very sharp; its radiated intensity has a directivity of only 3/2 and a FWHM of 90°. Higher directivity, however, can be reached with the half-wave dipole ($L = \lambda_0/2$) and full-wave dipole antenna ($L = \lambda_0$) for which the quadrupole pattern cancels:

$$I(\theta) = \frac{Z_0 I_0}{8\pi^2} \frac{\cos^2\left(\frac{\pi}{2}\cos\theta\right)}{\sin^2\theta} \quad (6.2.48)$$

for $K_0 L = \pi$ and

$$I(\theta) = \frac{Z_0 I_0}{8\pi^2} \frac{4\cos^4\left(\frac{\pi}{2}\cos\theta\right)}{\sin^2\theta} \quad (6.2.49)$$

for $K_0 L = 2\pi$. The directivity and FWHM of their radiation patterns are 1.64 and 78° for the half-wave dipole and 2.4 and 48° for the full-wave dipole.

The EM wavelength for oceanographic HFR is typically of the order of 20 m. It is therefore practically difficult to devise half-wave and full-wave dipole antennas. However, one often uses antenna rods that have the physical size of a quarter wavelength ($L = \lambda_0/4$), referred to as *quarter-wave monopole*. When such a based-fed antenna is installed on a conducting ground plane, it can be shown [145] to radiate equivalently to a virtual half-wave center-fed dipole antenna by virtue of image theory (Fig. 6.3). In practice, the conducting plane is obtained by connecting the antenna to the ground by means of a few metallic wires referred to as *radials* (Fig. 6.4).

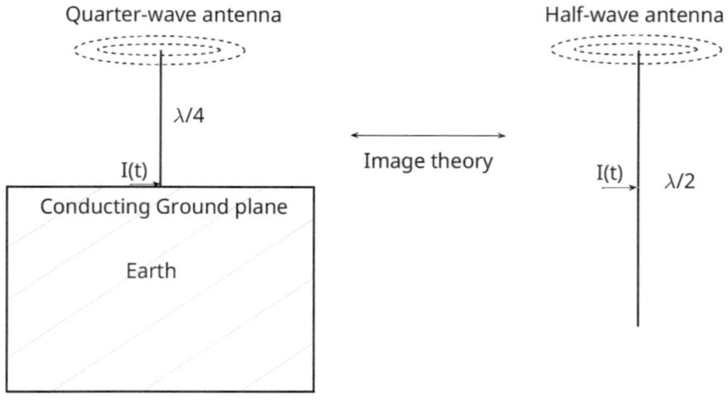

Fig. 6.3 A quarter-wave antenna radiating over a conducting ground plane is equivalent to a half-wave dipole antenna radiating in vacuum by virtue of image theory

Fig. 6.4 A conducting plane for the quarter-wave antenna is obtained by connecting it to the ground with a few metallic wires. Here, a view of one $\lambda_0/4$ antenna at Fort Peyras, Toulon

7 Antenna Radiation Over a Flat Ocean

Before we consider the scattering from a rough sea surface, we start with the case of a plane interface, which serves as a reference for the latter. We consider Maxwell's equation in a conducting medium instead of the vacuum, whose electrical properties are described by the complex permittivity and impedance of seawater. We introduce the canonical horizontal and vertical EM polarization, which describes the orientation of the EM field relative to the reference plane and plays a major role in the reflective properties of the surface. We recall the definition and properties of the Fresnel reflection coefficients, which gives a first understanding of why HFR use only vertically polarized electric fields.

7.1 The Electromagnetic Field in Seawater

Seawater at radio frequencies can be considered as a non-magnetic ($\mu_r = 1$), homogeneous medium with a relative dielectric constant ϵ_r that depends on the operating frequency. Due to its strong concentration of ions, salted water is a conducting medium so that any exciting electric field gives rise to an ohmic current:

$$\mathbf{j} = \gamma \mathbf{E} \tag{7.1.1}$$

The constant γ is the *conductivity* and is measured in Siemens per meter (S/m). When specifying the harmonic Maxwell equations (6.1.8) to seawater we obtain a set of simplified equations after elimination of the auxiliary fields:

$$\begin{aligned}(\nabla \cdot \mathbf{E}) &= 0\\ (\nabla \cdot \mathbf{H}) &= 0\\ (\nabla \times \mathbf{E}) &= i\omega_0\mu_0\mathbf{H}\\ (\nabla \times \mathbf{H}) &= -i\omega_0\epsilon_0\epsilon_r\mathbf{E}\end{aligned} \tag{7.1.2}$$

C.-A. Guérin, *High-Frequency Radars for the Oceanographer*, Geophysics and Environmental Physics, https://doi.org/10.1007/978-3-031-95642-3_7

where the *complex relative permittivity* has been introduced to take into account the effect of conductivity:

$$\epsilon_r = \epsilon_r' + i\frac{\gamma}{\epsilon_0\omega_0} \tag{7.1.3}$$

Following the same mathematical manipulations as previously we obtain the usual Helmholtz equation for the electric (and magnetic) field inside water:

$$\nabla^2\mathbf{E} + \epsilon_r K_0^2\mathbf{E} = 0 \tag{7.1.4}$$

Again, the solutions of this equation are the transverse plane waves:

$$\mathbf{E} = \mathrm{e}^{\mathrm{i}\mathbf{K}\cdot\mathbf{R}}\mathbf{E}_0,\ \mathbf{H} = \mathrm{e}^{\mathrm{i}\mathbf{K}\cdot\mathbf{R}}\mathbf{H}_0, \tag{7.1.5}$$

with $\mathbf{K}\cdot\mathbf{E} = 0$ and $\mathbf{K}\cdot\mathbf{H} = 0$. However, $\mathbf{K}$ is now a complex vector such that:

$$\mathbf{K}\cdot\mathbf{K} = \epsilon_r K_0^2 \tag{7.1.6}$$

and can thus be written as $\mathbf{K} = (K' + \mathrm{i}K'')\mathbf{u}$ for some unit vector $\mathbf{u}$ and some real numbers K' and K''. Since $K'' \neq 0$, this means that the plane wave (7.1.5) is exponentially damped or amplified when propagating in direction $\mathbf{u}$; such kind of solution can only exist in a medium that is bounded on one side as it is the case for the ocean volume which is limited on the top. To have physical solutions propagating in the direction of the wave vector when restoring the $e^{-\mathrm{i}\omega_0 t}$ time dependence, one must ensure that the complex wavenumber has an non-negative imaginary part ($K'' > 0$), which specifies it from (7.1.6) with no ambiguity. With this convention the field is exponentially damped in the medium along the direction of propagation:

$$E \sim e^{-K''\mathbf{u}\cdot\mathbf{R}} \tag{7.1.7}$$

For a good conductor the imaginary part of the complex permittivity is much larger than its real part ($\gamma >> \epsilon_r'\epsilon_0\omega_0$), so that:

$$K' = K'' = \sqrt{\frac{\gamma}{2\epsilon_0\omega_0}}K_0 \tag{7.1.8}$$

It follows that a propagation over a distance l results in an attenuation factor $\mathrm{e}^{-l/d}$ for the field amplitude E, with

$$d = \frac{1}{\sqrt{\pi\mu_0\gamma f_0}}, \tag{7.1.9}$$

where as usual $f_0 = \omega_0/(2\pi)$ is the EM frequency. This defines a *penetration depth* or *skin depth* d in water which characterizes the depth the EM field can "reach" when propagating down to the surface (see Fig. 7.1).

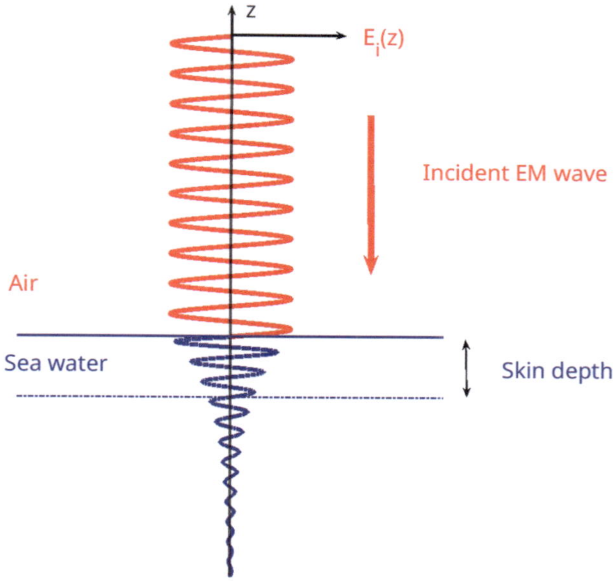

Fig. 7.1 The EM field incident from the air is attenuated exponentially in a lossy medium (seawater) over a distance called the skin depth

The real part of the complex permittivity of seawater is given by the relative permittivity of freshwater, which takes a constant value for radio waves $\epsilon_r \simeq 80$. Seawater conductivity is a function of temperature T and salinity S. Its reference value ($T = 15\,°\text{C}$, $S \simeq 35$ PSU) is

$$\gamma \simeq 4.3\ \text{S/m} \tag{7.1.10}$$

leading to the following expression for the complex permittivity of seawater as a function of the radar frequency:

$$\epsilon_r \simeq 80 + \frac{7.73\ 10^4}{f_0^{MHz}}\text{i}, \tag{7.1.11}$$

where $f_0^{MHz} = 10^6 f_0$ is the EM frequency given in MHz. For the penetration depth we obtain:

$$d \simeq \frac{0.24}{\sqrt{f_0^{MHz}}} \tag{7.1.12}$$

whose values range from 4 to 14 cm in the HF band ($f_0 = 3{-}30$ MHz) (Fig. 7.2).

A key concept in describing the propagation of EM waves over the sea surface is the *intrinsic impedance* of seawater, $Z = \sqrt{\mu/\epsilon}$, which the ratio of the electric to the magnetic field amplitude inside the medium. It is most commonly measured

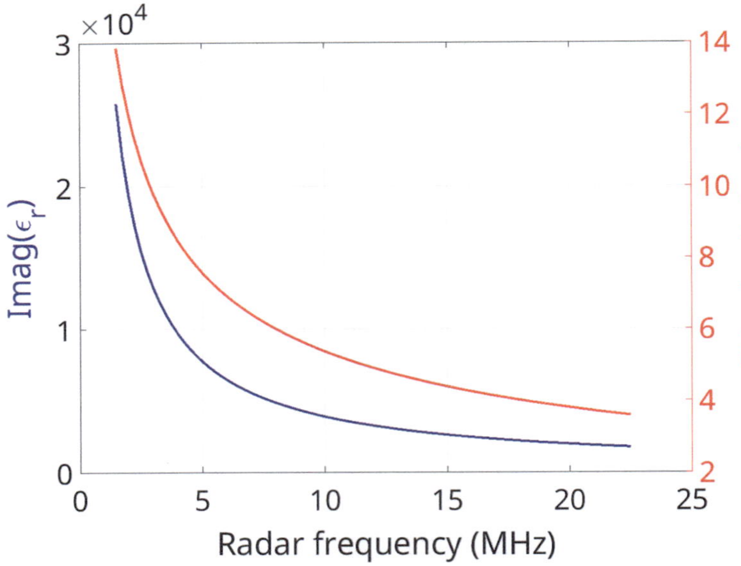

Fig. 7.2 Imaginary part of the complex relative permittivity and skin depth (in cm) in seawater as a function of the radar frequency (in MHz)

relative to the vacuum impedance by the dimensionless quantity $Z_r = Z/Z_0$. For a non-magnetic medium such as seawater, it is given by:

$$Z_r = \frac{1}{\sqrt{\epsilon_r}}, \tag{7.1.13}$$

which is the inverse square root of the complex relative permittivity and is equivalent to the inverse of the optical index ($Z = 1/n$). The impedance controls the boundary condition of the EM field at the air-sea interface (see the Appendix C) and is therefore called the *surface impedance*. It can be obtained from the expression (7.1.3) of the relative complex permittivity. It is on the order of:

$$Z_r \simeq 0.01 - 0.01\text{i} \tag{7.1.14}$$

in the middle of the HF frequency band and has a slight dependence on the radar frequency (Fig. 7.3).

7.2 The Half-Space Problem

The reflection of EM wave from a flat sea surface is mathematically described by the so-called half-space problem. In this idealized situation, two semi-infinite homogeneous media are separated by a flat horizontal interface, say $z = 0$. The upper medium is the air, which we assimilate to vacuum from an EM point of view, and the

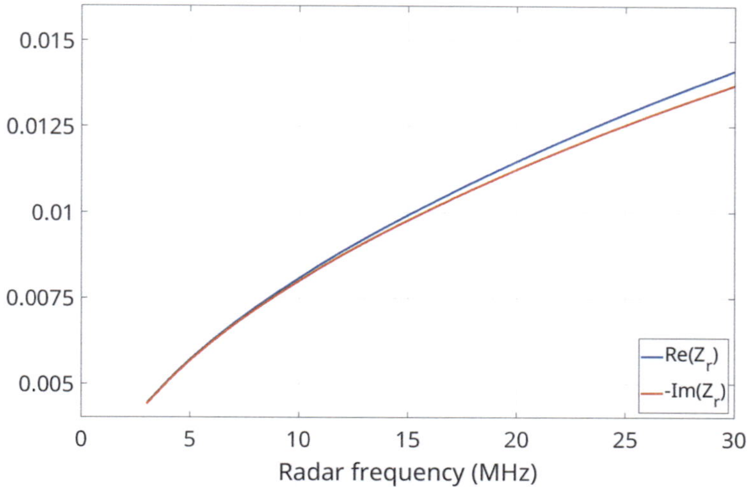

Fig. 7.3 Real and imaginary part of the relative impedance Z_r of the sea surface in the HF regime

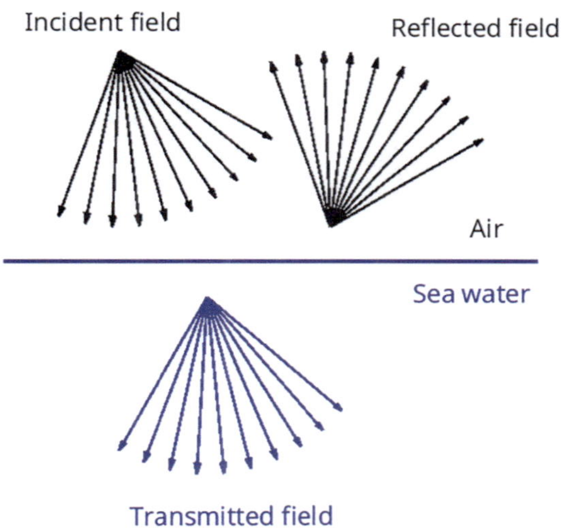

Fig. 7.4 The total electric field produced by a radiating source in free space is modified by the presence of a lower medium, delimited by a flat interface. The extra field produced in the upper medium is the reflected field and the field inside the lower medium is the transmitted field

lower medium is a lossy medium with complex permittivity representing the sea. A radiating source in the air is creating an EM field that impinges on the lower medium, called the *incident field*. As is obvious to everyone, the incident field is perturbed by the presence of the interface. In the upper medium there is an additional term, referred to as the *reflected field*, which propagates upwards and interferes with the incident field to form the *total field*. The field in the lower medium is also different from the incident field in free space and is called the *transmitted field* (Fig. 7.4). The half-space problem consists in finding the reflected and transmitted fields for a given source of radiation. It is solved using the boundary conditions at the interface.

To simplify the problem, suppose that the incident electromagnetic field is independent on one space coordinate, say y, so that its direction of propagation belongs to the vertical plane (x, z):

$$\mathbf{E}_i = \mathbf{E}_i(x, z), \ \mathbf{H}_i = \mathbf{H}_i(x, z). \tag{7.2.15}$$

This the case, for example, for an incident plane wave with a wave vector in a vertical plane. Since the complex permittivity does not depend on the y coordinate, the problem is invariant to translation along this coordinate. Therefore, the total EM field in space is independent of the variable y, so that

$$\mathbf{E} = \mathbf{E}(x, z), \ \mathbf{H} = \mathbf{H}(x, z). \tag{7.2.16}$$

It follows that in the harmonic Maxwell equations (7.1.2), one can replace any derivative with respect to the variable y by zero. Denoting by the corresponding superscripts x, y or z the components of the fields along each axis, we can write them in expanded form:

$$\begin{aligned} \partial_x E_x + \partial_z E_z &= 0 \\ \partial_x H_x + \partial_z H_z &= 0 \end{aligned} \tag{7.2.17}$$

for the divergence equations and:

$$\begin{aligned} -\partial_z E_y &= \mathrm{i}\omega_0 \mu_0 H_x \\ \partial_z E_x - \partial_x E_z &= \mathrm{i}\omega_0 \mu_0 H_y \\ \partial_x E_y &= \mathrm{i}\omega_0 \mu_0 H_z \end{aligned} \tag{7.2.18}$$

as well as:

$$\begin{aligned} -\partial_z H_y &= \mathrm{i}\omega_0 \epsilon(z) E_x \\ \partial_z H_x - \partial_x H_z &= -\mathrm{i}\omega_0 \epsilon(z) E_y \\ \partial_x H_y &= -\mathrm{i}\omega_0 \epsilon(z) E_z \end{aligned} \tag{7.2.19}$$

for the curl equations, where the complex permittivity $\epsilon(z)$ takes a different constant value in the upper $(\epsilon(z) = \epsilon_0)$ and the lower $(\epsilon(z) = \epsilon_0 \epsilon_r)$ medium. As can be seen from Eqs. (7.2.17) and (7.2.18), the three components (E_x, E_z, H_y) and (H_x, H_z, E_y) can be separated in two uncoupled subsystems of four equations. Therefore, an incident EM field whose non-zero components belong to one of these subsystems cannot excite the components of the other subsystem. In other words, if the incident field is of the form (E_{xi}, E_{zi}, H_{yi}) (or (H_{xi}, H_{zi}, E_{yi})), this remains true for the total field below and above the flat interface. This naturally leads to the definition of two canonical polarization states as follows. When the incident electric field $\mathbf{E}_i$ has a unique non-zero component along one horizontal axis (here, the $\mathbf{y}$ direction) and propagates in the vertical plane (here, (x, z)), this remains true for the total electric field, which is said to be in *horizontal polarization* ; if it is the incident magnetic field $\mathbf{H}_i$ which has a unique non-zero component along the horizontal axis and propagates in the vertical plane, the EM field is said to be in *vertical polarization*.

7.3 The Canonical Polarization Basis

As stated in Chap. 6, the incident field can be written as a sum of linearly polarized plane waves of the form:

$$\mathbf{E}_i(\mathbf{R}) = e^{\mathrm{i}\mathbf{K_0}\cdot\mathbf{R}}\mathbf{p} = e^{\mathrm{i}(\mathbf{k_0}\cdot\mathbf{r} - q_0 z)}\mathbf{p}, \tag{7.3.20}$$

where $\mathbf{p}$ is some transverse constant vector ($\mathbf{K_0}\cdot\mathbf{p} = 0$). The incident wave vector $\mathbf{K_0} = \mathbf{k_0} - q_0\mathbf{z}$ is decomposed into an horizontal ($\mathbf{k_0}$) and vertical (q_0) component, with $k_0^2 + q_0^2 = K_0^2$ and $q_0 > 0$. In the case of the horizontal polarization, the electric field is perpendicular to both the vertical vector $\mathbf{z}$ and the incident wave vector $\mathbf{K_0}$, so it is directed along the unit vector:

$$\mathbf{p}_H^-(\mathbf{k_0}) = \mathbf{z}\times\widehat{\mathbf{k_0}}, \tag{7.3.21}$$

where the minus supscript indicates that the plane wave propagates downwards. In vertical polarization, it is the magnetic field that is horizontal; the electric field is perpendicular both to this horizontal vector and to the wave vector in the incidence plane, so that it is directed along the unit vector:

$$\mathbf{p}_V^-(\mathbf{k_0}) = \widehat{\mathbf{K_0}}\times(\mathbf{z}\times\widehat{\mathbf{k_0}}) = K_0^{-1}(k_0\mathbf{z} + q_0\widehat{\mathbf{k_0}}) \tag{7.3.22}$$

Likewise, the reflected field can be decomposed into a sum of upwards propagating plane waves:

$$\mathbf{E}_r(\mathbf{R}) = e^{\mathrm{i}\mathbf{K}\cdot\mathbf{R}}\mathbf{p} = e^{\mathrm{i}(\mathbf{k}\cdot\mathbf{r} + qz)}\mathbf{p}, \tag{7.3.23}$$

where $\mathbf{K} = \mathbf{k} + q\mathbf{z}$ and $k^2 + q^2 = K_0^2$. It is directed along the unit vectors:

$$\mathbf{p}_H^+(\mathbf{k}) = \mathbf{z}\times\widehat{\mathbf{k}} \tag{7.3.24}$$

and

$$\mathbf{p}_V^+(\mathbf{k}) = \widehat{\mathbf{K}}\times(\mathbf{z}\times\widehat{\mathbf{k}}) = K_0^{-1}(k\mathbf{z} - q\widehat{\mathbf{k}}), \tag{7.3.25}$$

in horizontal and vertical polarization, respectively, where the plus subscripts indicate that waves are propagating upwards. The 2 vectors ($\mathbf{p}_H^\pm(\mathbf{k}), \mathbf{K}\times\mathbf{p}_H^\pm(\mathbf{k})$) form the canonical basis for upwards/downwards propagating EM plane waves in horizontal polarization; similarly, ($\mathbf{p}_V^\pm(\mathbf{k}), \mathbf{K}\times\mathbf{p}_V^\pm(\mathbf{k})$) is the canonical basis for the vertical polarization. When solving the half-space problem each case of polarization can be treated separately and in fact it is sufficient to find the unique non-zero component of the electric or magnetic field depending on whether the polarization is horizontal or vertical (Fig. 7.5).

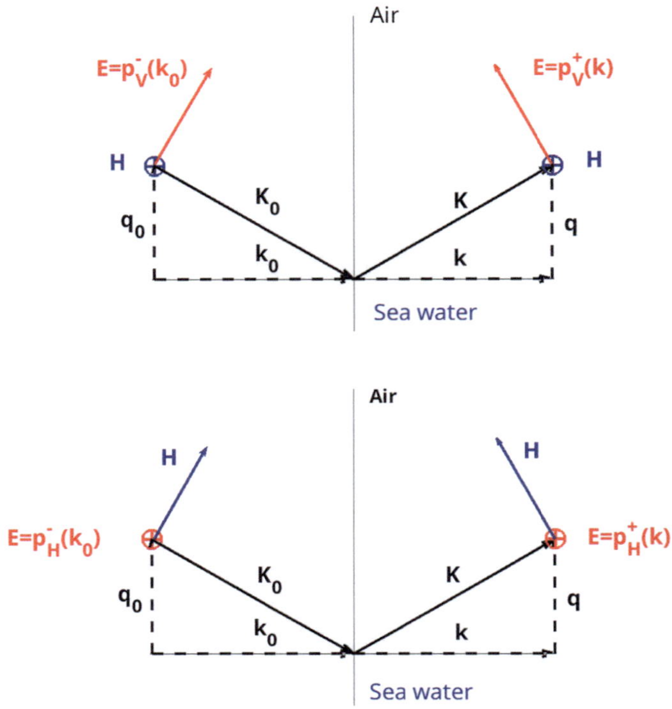

Fig. 7.5 The canonical polarization basis. In the horizontal polarization case (top), the electric field has a unique nonzero component along an horizontal axis (by convention, the **y** axis, shown with blue crosses); the wave vector and the magnetic field both lie in a vertical plane (by convention, the (**x**, **z**) plane). In the vertical polarization case (bottom), the same holds by swapping the role of the electric and magnetic field

7.4 Reflection from a Flat Interface

It is well known that EM plane waves are reflected and transmitted across a plane interface according to the Snell-Descartes laws, but their reflection and transmission coefficients depend on the incident polarization state. This is a consequence of the different boundary conditions depending on the polarization and is described in more detail in the Appendix D. Consider a plane boundary $z = 0$ separating the vacuum ($z > 0$) from the sea surface ($z < 0$), illuminated from above by an incident linearly polarized plane wave with wave vector $\mathbf{K_0} = \mathbf{k_0} - q_0\mathbf{z}$. The total electric field over the surface combines the incident and the reflected field. For an incident electric field with unit amplitude, it can be written as:

$$\begin{aligned}\mathbf{E}(\mathbf{R}) &= \mathbf{E}(\mathbf{r}, z) = e^{\mathrm{i}\mathbf{k_0}\cdot\mathbf{r}}\left[e^{-\mathrm{i}q_0 z}\mathbf{p}_H^-(\mathbf{k}_0) + \mathcal{R}_H e^{\mathrm{i}q_0 z}\mathbf{p}_H^+(\mathbf{k}_0)\right] \\ \mathbf{H}(\mathbf{R}) &= \mathbf{H}(\mathbf{r}, z) = Z_0^{-1}\left[e^{-\mathrm{i}q_0 z}\mathbf{p}_V^-(\mathbf{k}_0) + \mathcal{R}_H e^{\mathrm{i}q_0 z}\mathbf{p}_V^+(\mathbf{k}_0)\right]\end{aligned} \tag{7.4.26}$$

in horizontal polarization and

$$\begin{aligned}\mathbf{E}(\mathbf{R}) &= \mathbf{E}(\mathbf{r}, z) = e^{\mathrm{i}\mathbf{k}_0\cdot\mathbf{r}}\left[e^{-\mathrm{i}q_0 z}\mathbf{p}_V^-(\mathbf{k}_0) + \mathcal{R}_V e^{\mathrm{i}q_0 z}\mathbf{p}_V^+(\mathbf{k}_0)\right]\\ \mathbf{H}(\mathbf{R}) &= \mathbf{H}(\mathbf{r}, z) = -Z_0^{-1}e^{\mathrm{i}\mathbf{k}_0\cdot\mathbf{r}}\left[e^{-\mathrm{i}q_0 z}\mathbf{p}_H^-(\mathbf{k}_0) + \mathcal{R}_V e^{\mathrm{i}q_0 z}\mathbf{p}_H^+(\mathbf{k}_0)\right]\end{aligned} \tag{7.4.27}$$

in vertical polarization. The reflection coefficients $\mathcal{R}_H$ and $\mathcal{R}_V$ are called the *Fresnel reflection coefficients* (see Appendix D). They are given by:

$$\mathcal{R}_H(k_0) = \frac{q_0 - q_0'}{q_0 + q_0'}, \quad \mathcal{R}_V(k_0) = \frac{\epsilon_r q_0 - q_0'}{\epsilon_r q_0 + q_0'}, \tag{7.4.28}$$

where

$$q' = \sqrt{\epsilon_r K_0^2 - k_0^2} \tag{7.4.29}$$

with $\mathrm{Im}[q'] \geq 0$. As we shall see, only the vertical polarization is of interest in the context of HFR. For real values of ϵ_r, the reflection coefficient undergoes a change of sign whenever $\epsilon_r q_0 - q_0' = 0$. This corresponds to an angle of incidence:

$$\theta_B = \mathrm{atan}(\sqrt{\epsilon_r}), \tag{7.4.30}$$

referred to as the *Brewster angle*. For complex values of ϵ_r, the reflection coefficient is never exactly zero. However, its absolute value passes through a minimum, which defines a pseudo-Brewster angle. Figure 7.6 shows the variation of the Fresnel coefficients with the angle of incidence for pure water ($\epsilon_r = 80$) and seawater ($\epsilon_r = 80 + 5000\mathrm{i}$). Due to its high concentration of dissolved ions, seawater has a large conductivity resulting in a large imaginary part for its complex permittivity. Pure water, on the other hand, is a poor conductor and has zero conductivity. The Brewster angle for radio waves over fresh water is close to grazing ($\theta_B = 83.6°$) and the pseudo-Brewster angle for seawater is at about 1° away from grazing ($\theta_B = 88.9°$).

For strongly conductive media we may approximate $q_0' \simeq \sqrt{\epsilon_r}K_0$, which allows to express the Fresnel coefficients in very simple form in terms of the surface impedance Z_r:

$$\mathcal{R}_H(k_0) = \frac{Z_r\cos\theta - 1}{Z_r\cos\theta + 1}, \quad \mathcal{R}_V(k_0) = \frac{\cos\theta - Z_r}{\cos\theta + Z_r}. \tag{7.4.31}$$

Thee Brewster angle under this approximation is seen to be:

$$\theta_B = \mathrm{acos}(\mathrm{Re}[Z_r]) \tag{7.4.32}$$

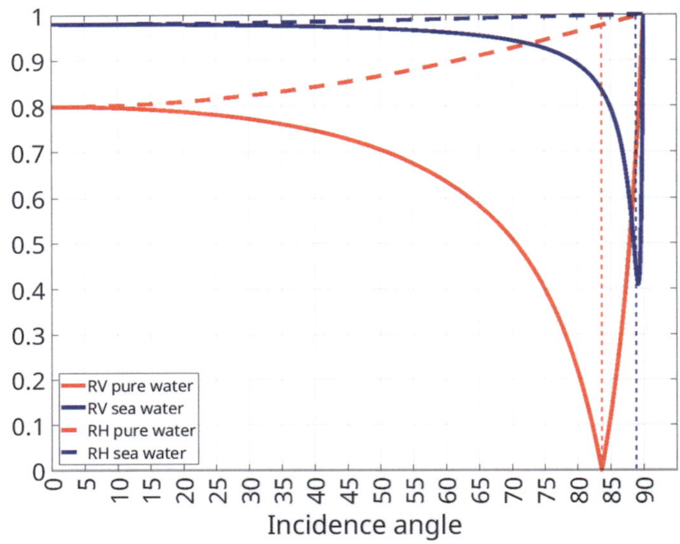

Fig. 7.6 Fresnel reflection coefficients (shown in absolute value) for the vertical (solid lines) and horizontal (dashed lines) polarization in the HF frequencies for pure (red lines) and seawater (blue lines). The vertical dashed line shows the location of the Brewster ($\theta_B = 83.6°$) and pseudo-Brewster angle ($\theta_B = 88.9°$)

7.5 Why Is Vertical Polarization Better Than Horizontal?

HFR use only vertical polarization for both transmitting and receiving arrays. Why is this? The first part of the answer was given in Chap. 6 where we saw that the radiation of a vertical dipole antenna is maximal in a direction perpendicular to the axis of the antenna. A vertical antenna therefore radiates dominantly towards the sea while its emission towards the sky is reduced. On the contrary, a horizontal antenna has its main radiation lobe in a vertical plane, which is not suitable for this purpose. The second part of the answer lies in the Fresnel reflection coefficient $\mathcal{R}$ (which can be either $\mathcal{R}_V$ or $\mathcal{R}_H$). A simple and fair picture of what happens in the far field at non-grazing angles can be obtained by identifying the local electric field with a plane wave of amplitude $\sim 1/R$ and the angle of incidence $\cos\theta = H/R$, where H is the height of the transmitting antenna above the sea level. The total electric field at the surface is therefore:

$$E \sim \frac{1+\mathcal{R}}{R}. \tag{7.5.33}$$

It turns out that the limit $\mathcal{R} \to -1$ is reached very rapidly in H polarization as the distance R increases:

$$1 + \mathcal{R}_H \simeq 2Z_r \cos\theta \tag{7.5.34}$$

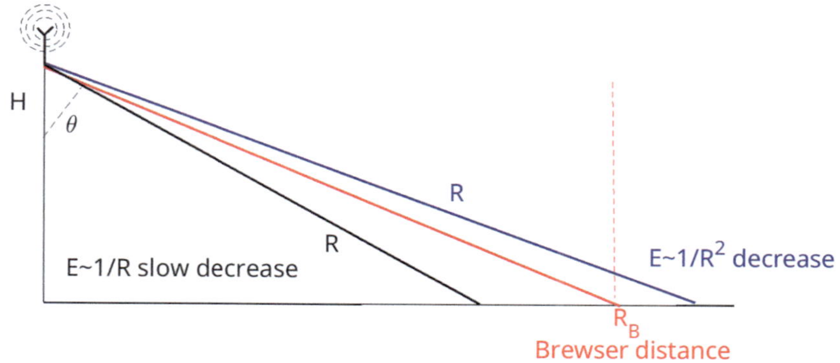

Fig. 7.7 In a simple plane wave approximation, the total electric field at the sea surface produced by a vertical dipole antenna at an altitude H has a slow $1/R$ decrease until it reaches the Brewster angle, past which it decreases like $1/R^2$

Fig. 7.8 Transmit antenna array at the site of Fort Peyras, Toulon, France. A vertical antenna rod produces a vertically polarized electric field. When the same antenna lies on the ground (here accidentally after a violent wind event), the transmitted electric field turns to horizontal (but this is not very useful!)

so that $E \sim 1/R^2$ even at short distances. On the contrary, the term:

$$1 + \mathcal{R}_V \simeq \frac{2\cos\theta}{\cos\theta + Z_r} \tag{7.5.35}$$

is very slow to decrease in vertical polarization and, in fact, can only become small after the pseudo Brewster angle ($\cos\theta << |Z_r|$) where it follows a $1/R$ decay. We therefore have 2 distinct regions for the field attenuation in vertical polarization (Fig. 7.7), namely $E \sim 1/R$ for $R < R_B$ and $E \sim 1/R^2$ for $R > R_B$, where $R_B = H/\cos\theta_B$ is the distance at which the pseudo-Brewster angle is reached (a distance which is quite large, given the extreme value of the Brewster angle).

A vertically polarized electric field can be produced very simply with a vertical antenna rod such as the $\lambda_0/4$ dipole antenna that are used for phased array radars An horizontally polarized electric field is obtained by laying the same antenna on the ground (Fig. 7.8).

Why Do HFRs See "Over the Horizon"? 8

The operating range of HFR goes well beyond the line of horizon, a remarkable property that has been known and used for decades in defense applications and justifies the common denomination of *Over-The-Horizon* (OTH) radars. The physical origin of the long distance propagation is the presence of an electromagnetic *ground wave*, also called *surface wave* (not to be confused with the gravity waves !). This is a component of the EM field which remains localized close to the interface and undergoes little attenuation as it propagates away from the source, following the Earth curvature. It adds up with the ordinary EM waves, called the *space waves*, which propagate in the atmosphere in direct path from the transmitter or after reflection by the sea surface. Surface waves can propagate two ways after scattering from an object or from the gravity waves (Fig. 8.1), which makes them very useful for many remote sensing applications. There has been a longstanding controversy in the literature as to the nature of surface waves. We refer to [48] for an excellent historical survey and a clarification of the concept. Today, the mechanism of ground wave propagation is well understood. In this chapter, we review briefly the main steps that lead to its correct formulation. A more detailed analysis can be found in classical papers [149,200–202,272] and textbooks [145,165,190,271]. This allows us to estimate the basic transmission loss, which quantifies the attenuation of the electric field as it propagates away from the transmitting source.

8.1 Ground Waves

The first model used to explain ground wave propagation is that of a "trapped wave" or "guided wave" by analogy with the electric current propagating in a wire. In vertical polarization and for highly conducting media, there exists an exact and simple solution to Maxwell's equations over a flat conducting interface where the electric

C.-A. Guérin, *High-Frequency Radars for the Oceanographer*, Geophysics and Environmental Physics, https://doi.org/10.1007/978-3-031-95642-3_8

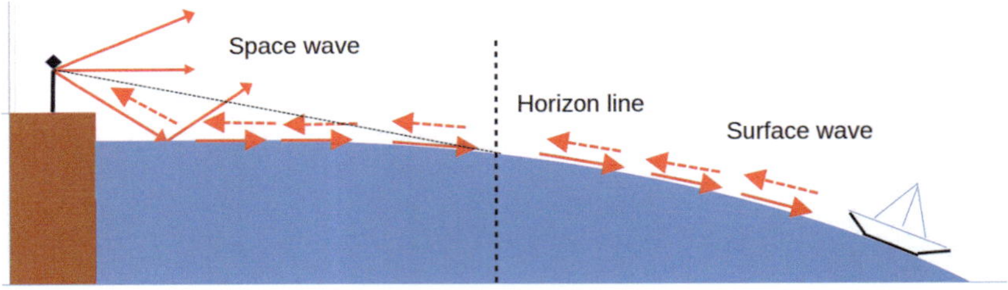

Fig. 8.1 Transmitted radio waves in the HF band can propagate not only in the atmosphere (space waves) but also along the sea surface beyond the horizon, following the Earth curvature (surface waves)

field is confined vertically close to the interface (in the sense that it is exponentially attenuated on both side) but can propagate with little attenuation along the horizontal direction. This solution was first proposed by Zenneck [297], two years before Sommerfeld's solution, and can be found in a modern presentation in the Appendix E. Zenneck's model was commonly accepted at first because it seemed to explain very well propagation over great distances. However, there was no indication as how one could excite such a propagation mode with an antenna. Two years later, Sommerfeld in a famous paper [248] solved the electromagnetic field generated by a vertical dipole on a flat conducting surface, a canonical problem which we will in the following refer to as the *Sommerfeld problem* . For this, he had to evaluate a Weyl integral by using an integration in the complex plane involving the Fresnel reflection coefficient (D.0.16), which due to the complex permittivity has a pole near the real axis. One component of his solution, corresponding to the residue at this pole, was identified with a cylindrical form of the Zenneck surface wave, thereby given more credence to the latter solution. However, the physical reality of the Zenneck wave was questioned later by Weyl [284], Wise [285], and Norton [201] (among others) who showed with more accurate calculations that a Zenneck like wave did not exist in the solution nor complied to the experimental observations. This led some authors to think that Sommerfeld initial solution was flawed with a sign error [199]. However, a very cautious investigation led by Collin almost one century later [48] put an end to the controversy and showed that Sommerfeld initial solution is actually correct; the apparent Zenneck term is implicitly canceled out by another contribution of the solution that can be made explicit after a power series expansion. Today, the accepted modern solution to the Sommerfeld problem is the more complete description by Norton, which even takes into account the curvature of a spherical earth. In the next section we present a simplified derivation of this solution for a smooth interface on a flat earth. No one has ever observed a Zenneck wave for radio wave propagating on the sea surface, because there is no practical way to "excite" it with the incident field. However, because of its simplicity, the Zenneck model is sometimes used, in particular for the search of an effective surface impedance [13], an issue that we will discuss in the following chapter.

8.2 Sommerfeld's Solution

The failure with the Zenneck solution is that it does not consider the way the electric field is produced by the antenna. A more realistic model consists in calculating the EM field radiated by a dipole antenna above a conducting interface. Consider a vertical dipole current source $\mathbf{j}_0 = j_0\mathbf{z}$ located at $(\mathbf{0}, H)$ above a flat interface $z = 0$ separating the air $(z > 0)$ from a strongly conducting medium with impedance $Z_r = 1/\sqrt{\epsilon_r} = 1/n$ and let us calculate the total electromagnetic field produced at any observation point $\mathbf{R} = (\mathbf{r}, z)$ above the interface (Fig. 8.2). For this, it is advantageous to use the potential vector, by noting that the total electric field produced by an incident vertically polarized plane wave is associated to an incident and reflected potential vector along the $\mathbf{z}$ direction, with the same Fresnel reflection coefficient. Precisely, it comes from (7.4.26) and (6.2.34):

$$\begin{aligned} \mathbf{E}(\mathbf{r}, z) &= e^{\mathrm{i}\mathbf{k}_0\cdot\mathbf{r}}\left[e^{-\mathrm{i}q_0 z}\mathbf{p}_V^-(\mathbf{k}_0) + \mathcal{R}_V e^{\mathrm{i}q_0 z}\mathbf{p}_V^+(\mathbf{k}_0)\right] \\ \Leftrightarrow \mathbf{A}(\mathbf{r}, z) &= -\frac{\mathrm{i}\sqrt{\epsilon_0\mu_0}}{k_0}e^{\mathrm{i}\mathbf{k}_0\cdot\mathbf{r}}\left[e^{-\mathrm{i}q_0 z} + \mathcal{R}_V e^{\mathrm{i}q_0 z}\right]\mathbf{z}, \end{aligned} \tag{8.2.1}$$

Therefore, using the expression (6.2.26) of the incident potential vector generated by a dipole source and the Weyl representation (B.0.7) of the free space Green's function, we may express the potential vector generated by the dipole antenna as the sum of 2 Weyl integrals:

$$\mathbf{A}(\mathbf{r}, z) = \mu_0\frac{\mathrm{i}}{8\pi^2}\left(\int\frac{d\mathbf{k}}{q}\,\mathrm{e}^{i(\mathbf{k}\cdot\mathbf{r}+q|z-H|)} + \int\frac{d\mathbf{k}}{q}\,\mathcal{R}_V(k)\mathrm{e}^{i(\mathbf{k}\cdot\mathbf{r}+q|z+H|)}\right)\mathbf{j}_0, \tag{8.2.2}$$

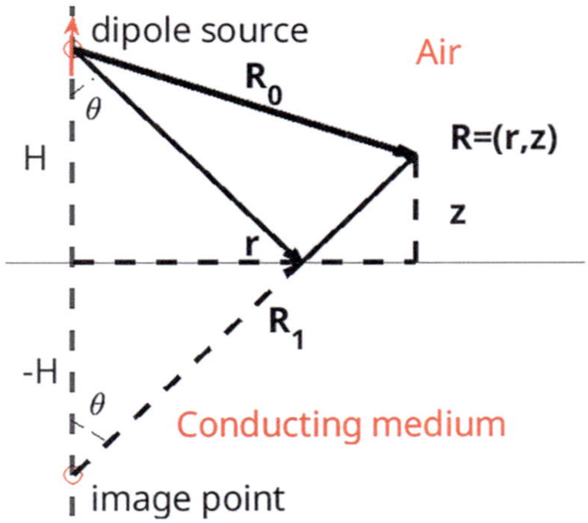

Fig. 8.2 The total field at an observation point $\mathbf{R}$ above the surface is the sum of the incident field in direct path (distance R_0) and the reflected field that is in direct path from an image source below the surface (distance R_1)

which run over $\mathbb{R}^2$ and involve the Fresnel reflection coefficient $\mathcal{R}_V(k)$ for propagating ($k \leq K_0$) as well as evanescent ($k > K_0$) wave vectors. In the following we will use the following convenient notation to express a Weyl integral with an arbitrary weight $f(\mathbf{k})$:

$$\mathcal{W}[f](r,z) = -\frac{\mathrm{i}}{8\pi^2}\int \frac{d\mathbf{k}}{q} f(\mathbf{k}) \mathrm{e}^{i(\mathbf{k}\cdot\mathbf{r}+qz)}, \tag{8.2.3}$$

where it is assumed that $z \geq 0$. We may thus rewrite in a more compact form:

$$\mathbf{A}(\mathbf{r},z) = -\mu_0\left[\mathcal{W}[1](\mathbf{r},|z-H|) + \mathcal{W}[\mathcal{R}_V](\mathbf{r},z+H)\right]\mathbf{j}_0 \tag{8.2.4}$$

Let us denote $\mathbf{R}_0 = (\mathbf{r}, z - H)$ the position vector of the observation point with respect to the source of current and let us introduce the image of the source point, which is symmetrically located underneath the horizontal plane. It corresponds to a position vector $\mathbf{R}_1 = (\mathbf{r}, z + H)$. With these notations we have simply:

$$\begin{aligned} \mathcal{W}[1](\mathbf{r},|z-H|) &= G(R_0) \\ \mathcal{W}[1](\mathbf{r},z+H) &= G(R_1) \end{aligned} \tag{8.2.5}$$

and thus we may decompose:

$$\mathbf{A}(\mathbf{r},z) = -\mu_0\left[G(R_0) + G(R_1) + \mathcal{W}[\mathcal{R}_V - 1](\mathbf{r},z+H)\right]\mathbf{j}_0 \tag{8.2.6}$$

In the limit of a perfectly conducting medium, $\mathcal{R}_V \to 1$, so that the third term in (8.2.6) is the extra contribution of finite conductivity. The evaluation of the Weyl integral is a delicate matter. It can be obtained through an integration in the complex plane which, however, requires much caution [48]. It can also be obtained through a stationary phase argument, which we present in the Appendix F in a simplified way following Norton's derivation [200,202]. In the end, the potential vector in vertical polarization is obtained as:

$$\begin{aligned} \mathbf{A}(\mathbf{r},z) = &-\mu_0 G(R_0)\mathbf{j}_0 - \mu_0\frac{\cos\theta - Z_r}{\cos\theta + Z_r}G(R_1)\mathbf{j}_0 \\ &-\mu_0\frac{2Z_r}{Z_r+\cos\theta}\mathrm{SO}\left(\frac{\mathrm{i}K_0R}{2}\frac{(Z_r+\cos\theta)^2}{1+\cos^2\theta}\right)G(R_1)\mathbf{j}_0, \end{aligned} \tag{8.2.7}$$

where:

$$\cos\theta = \frac{z+H}{R_1} \tag{8.2.8}$$

is the observation angle seen from the image source point and SO is the so-called *Sommerfeld attenuation function* :

$$\mathrm{SO}(u) = 1 + \mathrm{i}\sqrt{\pi u}\,\mathrm{e}^{-u}\mathrm{erfc}(-\mathrm{i}\sqrt{u}), \tag{8.2.9}$$

where erfc is the complementary error function. There is no simple analytical expression for this function, which can be obtained as a series expansion. It can be approximated by:

$$\mathrm{SO}(u) \simeq 1 - \mathrm{i}\sqrt{\pi}\mathrm{e}^{-u} \tag{8.2.10}$$

for small $|u|$ and by

$$\mathrm{SO}(u) \simeq -\frac{1}{2u} \tag{8.2.11}$$

for large values of $|u|$ ($|u| > 20$).

8.3 Space Wave and Surface Wave

In the expression (8.2.7) of the vector potential we can identify the Fresnel reflection coefficient (7.4.31) at the incidence angle θ for a strongly conductive medium ($|Z_r| << 1$):

$$\frac{\cos\theta - Z_r}{\cos\theta + Z_r} = \mathcal{R}_V(\theta), \quad \frac{2Z_r}{Z_r + \cos\theta} = 1 - \mathcal{R}_V(\theta), \tag{8.3.12}$$

and we may thus rewrite:

$$\mathbf{A}(\mathbf{r}, z) = \mathbf{A_{sp}}(\mathbf{r}, z) + \mathbf{A_{su}}(\mathbf{r}, z), \tag{8.3.13}$$

with

$$\mathbf{A_{sp}}(\mathbf{r}, z) = -\mu_0[G(R_0) + \mathcal{R}_V G(R_1)] \tag{8.3.14}$$

and

$$\mathbf{A_{su}}(\mathbf{r}, z) = -\mu_0\,[1 - \mathcal{R}_V)\mathrm{SO}(u)G(R_1)] \tag{8.3.15}$$

At large horizontal distance ($K_0 r >> 1$, $r/H >> 1$ and $r/z >> 1$) beyond the Brewster angle ($\theta \to \pi/2$), we have $\mathcal{R}_V \to -1$ and $R_0 - R_1 \to 0$, so that the combination of the first two terms in (8.3.14) cancels out. The component $\mathbf{A_{sp}}$ is therefore referred to as the *space wave* , because it eventually vanishes on the surface. The other component $\mathbf{A_{su}}$, however, has a finite limit at grazing angle since $1 - \mathcal{R}_V \to 2$:

$$\mathbf{A_{su}}(\mathbf{r}, z) \to -2\mu_0 F(R)G(R_1)\mathbf{j}_0, \tag{8.3.16}$$

where:

$$F(R) = \mathrm{SO}\left(\mathrm{i}\frac{1}{2}K_0 R Z_r^2\right) \tag{8.3.17}$$

is called the *Norton attenuation factor* . The dimensionless parameter:

$$p = \mathrm{i}\frac{1}{2}K_0 R Z_r^2 = \mathrm{i}\frac{\pi R}{\lambda_0 \epsilon_r} \tag{8.3.18}$$

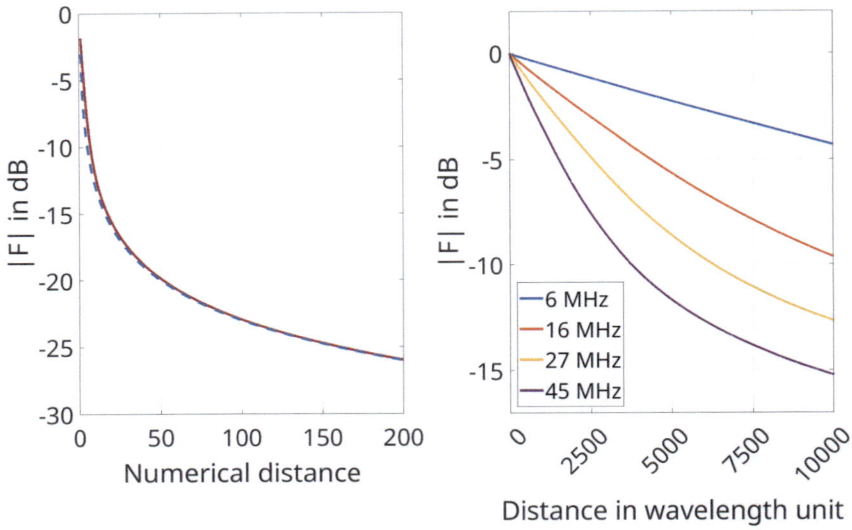

Fig. 8.3 Norton attenuation factor $|F|$ shown as a function of the numerical distance (left) and as a function of the distance in wavelength for a few radar frequencies (right). The dashed line is the asymptotic approximation $|F(p)| \simeq 1/2p$

that enters in the Sommerfeld function is defined as the *numerical distance* and the component $\mathbf{A_{su}}$ is referred to as the *surface wave* . Figure 8.3 shows the evolution of the magnitude $|F(p)|$ of the Norton attenuation factor as a function of the numerical distance and also as a function of the distance in wavelength unit. As seen, the $1/(2p)$ asymptotic regime (8.2.11) is reached after a few units of numerical distance. However, when expressed in wavelength units, it is the small p asymptotic regime (8.2.10) which holds for at least a few thousands units, depending on the radar frequency. The electric field can be obtained from the vector potential $\mathbf{A}$ by taking its double curl (6.2.34). After some calculation this yields, retaining the dominant negative powers of R:

$$\mathbf{E_{sp}}(\mathbf{r}, z) = -\mathrm{i}K_0 Z_0 (1 + \mathcal{R}_V) G(R_0) j_0 \widehat{\mathbf{R}} \times (\widehat{\mathbf{R}} \times \mathbf{z}) \tag{8.3.19}$$

for the space wave component and

$$\mathbf{E_{su}}(\mathbf{r}, z) = -\mathrm{i}K_0 Z_0 (1 - \mathcal{R}_V) F(R) G(R_0) j_0 (\mathbf{z} + Z_r \widehat{\mathbf{r}}) \tag{8.3.20}$$

for the surface wave component. Denoting E^0_{sp} and E^0_{su} the respective field strengths at a small reference distance $R = 1$, we have at long distance R (where $\mathcal{R}_V \to -1$):

$$\begin{aligned} E_{\mathrm{sp}} &= \frac{|1 + \mathcal{R}_V|}{R} E^0_{\mathrm{sp}} \\ E_{\mathrm{su}} &= 2\frac{F(R)}{R} E^0_{\mathrm{su}} \end{aligned} \tag{8.3.21}$$

The decrease of the space wave has been anticipated in Sect. 7.5 with a simple plane wave argument, with a transition from a $1/R$ to $1/R^2$ decay when passing the Brewster angle. The surface wave, on the other hand, decays like $F(R)/R$. However, as explained above, the Norton attenuation factor is a decreasing function of $|R|$ with $F \simeq 1$ for small $|p|$ and $|F| \simeq -1/2\,|p|$ for $|p| > 20$. Since $|Z_r^2| \sim 2.10^{-4} <<$ 1, the numerical distance is strongly reduced by the factor Z_r^2 and F is a slowly decreasing function of $K_0 R$. It thus takes thousands of radar wavelengths before the surface wave is significantly damped by the Norton attenuation factor, as shown in Fig. 8.3. Hence, the surface wave term evolves slowly from a $1/R$ decrease (small p) to a $1/R^2$ decrease (large p) but the $1/R^2$ is never reached in practice. In the end, the overall decay of the surface wave term $\mathbf{E}_{su}$ is $1/R$ on physically reasonable distances. As seen in Fig. 8.3, the Norton attenuation factor in the HF domain $3-30$ MHz decreases by less than 10 dB after 5000 radar wavelengths, which implies distance of a few hundreds of kilometers. If one neglects the variation of the Norton attenuation factor over the first kilometers ($F \simeq 1$), then one can express the ratio of the space wave to surface wave component with the simple formula:

$$\frac{E_{sp}}{E_{su}} \simeq \left|\frac{1+\mathcal{R}_V}{1-\mathcal{R}_V}\right|. \tag{8.3.22}$$

This means that the space wave is dominant until the pseudo Brewster angle is reached, after which it becomes negligible compared to the surface wave (see Eq. 8.3.22) (Fig. 8.4).

A particular feature of the surface wave is that it is not a purely transverse EM wave, as is the space wave. From the expression (8.3.20) we see that in the grazing limit $\theta \rightarrow \pi/2$, the electric field still has a longitudinal component which makes its polarization slightly tilted with respect to the transverse direction $\mathbf{z}$. The tilt angle ψ is

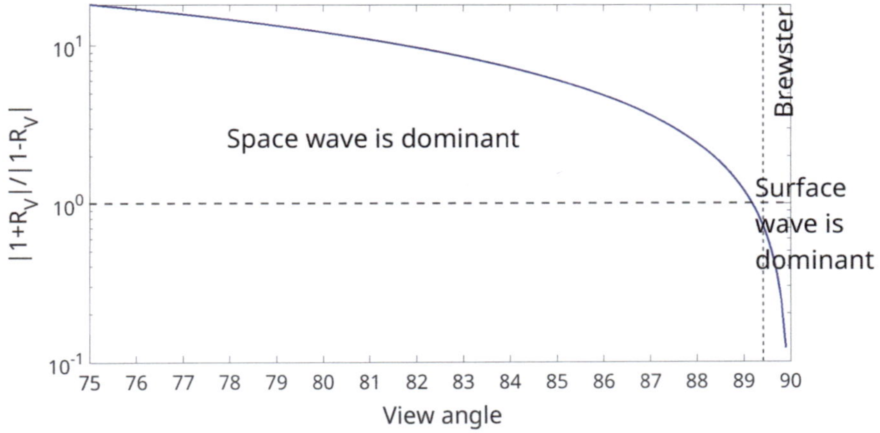

Fig. 8.4 Ratio (8.3.22) of the space wave to surface wave electric field component as a function of the view angle θ. The transition between the space wave and surface wave regime occurs around the pseudo Brewster angle

found to be the complementary angle to the pseuso Brewster angle ($\psi = \pi/2 - \theta_B$):

$$\tan \psi = \mathrm{Re}[Z_r] \tag{8.3.23}$$

Even though it is small, the tilt of the electric field from the vertical direction has important consequences, as it allows a small vertical component for the Poynting vector $\mathbf{P} = \frac{1}{2}\mathbf{E} \times \mathbf{H}^*$. This implies that a small fraction of electromagnetic power is transferred downwards to seawater and explains why the ground wave cannot propagate for ever. Note that the Norton and the Zenneck wave share the same polarization tilt. This is why the simple Zenneck model is sometimes employed as a "local" version of the Norton wave [13].

The Sommerfeld problem can also be solved for an horizontal dipole source by a simple adaptation of the previous technique. The same formula (8.2.7) is obtained for the vector potential with another numerical distance:

$$p' = \frac{\mathrm{i}K_0 R}{2Z_r^2} = \mathrm{i}\frac{\pi \epsilon_r R}{\lambda_0} \tag{8.3.24}$$

In contrast to the vertical polarization case, the Norton attenuation factor in horizontal polarization decreases rapidly with $K_0 R$ because the numerical distance is made much larger by the multiplication factor $1/Z_r^2$, so that the $1/R^2$ regime is reached at the very first wavelength.

8.4 Basic Transmission Loss

The attenuation of the EM field as it propagates over the sea surface is usually described in terms of the *basic transmission loss*, a concept that was popularized by Norton [204,205]. It is defined as the ratio (in dB) of the power radiated from an isotropic transmitting antenna to the power received by an equivalent loss-free antenna at a distance R (Fig. 8.5). In the context of ground wave propagation it is given by:

$$L_b = -20 \log_{10} \left(\frac{\lambda_0 |F(R)|}{4\pi R} \right), \tag{8.4.25}$$

where there is an additional factor $F(R)$ compared to the transmission loss in free space. This is a function of the distance and radar frequency. In the radio engineering community, this quantity is often expressed in terms of the radar frequency f_0 and the relative field strength $I_{dB} = 20 \log_{10} |E(R)/E(1)|$, which for the ground wave is approximately $I_{dB} = 20 \log_{10}(2 |F(R)| / R)$ in view of Eq. (8.3.20) (Fig. 8.6):

$$L_b = 142 \text{ dB} - 20 \log_{10} f_0 - I_{dB} \tag{8.4.26}$$

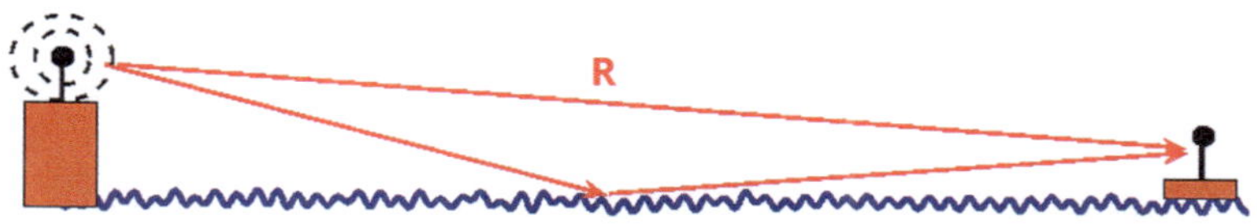

Fig. 8.5 The basic transmission loss is defined as the fraction of the power transmitted from an isotropic antenna to another isotropic receive antenna at a distance R over the sea surface

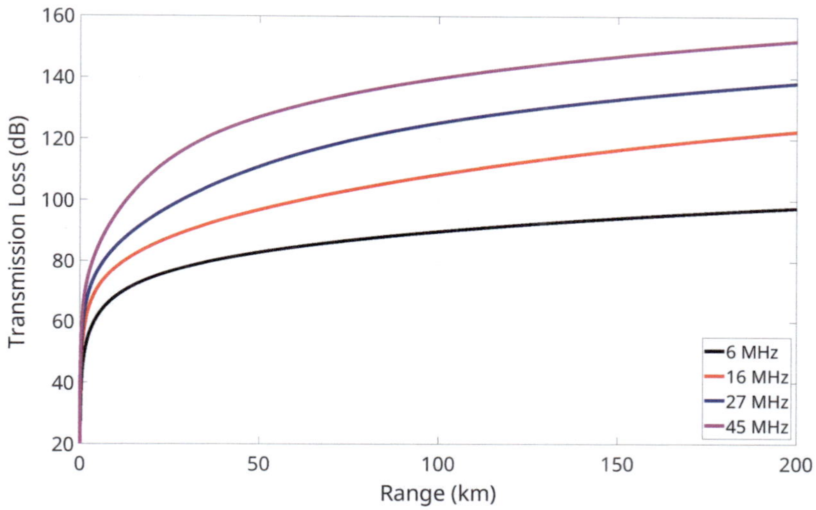

Fig. 8.6 Basic transmission loss as a function of the propagation distance over a flat sea surface for various radar frequencies

For very large distances, the planar Earth approximation used in the Sommerfeld problem is no longer valid and corrections for the Earth's curvature must be introduced, where the distance is now defined as the arc length over the Earth's surface. The ground wave solution for a spherical Earth is more much complicated than the planar one and we will not go into its description, referring the interested reader to specialized papers and books on the subject (e.g. [165,190,203,269]. The ground wave propagating on a spherical surface is slightly more attenuated than on a flat surface and departs more quickly from the $1/R$ decay. At the usual radar frequencies used for radio oceanography (13–16 MHz), the difference is only a few dB within the operational range (say, about 100 km). Figure 8.7 shows the additional transmission loss resulting from a spherical earth model compared to a planar earth model. It has been computed using the technique presented in Chap. 9 of [165] (the codes have been shared by Joel Johnson (Ohio State University)).

Another effect not considered in Sommerfeld's original formulation is the presence of islands or mixed sea-land paths along the propagation direction of the ground waves. It is obvious that the crossing of land obstacles such as islands reduces the local electric field strength. However, ground waves have another remarkable property, which is that they can recover in strength after passing the obstacle, an effect

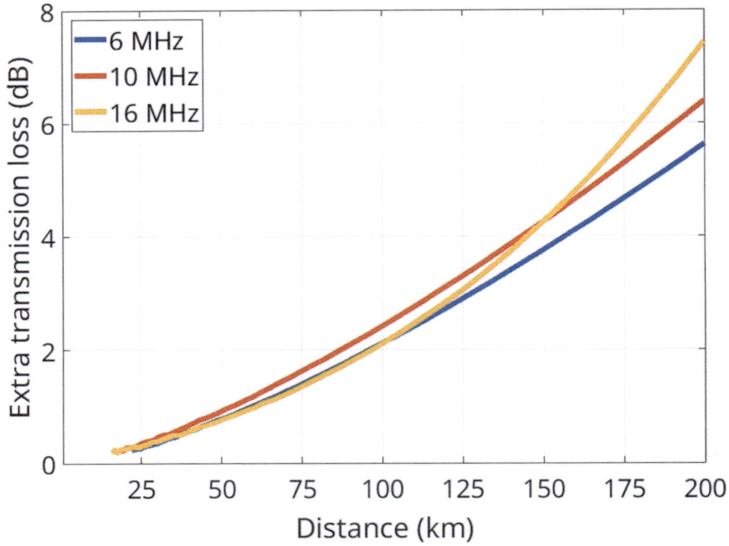

Fig. 8.7 Additional transmission loss between the spherical and the flat Earth solution

known as the Millington effect [194, 195]. The effect of the varying permittivity of a flat terrain has been treated analytically by Clemmov and Bremmer [36,46] and numerical solutions have been found to include the effect of roughness and topography [4,35,235]. Today, the propagation of ground waves over land and sea can be accurately modelled. However, for the purposes of this book, only the original historical solution for flat oceans by Sommerfeld and Norton will be considered.

Antenna Radiation Over a Moving Sea Surface

9

So far we have only considered the propagation of EM waves above a plane air/seawater interface, which is governed by the specular reflection and the interference with the incident field. In presence of surface roughness, such as that induced by the gravity waves, the electromagnetic field is scattered in various directions of space, in particular in the backscattering direction where it can be measured by the radar. The description of the scattered field in the context of HFR is a special case of rough surface scattering, a broad field of research that includes optical, acoustic or radio wave scattering and which can be described within a common mathematical framework. In this chapter, we recall the main notions of the theory of rough surface scattering and we introduce the perturbation expansion which gives the scattered field as a function of the sea surface roughness in the case where the radar wavelength is much larger than the wave height. We introduce the radar and Doppler cross sections, which are the primary observables of an HFR experiment, and relate them to the ocean wave spectrum. We introduce the notion of effective surface impedance, which allows to quantify the attenuation by surface roughness in the ground wave propagation, and obtain some simple estimates for the backscattered power as a function of the distance to the radar.

9.1 The EM Field Scattered from a Rough Surface

In presence of a rough interface, the incident EM wave is reflected not only in the specular direction but also in various directions of space. These additional waves are called the scattered field $\mathbf{E_s}$ (Fig. 9.1). They interfere with the incident field $\mathbf{E_i}$ over the sea surface to form the *total electric field* $\mathbf{E}$. The scattered field can be mathematically described as the complementary field of the incident field, which is in general known and controlled:

$$\mathbf{E_s} = \mathbf{E} - \mathbf{E_i} \tag{9.1.1}$$

C.-A. Guérin, *High-Frequency Radars for the Oceanographer*, Geophysics and Environmental Physics, https://doi.org/10.1007/978-3-031-95642-3_9

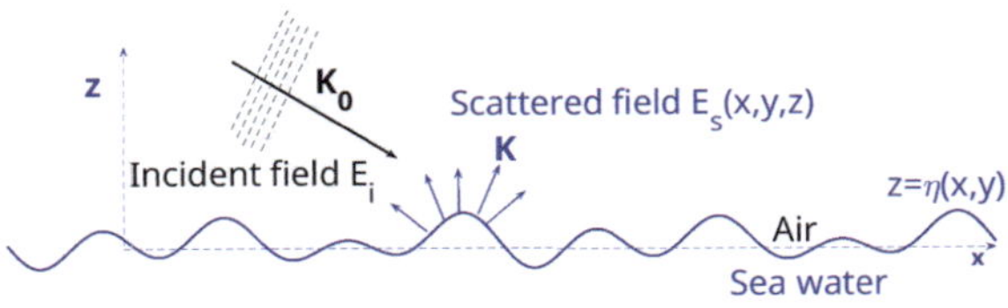

Fig. 9.1 An incident EM field on a rough surface gives rise to upwards scattered waves in all directions of space

The scattered field carries information on the surface topography and it is the purpose of remote sensing to analyze it. In a practical radar experiment, the incident field and the scattered field are separated, either by the alternate transmit and receive modes (IFMCW systems) or by their Doppler frequency shift on the fast time (FMCW systems). The voltage which is recorded on the antenna is therefore a direct measurement, apart from a multiplicative factor, of the scattered electric field at the antenna location. The natural system of coordinates to describe wave scattering from a rough surface is attached to the mean plane (x, y) of elevations, which is horizontal, and to the upward vertical axis $\mathbf{z}$. The incident field is usually assumed to be a monochromatic plane wave of unit amplitude in the canonical polarization basis:

$$\mathbf{E_i}(\mathbf{R}) = e^{i\mathbf{K_0}\cdot\mathbf{R}}\mathbf{E_0} \tag{9.1.2}$$

where the sinusoidal $e^{-i\omega_0 t}$ is implicit and $\mathbf{R} = (x, y, z) = (\mathbf{r}, z)$ is the position vector of observation. Here, $\mathbf{E_0} = \mathbf{p}_V^-(\mathbf{k_0})$ or $\mathbf{E_0} = \mathbf{p}_H^-(\mathbf{k_0})$ depending on whether the incident field is vertically or horizontally polarized (see Chap. 7). Due to the linearity of Maxwell's equations, the scattered field associated with any incident field can be obtained by superposition of these solutions and thus there is no loss of generality in assuming this particular incident field. The scattered field from an infinite surface (which we assume at this stage to be "frozen") can be written as a Rayleigh expansion (6.1.20) with upwards propagating plane waves only:

$$\mathbf{E_s}(\mathbf{R}) = \int d\mathbf{k} e^{i(\mathbf{k}\cdot\mathbf{r}+qz)}\mathsf{A}(\mathbf{k}, \mathbf{k_0})\mathbf{E_0}, \tag{9.1.3}$$

where $\mathbf{k_0}$ is the horizontal projection of the three-dimensional incident wave vector $\mathbf{k_0}$ while the two-dimensional vector $\mathbf{k}$ runs over $\mathbb{R}^2$ with the convention $q = \sqrt{K_0^2 - k^2}$ and $\text{Im}[q] \geq 0$. For $k \leq K_0$, $\mathbf{k}$ and $q > 0$ can be seen as the horizontal and vertical projection of the wave vector $\mathbf{K} = \mathbf{k} + q\mathbf{z}$ of an upwards propagating plane wave (Fig. 9.2). The components with $k \leq K_0$ correspond to propagating plane waves while the component $k > K_0$ in the integral correspond to evanescent plane waves. With the convention that q is the positive imaginary square root, these waves are exponentially attenuated when z increases along the positive vertical axis. The operator $\mathsf{A}(\mathbf{k}, \mathbf{k_0})$ is called the *scattering amplitude* and relates any pair of incoming and outgoing plane waves. It can be described by four coefficients using the canonical

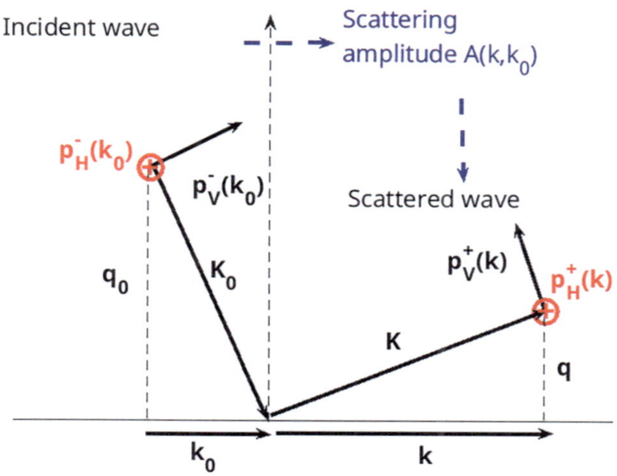

Fig. 9.2 Geometry of wave vectors. The incident and scattered plane waves are labeled through the horizontal projection of their wave vector, which are the integration variables in the Rayleigh expansion of the scattered field

polarization basis of vertically/horizontally polarized plane waves. By decomposing the incident polarization onto the canonical basis:

$$\mathbf{E_i}(\mathbf{R}) = e^{\mathrm{i}\mathbf{K_0}\cdot\mathbf{R}}\left[\alpha_V \mathbf{p}_V^-(\mathbf{k_0}) + \alpha_H \mathbf{p}_H^-(\mathbf{k_0})\right], \tag{9.1.4}$$

we obtain the expression for the scattered field with (9.1.3) and:

$$\begin{aligned}\mathsf{A}(\mathbf{k}, \mathbf{k_0})\mathbf{E_0} &= \alpha_V A_{VV}(\mathbf{k}, \mathbf{k_0})\mathbf{p}_V^+(\mathbf{k}) + \alpha_V A_{VH}(\mathbf{k}, \mathbf{k_0})\mathbf{p}_H^+(\mathbf{k}) \\ &\quad + \alpha_H A_{HV}(\mathbf{k}, \mathbf{k_0})\mathbf{p}_V^+(\mathbf{k}) + \alpha_H A_{HH}(\mathbf{k}, \mathbf{k_0})\mathbf{p}_H^+(\mathbf{k})\end{aligned} \tag{9.1.5}$$

for some complex coefficients A_{VV}, A_{VH}, A_{HV} and A_{HH}. These coefficients express the transfer from one given incident polarization component to any other scattered field polarization component; they form a 2×2 matrix called the *scattering matrix* :

$$\mathsf{A}(\mathbf{k}, \mathbf{k_0}) = \begin{pmatrix} A_{VV}(\mathbf{k}, \mathbf{k_0}) & A_{VH}(\mathbf{k}, \mathbf{k_0}) \\ A_{HV}(\mathbf{k}, \mathbf{k_0}) & A_{HH}(\mathbf{k}, \mathbf{k_0}) \end{pmatrix} \tag{9.1.6}$$

Its diagonal coefficients (A_{VV}, A_{HH}) are the *co-polarized* scattering amplitudes while its off-diagonal coefficients (A_{VH}, A_{HV}) are the *cross-copolarized* scattering amplitudes. By taking R to infinity in a prescribed direction $\mathbf{K}$ and using the Weyl integral representation (B.0.7) of the free space Green's function, we obtain the far-field expression:

$$\mathbf{E_s}(\mathbf{R}) \simeq -2\mathrm{i}\pi \frac{\mathrm{e}^{\mathrm{i}KR}}{R} q\, \mathsf{A}(\mathbf{k}, \mathbf{k_0})\mathbf{E_0} \tag{9.1.7}$$

This states that the scattered field is proportional in every direction to an outgoing spherical wave, a condition which is sometimes expressed through the additional *Sommerfeld radiation condition* for scalar fields and *Silver-Müller* radiation condition for EM waves. There is, however, a noticeable difference between the plane wave Rayleigh expansion (9.1.3) and the spherical wave asymptotic (9.1.7): the former representation is exact in the half-space domain which is above the maximal surface elevation ($z > z_{\max}$) while the latter is only valid at far distance. Note that the scattering amplitude has the dimension of a squared distance ($[M^2]$). There exist other conventions where the q factor is absorbed in the definition of $\mathsf{A}(\mathbf{k}, \mathbf{k_0})$, in such a way that the scattering amplitude has the dimension of a distance (see e.g. the review paper [73]).

9.2 How Does the Scattered Field Relate to Surface Roughness?

There is in general no exact analytical solution for the scattering amplitude as a function of surface roughness. However, in the HF and VHF regime, where the radar wavelength is much greater than the wave height, the scattered field can be described very accurately by a perturbation expansion referred to as *Rayleigh-Rice theory* or the *Small Perturbation Method* [73,267]. If we denote $z = \eta(\mathbf{r}) = \eta(x, y)$ the vertical deviation of the rough interface from its mean plane and $\langle \eta^2 \rangle^{1/2}$ its root-mean-square value, we may expand the scattered field in orders of the small dimensionless parameter $K_0 \langle \eta^2 \rangle^{1/2}$, where $K_0 = 2\pi/\lambda_0$ is as usual the radar wavenumber. We may thus seek the scattering amplitude in the form:

$$\mathsf{A}(\mathbf{k}, \mathbf{k_0}) = \mathsf{A}_0(\mathbf{k}, \mathbf{k_0}) + \mathsf{A}_1(\mathbf{k}, \mathbf{k_0}) + \mathsf{A}_2(\mathbf{k}, \mathbf{k_0}) + \cdots , \tag{9.2.8}$$

where the nth term $\mathsf{A}_n(\mathbf{k}, \mathbf{k_0})$ is of order $(K_0 \langle \eta^2 \rangle^{1/2})^n$. The time-dependence is ignored, which amounts to consider the instantaneous scattering amplitude of the sea surface at some instant. The most general mathematical expression of this perturbation series is obtained through a functional expansion with respect to surface roughness η. For an infinite rough surface, it can be shown [267] that the scattering amplitude under a second-order perturbation expansion takes the universal functional form:

$$\begin{aligned}
\mathsf{A}_0(\mathbf{k}, \mathbf{k_0}) &= \mathsf{B}_0(\mathbf{k_0})\delta(\mathbf{k} - \mathbf{k_0}) \\
\mathsf{A}_1(\mathbf{k}, \mathbf{k_0}) &= -\mathrm{i}\mathsf{B}_1(\mathbf{k}, \mathbf{k_0})\widehat{\eta}(\mathbf{k_B}) \\
\mathsf{A}_2(\mathbf{k}, \mathbf{k_0}) &= \int \mathsf{B}_2(\mathbf{k}, \mathbf{k_0}, \mathbf{k_0} + \mathbf{k_1})\widehat{\eta}(\mathbf{k_1})\widehat{\eta}(\mathbf{k_B} - \mathbf{k_1})d\mathbf{k_1} \\
\mathsf{A}_3(\mathbf{k}, \mathbf{k_0}) &= \mathrm{i} \int \mathsf{B}_3(\mathbf{k}, \mathbf{k_0}, \mathbf{k_0} + \mathbf{k_1}, \mathbf{k_0} + \mathbf{k_2})\widehat{\eta}(\mathbf{k_1})\widehat{\eta}(\mathbf{k_2} - \mathbf{k_1})\widehat{\eta}(\mathbf{k_2} - \mathbf{k_1})d\mathbf{k_1} \\
&\cdots
\end{aligned} \tag{9.2.9}$$

where $\mathbf{k_B} = \mathbf{k} - \mathbf{k_0}$ is the *bistatic Bragg wave vector*. The operators $\mathsf{B}_0, \mathsf{B}_1, \mathsf{B}_2, \ldots,$ do not depend on roughness but only on the scattering geometry and the dielectric constant of the medium; they are referred to as the *kernels* of the expansion at various orders. To simplify the notations, we will omit their dependence on the wave vectors unless necessary, that is we will simply write $\mathsf{B}_1 = \mathsf{B}_1(\mathbf{k}, \mathbf{k_0})$ etc. The function $\widehat{\eta}$ is the two-dimensional spatial Fourier Transform of elevations:

$$\widehat{\eta}(\mathbf{k}) = \frac{1}{4\pi^2} \int_{\mathbf{L}} e^{-i\mathbf{k}\cdot\mathbf{r}} \eta(\mathbf{r}) d\mathbf{r}, \tag{9.2.10}$$

where the conventional factor $1/4\pi^2$ is chosen for convenience and the symbol $\int_{\mathbf{L}}$ indicates that the roughness is implicitly assumed of finite extent L in each dimension, that is the actual range of integration is a square domain $|x| < L/2$ and $|y| < L/2$. A physical interpretation of this expansion can be given in terms of the number of interactions of the EM wave with the surface. The zeroth-order term is the scattered field in the absence of roughness. It is limited to the specular reflection, which explains the Dirac function at $\mathbf{k} = \mathbf{k_0}$. The corresponding matrix is a diagonal composed of the Fresnel reflection coefficients in vertical and horizontal polarization, $\mathcal{R}_V(\mathbf{k_0})$ and $\mathcal{R}_H(\mathbf{k_0})$, respectively. The first-order term is the single-scattering term that is responsible for the dominant contribution of the scattered field in every direction but specular. It is proportional to the spatial Fourier transform of roughness at the Bragg wavevector $\mathbf{k_B}$ and confirms the observation of Crombie that the EM scattering mechanism at the sea surface involves primarily the Bragg wave (see Chap. 2). The second-order term involves a double electromagnetic interaction with the surface through the coupling of two spatial frequencies. Contrarily to the first-order, it implies all wavevectors that compose the gravity wave field at the sea surface, with various weights, however, given by the kernel B_2. This last kernel is not uniquely defined, as several different expressions can lead to the same integral. The third-order and higher scattering orders are of seldom use and are mentioned for completeness.

9.3 The Radar Cross Section

9.3.1 Definition

As seen, the scattered field carries the signature of the rough surface and can be used for the characterization of the latter. Once the received signal has been processed in range and azimuth (see Chap. 3), it can be identified with the complex demodulated field $E_s(R, \phi)$ scattered from a given radar cell at distance R and bearing ϕ from the receiver (Fig. 9.3). Because of the continuous motion of waves at the sea surface, the received intensity undergoes random fluctuations on a slow time and cannot be treated in a deterministic way. The only meaningful way to characterize this signal is to perform averages over the integration time, which is much longer than the typical time scale of the wave motion (the gravity waves field is completely renewed within a

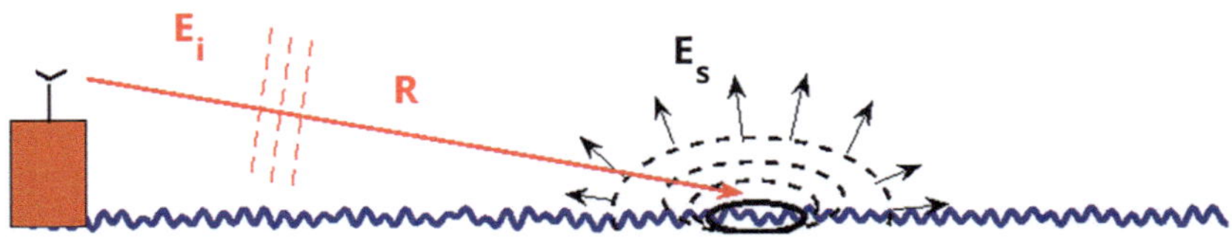

Fig. 9.3 After range and azimuth processing, the radar signal can be identified with the complex scattered field emanating from a scattering patch at a given radar cell

few seconds while the integration time is a few minutes). The most obvious averages are the so-called *coherent field*:

$$\langle E_s \rangle = \frac{1}{T} \int_0^T E_s(t) dt \qquad (9.3.11)$$

and the mean scattered intensity:

$$\langle |E_s|^2 \rangle = \frac{1}{T} \int_0^T |E_s|^2 (t) dt \qquad (9.3.12)$$

While these definitions are well suited to what can be experimentally measured, on the theoretical level it is mathematically more convenient to handle such time averages as ensemble averages of random processes. For this, one usually assumes that the sea roughness is described by a spatio-temporal random process $\eta(\mathbf{r}, t)$, which is stationary both in space and time, with the consequence that the scattered field is also a stationary random processes. If one further assume the ergodicity of this process, the above time average can be considered as an ensemble average in the limit of large integration time. It is quite intuitive that, due to its random fluctuations, the complex electric field averages to zero, unless it contains the specular reflection of the mean plane of the sea surface. Hence the coherent field is limited to the specular direction of observation ($\mathbf{k} = \mathbf{k_0}$). The mean scattered intensity $\langle |E_s|^2 \rangle$ is a more informative quantity. In order to characterize the intrinsic properties of the sea surface, it must be corrected for its primary dependence on:

- the distance measured from the center of the radar cell, which decays as $1/R^2$
- the spatial resolution, which is proportional to the area of the radar cell
- the transmitted power, which is proportional to the square modulus of the incident field.

One therefore defines the so-called *normalized radar cross section* (NRCS) as[1]:

$$\sigma^0 = \frac{4\pi R^2 \langle |E_s|^2 \rangle}{\text{Area} \times |E_i|^2}, \qquad (9.3.13)$$

[1] The factor 4π is conventional and originates from the standard definition of the radar cross section of a bounded target, RCS=$4\pi R^2 \, |E_s|^2 \, / \, |E_i|^2$.

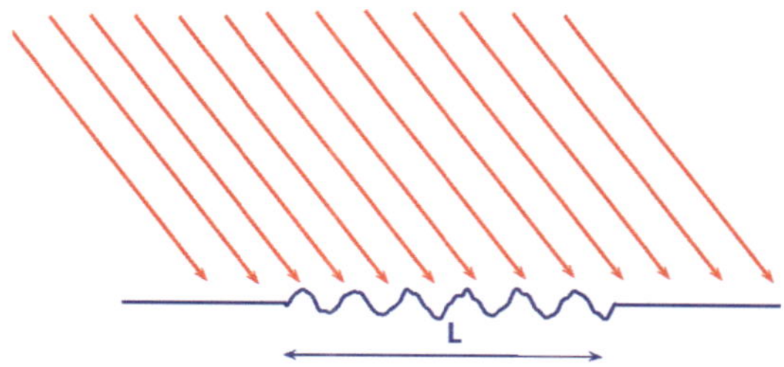

Fig. 9.4 A finite surface patch can be mathematically obtained with a local deformation of an infinite flat surface uniformly illuminated by a plane wave

where it is assumed that the distance R and the radar cell area are large enough so that the above quantity has converged to its limiting value. Even though this notion is more abstract than that of antenna voltage, which is the primary observed quantity, it is fundamental because it allows to interpret every single, particular experiment in universal terms with standardized definitions. To calculate the NRCS, it is convenient to assume that the incident field on the radar cell can be assimilated to a plane wave, which is a reasonable assumption if the scattering patch is sufficiently far from the transmitter (compared to both the radar wavelength and the size of the antenna, see Chap. 6). In order to limit the scattering area, one therefore assumes that the surface has a roughness of finite extent, say L in every dimension, and is otherwise smooth (Fig. 9.4). With these assumptions, it follows from (9.1.7) and (9.3.13) that:

$$\sigma^0 = 16\pi^3 L^{-2} \langle |q\mathsf{A}(\mathbf{k}, \mathbf{k_0})|^2 \rangle, \tag{9.3.14}$$

where the mean square amplitude is implicitely proportional to the surface patch ($|q\mathsf{A}|^2 \sim L^2$).

The average power of the scattered field can be decomposed into a coherent ($|\langle E_s \rangle|^2$ and an incoherent contribution ($\langle |E_s|^2 \rangle - |\langle E_s \rangle|^2$), resulting in an associated coherent and incoherent NRCS. As mentioned, the NRCS measured away from the specular direction is purely incoherent; it is most often measured in the *backscattering direction*, which is opposite to the incident directions ($\mathbf{k} = -\mathbf{k_0}$). Backscattering NRCS are obtained with *monostatic* systems, for which the same antenna serves for transmitting and receiving, or *pseudo monostatic* systems, in which the receive and transmit antennas are separated but located nearby (Fig. 9.5). When the receive antenna system is at a separate location from the transmit system, the radar is said to be in a *bistatic* configuration and it measures a bistatic NRCS. In the language of radar, the transmitter and receiver antenna systems are simply referred to as *TX* and *RX*, respectively.

By projecting the incident and scattered fields onto the canonical (V, H) polarization basis, one can define 4 NRCS coefficients, $\sigma^0_{VV}, \sigma^0_{HH}, \sigma^0_{VH}$ and σ^0_{HV}, corresponding to the various polarization components. In the monostatic case, each of these dimensionless coefficients is referred to as *sigma naught* or the *back-scattering coefficient* ; in the bistatic case it is simply referred to as a *normalized bistatic radar cross section* or *bistatic NRCS*.

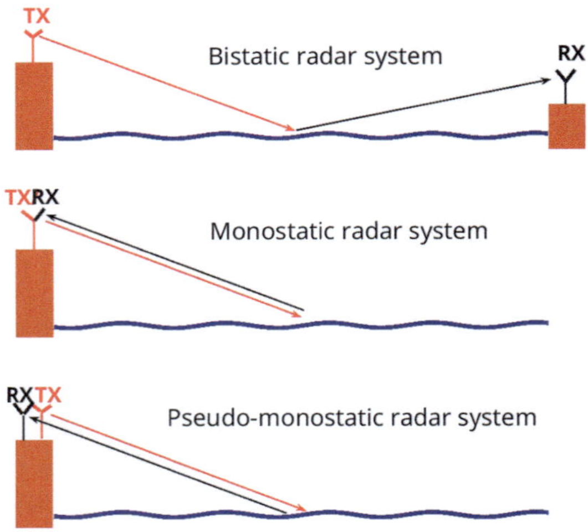

Fig. 9.5 Bistatic, monostatic and pseudo-monostatic configurations. The transmit and receive antenna are referred to as TX and RX, respectively

9.3.2 Calculation

We will now seek a perturbation expansion for the incoherent bistatic NRCS, away from the specular direction ($\mathbf{k_B} \neq \mathbf{0}$). At this stage it is necessary to make some basic assumption on the nature of the process describing the surface roughness. In a first step we will only consider the surface at given time, say $t = 0$, and assume that it is described by a wide-sense stationary random Gaussian process, $\eta(\mathbf{r}) = \eta(x, y)$, with a correlation function $C(\mathbf{r})$ and wavenumber power spectrum $S_0(\mathbf{k})$ as defined in (A.1.8) and (A.1.9). Upon changing the vertical reference coordinate, one can always assume that the process is centered, that is $\langle \eta(\mathbf{r}) \rangle = 0$. The roughness extension is limited to a patch of length L in each dimension, that is a sharp cut-off is implicitly applied to the process at the edges ($\eta(x, y) = 0$ for $|x| > L/2$ or $|y| > L/2$). The dominant contribution to the NRCS is obtained by taking the first-order amplitude A_1 in (9.3.14). It is of second-order in elevations:

$$\sigma_1^0 = 16\pi^3 \, |q\mathsf{B}_1|^2 \; L^{-2} \langle |\widehat{\eta}(\mathbf{k_B})|^2 \rangle \tag{9.3.15}$$

Now, the squared modulus of the Fourier transform of elevations can be rewritten as double integral:

$$|\widehat{\eta}(\mathbf{k_B})|^2 = \frac{1}{(2\pi)^4} \int_{\mathbf{L}} \int_{\mathbf{L}} \mathrm{e}^{-\mathrm{i}\mathbf{k_B}\cdot(\mathbf{r}-\mathbf{r}')} \eta(\mathbf{r})\eta(\mathbf{r}') d\mathbf{r} d\mathbf{r}', \tag{9.3.16}$$

which by taking the ensemble average leads to:

$$\langle |\widehat{\eta}(\mathbf{k_B})|^2 \rangle = \frac{1}{(2\pi)^4} \int_{\mathbf{L}} \int_{\mathbf{L}} \mathrm{e}^{-\mathrm{i}\mathbf{k_B}\cdot(\mathbf{r}-\mathbf{r}')} C(\mathbf{r}-\mathbf{r}')\rangle d\mathbf{r} d\mathbf{r}' \tag{9.3.17}$$

The integrand is therefore a function of the difference $\mathbf{r} - \mathbf{r}'$ only, so that a change of variables:

$$(\mathbf{r}, \mathbf{r}') \to (\mathbf{u} = \mathbf{r} - \mathbf{r}', \mathbf{r}') \tag{9.3.18}$$

reduces the double integral to a single one:

$$\langle|\widehat{\eta}(\mathbf{k_B})|^2\rangle = L^2\frac{1}{(2\pi)^4}\int_{\mathbf{L}} \mathrm{e}^{-i\mathbf{k_B}\cdot\mathbf{u}}C(\mathbf{u})d\mathbf{u} = \frac{L^2}{(2\pi)^2}S_0(\mathbf{k_B}) \tag{9.3.19}$$

It follows that radar cross section at the lowest order is proportional to the power spectrum of elevations:

$$\sigma_1^0 = 4\pi q^2\,|\mathsf{B}_1|^2\; S_0(\mathbf{k_B}), \tag{9.3.20}$$

The next term in the perturbation expansion of the NRCS is obtained with the cross-averages $\langle \mathsf{A}_1\mathsf{A}_2^*\rangle$, which involve 3-term correlations of the form $\langle\widehat{\eta}(\mathbf{k_1})\widehat{\eta}^*(\mathbf{k_2})\widehat{\eta}^*(\mathbf{k_3})\rangle$. These are null because the cross-average of an odd number of Fourier transforms of elevations is always zero. Hence, the next order term σ_2^0 in the NRCS expansion is of order 4 in elevations and appears through the products $\langle|\mathsf{A}_2|^2\rangle$ and $\langle \mathsf{A}_1\mathsf{A}_3^*\rangle$. We may therefore write:

$$\sigma_2^0 = (\sigma_2^0)_{2+2} + (\sigma_2^0)_{1+3} \tag{9.3.21}$$

with obvious notations. The first term is given by:

$$(\sigma_2^0)_{2+2} = 16\pi^3 q^2\,L^{-2}\left\langle\left|\int \mathsf{B}_2\,\widehat{\eta}(\mathbf{k_1})\widehat{\eta}(\mathbf{k_B}-\mathbf{k_1})d\mathbf{k_1}\right|^2\right\rangle \tag{9.3.22}$$

This expression involves a double integral with 4-term correlations of the form $\langle\widehat{\eta}(\mathbf{k_1})\widehat{\eta}(\mathbf{k_2})\widehat{\eta}(\mathbf{k_3})^*\widehat{\eta}(\mathbf{k_4})^*\rangle$ which, for Gaussian processes, can be decomposed in products of 2-term correlations:

$$\begin{aligned}\langle\widehat{\eta}(\mathbf{k_1})\widehat{\eta}(\mathbf{k_2})\widehat{\eta}^*(\mathbf{k_3})\widehat{\eta}^*(\mathbf{k_4})\rangle &= \langle\widehat{\eta}(\mathbf{k_1})\widehat{\eta}(\mathbf{k_2})\rangle\langle\widehat{\eta}^*(\mathbf{k_3})\widehat{\eta}^*(\mathbf{k_4})\rangle\\ &+ \langle\widehat{\eta}(\mathbf{k_1})\widehat{\eta}^*(\mathbf{k_3})\rangle\langle\widehat{\eta}(\mathbf{k_2})\widehat{\eta}^*(\mathbf{k_4})\rangle\\ &+ \langle\widehat{\eta}(\mathbf{k_1})\widehat{\eta}^*(\mathbf{k_4})\rangle\langle\widehat{\eta}(\mathbf{k_2})\widehat{\eta}^*(\mathbf{k_3})\rangle\end{aligned} \tag{9.3.23}$$

Developing again the expression of the Fourier transform and intertwining the correlation function in the 2-term products we obtain:

$$\begin{aligned}\langle\widehat{\eta}(\mathbf{k_1})\widehat{\eta}(\mathbf{k_B}-\mathbf{k_1})\rangle\langle\widehat{\eta}^*(\mathbf{k_2})\widehat{\eta}^*(\mathbf{k_B}-\mathbf{k_2})\rangle &= \frac{L^2}{(2\pi)^2}S_0(\mathbf{k_1})S_0(\mathbf{k_2})\delta(\mathbf{k_B})\\ \langle\widehat{\eta}(\mathbf{k_1})\widehat{\eta}^*(\mathbf{k_2})\rangle\langle\widehat{\eta}(\mathbf{k_B}-\mathbf{k_1})\widehat{\eta}^*(\mathbf{k_B}-\mathbf{k_2})\rangle &= \frac{L^2}{(2\pi)^2}S_0(\mathbf{k_1})S_0(\mathbf{k_B}-\mathbf{k_2})\delta(\mathbf{k_1}-\mathbf{k_2})\\ \langle\widehat{\eta}(\mathbf{k_1})\widehat{\eta}^*(\mathbf{k_B}-\mathbf{k_2})\rangle\langle\widehat{\eta}^*(\mathbf{k_2})\widehat{\eta}(\mathbf{k_B}-\mathbf{k_1})\rangle &= \frac{L^2}{(2\pi)^2}S_0(\mathbf{k_1})S_0(\mathbf{k_2})\delta(\mathbf{k_1}-(\mathbf{k_B}-\mathbf{k_2}))\end{aligned} \tag{9.3.24}$$

The first product contributes only to the coherent NRCS and is limited to the specular direction ($\mathbf{k_B}=\mathbf{0}$). The other two terms have an identical contribution to the NRCS (9.3.22) in view of the invariance property (G.2.2) of the kernel B_2, leading to:

$$(\sigma_2^0)_{22} = 8\pi q^2\int|\mathsf{B}_2|^2\;S_0(\mathbf{k_1})S_0(\mathbf{k_B}-\mathbf{k_1})d\mathbf{k_1} \tag{9.3.25}$$

The calculation of the term $(\sigma_2^0)_{1+3}$ is very similar. It implies the calculation of the following ensemble averages:

$$\begin{aligned}
&\langle\widehat{\eta}(\mathbf{k_B})\widehat{\eta}^*(\mathbf{k_1})\rangle\langle\widehat{\eta}^*(\mathbf{k_2}-\mathbf{k_1})\widehat{\eta}^*(\mathbf{k_B}-\mathbf{k_2})\rangle = 0\\
&\langle\widehat{\eta}(\mathbf{k_B})\widehat{\eta}^*(\mathbf{k_B}-\mathbf{k_2})\rangle\langle\widehat{\eta}^*(\mathbf{k_1})\widehat{\eta}^*(\mathbf{k_2}-\mathbf{k_1})\rangle = 0,\\
&\langle\widehat{\eta}(\mathbf{k_B})\widehat{\eta}^*(\mathbf{k_2}-\mathbf{k_1})\rangle\langle\widehat{\eta}^*(\mathbf{k_1})\widehat{\eta}^*(\mathbf{k_B}-\mathbf{k_2})\rangle = \frac{L^2}{(2\pi)^2}S_0(\mathbf{k_B})S_0(\mathbf{k_1})\delta(\mathbf{k_B}+\mathbf{k_1}-\mathbf{k_2})
\end{aligned} \tag{9.3.26}$$

where the first two averages are null because $S_0(\mathbf{0}) = 0$, as it is the case for an ocean wave spectrum. This leads to:

$$(\sigma_2^0)_{1+3} = 4\pi q^2 \mathsf{B}_1\ S_0(\mathbf{k_B}) \int \mathsf{B}_3^*\ S_0(\mathbf{k_1})d\mathbf{k_1} \tag{9.3.27}$$

In the domain of validity of the perturbation theory, the second-order NRCS is generally much smaller than the first-order one ($\sigma_2^0 << \sigma_1^0$) and is only considered when the former vanishes. This is the case for the cross-polarized component of the electric field (obtained by transmitting in vertical and receiving in horizontal polarization, or vice-versa), which can only be explained by multiple scattering interactions. However, in the context of HFR where only the co-polarized vertical polarization is used, it can be neglected. As we shall see, this is not the case for the Doppler cross section.

9.4 The Doppler Cross Section

9.4.1 Definition

The NRCS is based only on the intensity of the scattered field and is adapted to incoherent radars. A more complete description of the sea echo can be obtained with coherent radars using the temporal auto-correlation function:

$$\gamma(t) = \frac{4\pi R^2\langle E_s(t')E_s^*(t'+t)\rangle}{\text{Area}\times|E_i|^2}, \tag{9.4.28}$$

which has the same normalization as the NRCS. Here again, the ensemble average is defined though a time average:

$$\langle E_s(t')E_s^*(t'+t)\rangle = \frac{1}{T}\int_0^T E_s(t')E_s(t'+t)dt' \tag{9.4.29}$$

The cross section per unit bandwidth, also called the *Doppler cross section*, is defined as the Fourier Transform of this last function:

$$\sigma(\omega) = \frac{1}{2\pi}\int_{-\infty}^{+\infty} e^{-\mathrm{i}\omega t}\gamma(t)dt, \tag{9.4.30}$$

From (9.1.7) it can be related to the auto-correlation function of the scattering amplitude through:

$$\gamma(t) = 16\pi^3 L^{-2} |q|^2 \langle \mathsf{A}(\mathbf{k}, \mathbf{k_0}, t + t')\mathsf{A}^*(\mathbf{k}, \mathbf{k_0}, t')\rangle. \tag{9.4.31}$$

With this definition the NRCS is the integral of the Doppler cross section, i.e.:

$$\sigma^0 = \int_{-\infty}^{+\infty} \sigma(\omega)d\omega \tag{9.4.32}$$

The Doppler cross section admits the same decomposition as the radar cross section in terms of co-polarized ($\sigma_{VV}(\omega)$ or $\sigma_{HH}(\omega)$) or cross-polarized ($\sigma_{VH}(\omega)$ or $\sigma_{HV}(\omega)$) components. In the context of HFR, a vertical polarization of the incident electric field is needed to take advantage of the ground wave propagation mechanism (Chap. 8). The cross-polarized component of the backscattered field is in general negligible ($\sigma^0_{VH}(\omega) << \sigma^0_{VV}(\omega)$), it is only the co-polarized vertical component $\sigma^0_{VV}(\omega)$ that is of interest.

9.4.2 Calculation with Linear Sea Surfaces

The derivation of the Doppler cross section is very similar to that of the radar cross section. Again, it relies on the perturbation expansion (9.2.9) where a time dependence is now assumed for the scattering amplitude ($\mathsf{A}(\mathbf{k}, \mathbf{k_0}) \Rightarrow \mathsf{A}(\mathbf{k}, \mathbf{k_0}, t)$) and for the spatial Fourier transform of elevations ($\widehat{\eta}(\mathbf{k}) \Rightarrow \widehat{\eta}(\mathbf{k}, t)$). Again, the dominant contribution to the field auto-correlation function (9.4.28) is obtained by taking the first-order amplitude in (9.3.14):

$$\gamma_1(t) = 16\pi^3 |q\mathsf{B}_1|^2 \; L^{-2}\langle \widehat{\eta}(\mathbf{k_B}, t + t')\widehat{\eta}^*(\mathbf{k_B}, t')\rangle \tag{9.4.33}$$

The term under brackets can be estimated by introducing the space-time correlation function of elevations:

$$\langle \widehat{\eta}(\mathbf{k_B}, t + t')\widehat{\eta}^*(\mathbf{k_B}, t')\rangle = \frac{1}{(2\pi)^4} \int_{\mathbf{L}} \int_{\mathbf{L}} \mathrm{e}^{\mathrm{i}\mathbf{k_B}\cdot(\mathbf{r}-\mathbf{r}')} C(\mathbf{r} - \mathbf{r}', t)\rangle d\mathbf{r}d\mathbf{r}' \tag{9.4.34}$$

With the same change of variables (9.3.18) as before, this yields:

$$\int_{-\infty}^{+\infty} e^{-\mathrm{i}\omega t} \langle \widehat{\eta}(\mathbf{k_B}, t + t')\widehat{\eta}^*(\mathbf{k_B}, t')\rangle dt = \frac{L^2}{2\pi} S(\mathbf{k_B}, \omega) \tag{9.4.35}$$

This shows that the Doppler cross section at the lowest order in the perturbation expansion is proportional to the wavenumber-frequency spectrum (A.1.4):

$$\sigma_1(\omega) = 4\pi \; |q\mathsf{B}_1|^2 \; S(\mathbf{k_B}, \omega) \tag{9.4.36}$$

We now investigate the higher-order scattering contributions to the Doppler cross section. Again, the cross-averages $\langle \mathsf{A}_1 \mathsf{A}_2^* \rangle$ are null, so that the next nonzero contribution is given by the combinations $\mathsf{A}_2\mathsf{A}_2^*$ and $\mathsf{A}_1\mathsf{A}_3^*$. The $1+3$ combination is generally neglected because it is indeed smaller than the $2+2$ and (mostly) because the third-order kernel B_3 is prohibitively complex. At second order, the field auto-correlation function is therefore given by:

$$\gamma_2(t) = 16\pi^3 q^2 L^{-2} \int \mathsf{B}_2(\mathbf{k}, \mathbf{k_0}, \mathbf{k_0} + \mathbf{k_1}) \mathsf{B}_2^*(\mathbf{k}, \mathbf{k_0}, \mathbf{k_0} + \mathbf{k_2}) \times \langle \widehat{\eta}(\mathbf{k_1}, t') \widehat{\eta}(\mathbf{k_B} - \mathbf{k_1}, t') \widehat{\eta}^*(\mathbf{k_2}, t + t') \widehat{\eta}^*(\mathbf{k_B} - \mathbf{k_2}, t + t') \rangle d\mathbf{k_1} d\mathbf{k_2} \tag{9.4.37}$$

The same calculation rules that were employed in (9.3.24) to evaluate the NRCS apply to the field auto-correlation function, provided the power spectrum S_0 is replaced by the partial Fourier Transform of the surface correlation function:

$$S(\mathbf{k}, t) = \frac{1}{(2\pi)^2} \int_{\mathbb{R}^2} \mathrm{e}^{-\mathrm{i}\mathbf{k}\cdot\mathbf{r}} C(\mathbf{r}, t) d\mathbf{r} = \int e^{\mathrm{i}\omega t} S(\mathbf{k}, \omega) d\omega \tag{9.4.38}$$

yielding:

$$\gamma_2(t) = 4\pi q^2 \int |\mathsf{B}_2|^2 \; S(\mathbf{k_1}, t') S(\mathbf{k_B} - \mathbf{k_1}, t + t') d\mathbf{k_1} \tag{9.4.39}$$

To go further in the expression of the Doppler cross section it is necessary to specify how the sea surface moves, that is to make an assumption on the dispersion relation. *Linear surfaces* are composed of a superposition of independent waves traveling with the free gravity-wave dispersion relation $\omega = \omega(\mathbf{k}) = \sqrt{gk}$. The associated wavenumber-frequency spectrum is of the form:

$$S(\mathbf{k}, \omega) = \frac{1}{2}(S(\mathbf{k})\delta(\omega - \omega(\mathbf{k})) + S(-\mathbf{k})\delta(\omega + \omega(\mathbf{k}))) \tag{9.4.40}$$

where $S(\mathbf{k})$ is the directional ocean wavenumber spectrum defined in (2.5.28). It follows that the first-order Doppler cross section is composed of two peaks at plus/minus the Bragg frequency:

$$\sigma_1(\omega) = 2\pi q^2 \, |\mathsf{B}_1|^2 \; (S(\mathbf{k_B})\delta(\omega - \omega_B) + (S(-\mathbf{k_B})\delta(\omega + \omega_B)) \tag{9.4.41}$$

We now have an explicit formulation for the partial Fourier transform:

$$S(\mathbf{k}, t) = \frac{1}{2}(S(\mathbf{k})e^{\mathrm{i}\omega(\mathbf{k})t} + S(-\mathbf{k})e^{-\mathrm{i}\omega(\mathbf{k})t}) \tag{9.4.42}$$

which yields the following expression for the second-order Doppler cross section obtained with a linear sea surface:

$$\sigma_2^{\mathrm{lin}}(\omega) = 2\pi q^2 \sum_{n_1, n_2 = \pm 1} \int |\mathsf{B}_2|^2 \, S(n_1\mathbf{k_1}) S(n_2\mathbf{k_2}) \delta(\omega - n_1\omega_1 - n_2\omega_2) d\mathbf{k_1} \tag{9.4.43}$$

where $\mathbf{k_2} = \mathbf{k_B} - \mathbf{k_1}$, $\omega_1 = \omega(\mathbf{k_1})$ and $\omega_2 = \omega(\mathbf{k_2})$. Unlike the radar cross section, the Doppler cross section needs to be taken at second-order to be correctly described because the first-order contribution is limited to the Bragg frequency. Even though the second-order contribution is in general several orders of magnitude smaller than the first-order, it can be measured in the Doppler spectrum away from the Bragg lines.

9.4.3 Calculation with Non-linear Sea Surfaces

The linear assumption is sufficient to interpret the Doppler cross section at the lowest order. However, as we will see, it is insufficient to provide a complete description of the second-order. It turns out that the different waves forming the sea surface are not truly independent but undergo hydrodynamic interactions, which are a consequence of the non-linearity of the equations of Fluid Mechanics. For small wave steepness, an assumption which is in general verified for the decametric waves involved in HFR scattering, the Fourier components of the wave field are sought in the form of a perturbation expansion starting from a reference linear surface. The *nonlinear surface* is this written in the form:

$$\eta_{\text{nonlin}}(\mathbf{r}, t) = \eta(\mathbf{r}, t) + \eta_2(\mathbf{r}, t) + \eta_3(\mathbf{r}, t) + \cdots, \tag{9.4.44}$$

where $\eta(\mathbf{r}, t)$ is some reference linear surface, η_2 is a quadratic functional of η, etc. The solutions at each order of steepness are found iteratively by solving the classical boundary value problem for the velocity potential, assuming irrotational motion. This approach was developed by Barrick and Weber [27,283], who obtained an expansion of the nonlinear surface in terms of Fourier components. Considering the partial Fourier transform of roughness at fixed time, which is the quantity of interest in the perturbation expansion of the scattered field, we can rewrite the Barrick and Weber nonlinear correction of the sea surface as:

$$\widehat{\eta}_2(\mathbf{k_B}, t) = \int d\omega_1 d\omega_2 \, \mathrm{e}^{-\mathrm{i}(\omega_1+\omega_2)t} \int \mathbf{dk_1} \mathrm{i}\Gamma_H \, a(\mathbf{k_1}, \omega_1) a(\mathbf{k_B} - \mathbf{k_1}, \omega_2), \tag{9.4.45}$$

where Γ_H is the coupling coefficient (2.7.38) and

$$a(\mathbf{k}, \omega) = \frac{1}{(2\pi)^3} \int d\mathbf{r} d\omega \, \mathrm{e}^{-\mathrm{i}(\mathbf{k}\cdot\mathbf{r}-\omega t)} \eta(\mathbf{r}, t) \tag{9.4.46}$$

is the three-dimensional Fourier transform of the surface. This term is formally similar to the second-order amplitude A_2 in the perturbation expansion of the scattered field (9.2.9), since the latter can be rewritten as:

$$\mathsf{A}_2(\mathbf{k}, \mathbf{k_0}, t) = \int d\omega_1 d\omega_2 \, \mathrm{e}^{-\mathrm{i}(\omega_1+\omega_2)t} \int \mathbf{dk_1} \mathsf{B}_2 \, a(\mathbf{k_1}, \omega_2) a(\mathbf{k_B} - \mathbf{k_1}, \omega_2) \tag{9.4.47}$$

The quadratic nonlinear correction of the surface, when combined with the first-order scattering term $(-\mathrm{i}\mathsf{B}_1\widehat{\eta}_2(\mathbf{k_B}, t))$, therefore adds up to the second-order scattering term combined with the linear surface. Formally, this amounts to operate the following replacement over the second-order kernel:

$$\mathsf{B}_2 \to \mathsf{B}_2 + \Gamma_H \mathsf{B}_1 \tag{9.4.48}$$

Since the calculations are otherwise unchanged, this yields the following general expression for the second-order Doppler cross section of nonlinear sea surfaces:

$$\sigma_2^{\text{nonlin}}(\omega) = 2\pi q^2 \sum_{n_1,n_2=\pm 1} \int |\mathsf{B}_2 + \mathsf{B}_1\Gamma_H|^2 \, S(n_1\mathbf{k_1})S(n_2\mathbf{k_2})\delta(\omega - n_1\omega_1 - n_2\omega_2)d\mathbf{k_1} \tag{9.4.49}$$

where again $\mathbf{k_2} = \mathbf{k_B} - \mathbf{k_1}$, $\omega_1 = \omega(\mathbf{k_1})$, $\omega_2 = \omega(\mathbf{k_2})$ and Γ_H is given by (2.7.38) in the case of deep-water gravity waves. The complete expression of the Doppler cross section at second-order, once hydrodynamical interactions are taken into account, is thus:

$$\sigma(\omega) = \sigma_1(\omega) + \sigma_2^{\text{nonlin}}(\omega). \tag{9.4.50}$$

Note that we have ignored the third-order hydrodynamic term η_3 which, combined with the first-order scattering amplitude, also leads to a contribution that is of order 4 in elevations and is formally similar to the $\mathsf{A}_1\mathsf{A}_3^*$ contribution from third-order scattering interactions. Taking into account the complete expression of the second-order Doppler spectrum at fourth order in elevation implies replacing the kernel in (9.4.49) by:

$$|\mathsf{B}_2 + \mathsf{B}_1\Gamma_H|^2 + 2\mathrm{Re}[\mathsf{B}_1^*\Gamma_3] \tag{9.4.51}$$

for some third-order hydrodynamical coupling coefficient Γ_3. As noted by Creamer [52] (see also the discussion in [206]), the 1–3 hydrodynamical combination corrects the second-order ocean wave spectrum (2.7.39) from Barrick and Weber theory and prevents it from diverging at large wavenumbers. This correction has in general little impact in the frequency range of the ocean wave spectrum that contributes to the HFR Doppler cross section ($k_B = 0.1$–1 rad/m). However, it can become important if the roughness parameter is no longer small, which can occur even at moderate sea state in the upper part of the HF radio band. Besides this contribution to the second-order Doppler spectrum, third-order wave interactions are responsible for a change in the gravity wave dispersion relation [283], which modifies the celerity of the Bragg wave and has some implications in the interpretation of the surface current. This will be discussed in Sect. 12.1.2.

9.5 How Much Is Backscattered from the Sea?

9.5.1 The Radar Equation

The NRCS is a key factor for measuring the backscattered power. For a transmitting antenna of gain G_T and a co-located receiving antenna of gain G_R, the backscattered power from a sea surface patch of given area at a distance R is given by the so-called *radar equation*:

$$P_r = P_i \frac{G_T G_R \lambda_0^2}{(4\pi)^3 R^4} \sigma^0 \times \text{Area}, \tag{9.5.52}$$

where P_i is the transmitted power. At grazing angles where the illumination operates via a ground wave mechanism, the radar equation must be corrected by the round-trip Norton attenuation factor $F^4(R)$:

$$P_r = P_i \frac{G_T G_R \lambda_0^2}{(4\pi)^3 R^4} F^4(R) \sigma^0 \times \text{Area} \tag{9.5.53}$$

This additional factor $F^4(R)$ is equal to 1 at distances below the horizon (non-grazing angles) and quantifies the magnitude of the two-ways propagating surface wave beyond the horizon. The adaptation of the radar equation to the bistatic case is straightforward:

$$P_r = P_i \frac{G_T G_R \lambda_0^2}{(4\pi)^3 R_1^2 R_2^2} F^2(R_1) F^2(R_2) \sigma^0 \times \text{Area} \tag{9.5.54}$$

where R_1 and R_2 are the distance from the surface patch to the transmitter and receiver, respectively. Let us now apply the monostatic radar equation with Norton attenuation factor (9.5.53) to a radar cell, for which the area $R \Delta R \phi$ is proportional to the distance. If we neglect the variation of the sea surface radar cross section with distance along a given azimuth (this will be further commented in Chap. 10), when we obtain for the relative received power:

$$\frac{P_r(R)}{P_r(R_0)} = (R/R_0)^{-3} (F(R)/F(R_0))^4, \tag{9.5.55}$$

where R_0 is some reference distance at close range. It follows that the backscattered power decays approximately as R^{-3} due to the slow variation of the Norton attenuation factor.

9.5.2 Attenuation by Roughness

It is well known to radio-oceanographers that a heavy sea state can significantly diminish the received power compared to a calm sea and thus reduces the range of operation (see e.g. [155]). This dependence on sea state is not accounted for by the simple R^{-3} power law of (9.5.55). However, the effect of sea roughness on the basic transmission loss can be modelled by modifying the Norton attenuation factor in the radar equation. One way of doing this is to replace the seawater complex impedance $Z_r = \sqrt{\mu_0/\epsilon_r}$ with an effective impedance which accounts for the effect of surface roughness in the attenuation of the mean electric field. This was first done by Barrick [13,14] using the simple model of the one-dimensional Zenneck wave (Sect. 8.1) combined with Rayleigh-Rice perturbation approach. The weakness of this approach is that it is based on the Zenneck surface wave model, the physical reality of which has been denied in the literature. In the Appendix H, we propose an alternate and more general derivation of the effective impedance, based on the mean reflection coefficient. It is found that the effective surface impedance Z_e includes an additional term, which is proportional to the power spectrum of the surface:

$$Z_e = Z_r + \int P(\mathbf{k}_1) S_0(\mathbf{k}_1) d\mathbf{k}_1, \tag{9.5.56}$$

with:

$$P(\mathbf{k}_1) = \frac{(\mathbf{k_1} \cdot \mathbf{k_0})^2}{K_0 \sqrt{K_0^2 - \|\mathbf{k_0} + \mathbf{k_1}\|^2} + Z_r K_0^2} \tag{9.5.57}$$

The effective surface impedance can be computed at different wind speeds using a ocean wave spectrum model. Figure 9.6 shows a simulation with a Pierson-Moskowitz omnidirectional ocean spectrum combined with a symmetrized $\cos^4 \phi$ angular function (see Appendix A). As seen, both real and imaginary parts of the effective surface impedance follow a linear trend with wind speed and can increase up to about 30% as compared with a flat surface, with an effect more pronounced when the radar look direction is aligned with the wind direction. This induces an augmentation of the numerical distance and has an significant impact on the Norton attenuation factor and the basic transmission loss (8.4.25). Figure 9.7 shows the excess transmission loss of a rough sea surface as compared to a flat sea surface. The effective surface impedance has been calculated using the same directional spectral model and assuming the radar to be directed upwind. As expected, the excess transmission loss increases with sea state and ranges from about 1–6 dB depending on the radar frequency. Once the Norton factor has been corrected for by the effect of surface roughness, one can use (9.5.55) to evaluate the relative backscattered power in a radar cell. Figure 9.8 shows this attenuation as a function of range for a reference cell of $R_0 = 1$ km and a typical HFR frequency of 16 MHz. At 10 m/s wind speed, additional losses of the order of 10 dB are observed at large distances when compared to the flat ocean surface. Refined estimations of the effective surface impedance using directional spectral models and comparison with experimental measurements can be found in [14,85,86,113,155]. The decrease of the relative backscattered power sets the actual operating range of HFR. Taking a typical value of 60 dB for the maximum

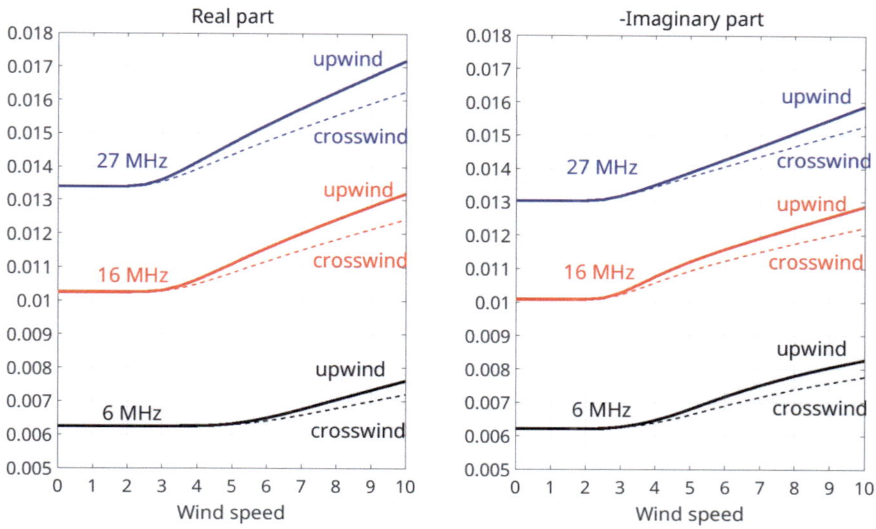

Fig. 9.6 Effective surface impedance (in absolute value) for different wind speeds and radar frequencies according to a Pierson-Moskowitz ocean wave spectrum with a $\cos^4 \phi$ angular spreading function. In this example, the radar look direction is either aligned with (solid lines) or perpendicular to (dashed lines) the wind direction.

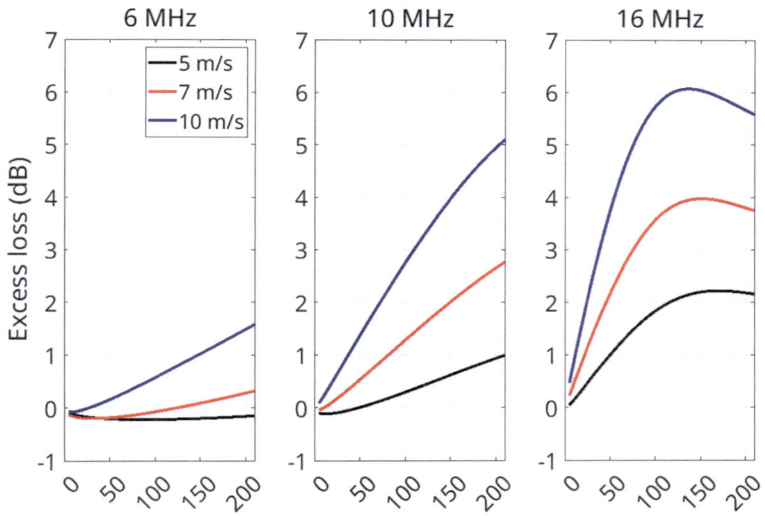

Fig. 9.7 Excess transmission loss due to sea roughness as a function of the propagation distance for various radar frequencies

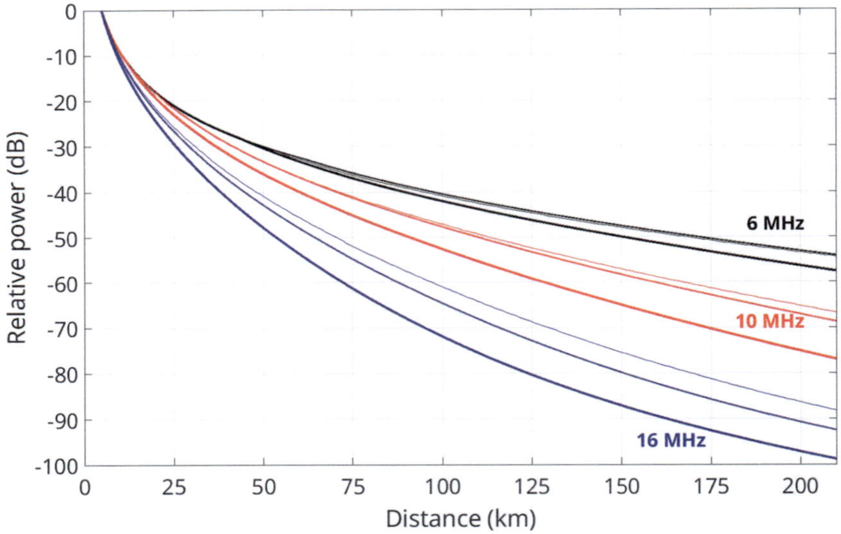

Fig. 9.8 Relative backscattered power at 3 radar frequencies (6, 10 and 16 MHz, sorted by color) as a function of range for 3 wind speeds (0, 5 and 10 m/s, sorted by increasing line thickness). The reference cell is located at $R = 1$ km

SNR (a value that also depends on the transmitted power) and a radar frequency of 16 MHz gives a range of 75–100 km depending on sea state. This is consistent with what is observed with the HFR station in Toulon described in Chap. 11.

Second-Order Doppler Spectra for HFR 10

We now come to what lies at the heart of the theoretical foundation of the HFR, namely the analytical derivation of the second-order ocean Doppler spectrum. For this we use the general description of the Doppler spectrum derived in the last chapter using the small perturbation approach and make it explicit in the case of HFR remote sensing. We recover the classical monostatic results known as Barrick's theory and also give, in a common framework, the much less classical bistatic results. We end up with Walsh and Gill theory, which is an alternative mathematical theory to describe the bistatic second-order Doppler spectra and is not fully equivalent to Barrick's theory.

10.1 Barrick's Theory

The classical theory of oceanographic HFR addresses a particular case of scattering theory. It assumes a vertically polarized electric field (both on transmit and receive), a monostatic configuration ($\mathbf{k} = -\mathbf{k_0}$), a grazing incidence ($\theta_i = \pi/2, k_0 = K_0$) and a strongly conducting surface ($|\epsilon_r| >> 1$ or $|Z_r| << 1$). The monostatic formulas for the second-order Doppler spectrum were first obtained by Barrick [15] and exploited by the same author in a series of seminal papers [17, 18, 173]. For this reason, this theoretical framework is often referred to as "Barrick's theory". In his famous 1972 paper, Barrick did not provide the details of his derivation and it seems that very few authors checked his formulas with their own derivation [114, 115, 227]. Here we rederive Barrick's theory as a limiting case of the second-order radar and Doppler cross sections that were obtained in the last chapter. However, the grazing limit of the radar cross section requires some caution because the kernel B_1 converges to

C.-A. Guérin, *High-Frequency Radars for the Oceanographer*, Geophysics and Environmental Physics, https://doi.org/10.1007/978-3-031-95642-3_10

different values according to whether the perfectly conducting limit is performed before or after the grazing angle limit. Indeed, we obtain from (G.3.17):

$$q_0 \mathsf{B}_1 \simeq -(4\epsilon_r - 2)q_0^2 \tag{10.1.1}$$

when taking the grazing limit ($q_0 \to 0$) for a given finite conductivity ϵ_r, while we obtain:

$$q_0 \mathsf{B}_1 \simeq -4K_0^2 \tag{10.1.2}$$

for a fixed incidence angle in the perfect conductor limit ($|\epsilon_r| \to \infty$), where as usual K_0 is the vacuum wavenumber and $q_0 = K_0 \cos\theta_i$ is the vertical component of the incident wave vector. It follows that in the former case, the cross section goes to zero as $\cos^4\theta_i$ (a theoretical behavior of the radar cross section that has been proven by Tatarskii [253] to hold universally for a plane wave illumination) while in the latter case it has a nonzero finite limit. Which limit should one consider? To answer this question, note that the transition point between the two asymptotic regimes takes place at $\epsilon_r q_0 \simeq q_0'$, that is around the pseudo-Breswter angle. In the HF band, the complex relative permittivity takes very large values (from about 4000 to 25000, see Fig. 7.2) and the pseudo-Brewster angle ranges from about 89.70 to 89.85°. Therefore the asymptotic regime (10.1.1) can only be attained at these extreme grazing angles. Now, it is practically impossible to produce a "pure" illuminating plane wave at such a grazing incidence.[1] This means that the incident field always contains incidences for which the limit (10.1.2) holds and prevails. As a result, we may take this last limit in the expressions (9.3.20) and (9.4.36). This yields:

$$\sigma_1^0 \simeq 2\mathcal{N} S_0(\mathbf{k_B}) \tag{10.1.3}$$

for the first-order radar cross section and:

$$\sigma_1(\omega) = \mathcal{N}\left(S(\mathbf{k_B})\delta(\omega - \omega_B) + S(-\mathbf{k_B})\delta(\omega + \omega_B)\right) \tag{10.1.4}$$

for the first-order Doppler cross section, with a normalization factor given by:

$$\mathcal{N} = 2^5 \pi K_0^4 \tag{10.1.5}$$

We recall that $\mathbf{k_B} = -2\mathbf{k_0}$ is the monostatic Bragg vector, $\omega_B = \sqrt{g k_B}$ is the radian Bragg frequency, $S(\mathbf{k})$ is the directional wavenumber ocean spectrum defined in

[1] As seen in Chap. 6, dipole antennas have a large angular width in the vertical plane, with a FWHM of 78° for the half-wavelength dipole. Shaping a very narrow beam at grazing incidence would require a very large vertical antenna. Now, for an antenna of length $L >> \lambda_0$, it is well-known from antenna theory (e.g. [8]) that the angular width of the transmitted beam is about $\Delta\theta = \lambda_0/L$. Hence, for a perfectly vertical antenna to create a beam at less than $1/200$ radian from grazing would imply a physical vertical length of about $400\lambda_0$, that is a few kilometers in the HF band !.

(2.5.28) and $S_0(\mathbf{k}) = \frac{1}{2}(S(\mathbf{k}) + S(-\mathbf{k}))$ the symmetrized wavenumber power spectrum (see Appendix A). Note that these relations are consistent with the radar cross section being the integral of the Doppler cross section, see (9.4.32). These well-known formulae for the first-order radar and Doppler cross section were first established by Barrick in his famous 1972 paper [15].[2]

Next we specify the second-order cross sections from (9.3.22) and (9.4.43). For this we use the expression for the second-order perturbation kernel in the back-scattering direction (G.3.24):

$$q_0 \mathsf{B}_2(-\mathbf{k_0}, \mathbf{k_0}, \mathbf{k_0} + \mathbf{k_1}) = \frac{2(\mathbf{k_0} \cdot \mathbf{k_1})^2 - 4K_0^2(\mathbf{k_0} \cdot \mathbf{k_1}) - 4K_0^2 k_1^2}{\sqrt{K_0^2 - \|\mathbf{k_0} + \mathbf{k_1}\|^2} + K_0 Z_r} \tag{10.1.6}$$

Note that in taking the strong conductor limit we have kept the $K_0 Z_r$ term at the denominator to prevent from a singularity arising at $\|\mathbf{k_0} + \mathbf{k_1}\| = K_0$. At grazing incidence ($k_0 = K_0$ and $q_0 = 0$), the expression (10.1.6) can be simplified by noting that:

$$K_0^2 - \|\mathbf{k_0} + \mathbf{k_1}\|^2 = \mathbf{k_1} \cdot \mathbf{k_2}, \tag{10.1.7}$$

with $\mathbf{k_2} = \mathbf{k_B} - \mathbf{k_1}$. After some rearrangement this leads to:

$$q_0 \mathsf{B}_2 = -4K_0^2 \Gamma_{EM} = q_0 \mathsf{B}_1 \Gamma_{EM} \tag{10.1.8}$$

where Γ_{EM} is Barrick's EM coupling coefficient [15]:

$$\Gamma_{EM}(\mathbf{k_1}, \mathbf{k_2}) = \frac{1}{2} \frac{k_B^{-2}(\mathbf{k_1} \cdot \mathbf{k_B})(\mathbf{k_2} \cdot \mathbf{k_B}) - 2\mathbf{k_1} \cdot \mathbf{k_2}}{\sqrt{\mathbf{k_1} \cdot \mathbf{k_2}} + K_0 Z_r} \tag{10.1.9}$$

As explained in Sect. 9.4.3, the second-order Doppler spectrum arising from nonlinear EM interactions must be complemented with nonlinear hydrodynamic interactions combined with a single scattering process, where the 2 separate contributions can be unified under a common coupling coefficient as in (9.4.49). In the end, this leads to the following formula for the second-order Doppler cross section:

$$\sigma_2(\omega) = \mathcal{N} \sum_{n_1 = \pm 1, n_2 = \pm 1} \int S(n_1 \mathbf{k_1}) S(n_2 \mathbf{k_2}) \, |\Gamma|^2 \, \delta(\omega - n_1 \omega_1 - n_2 \omega_2) d\mathbf{k_1} \tag{10.1.10}$$

where the different variables satisfy the usual relations:

$$\mathbf{k_1} + \mathbf{k_2} = -2\mathbf{k_0} = \mathbf{k_B}, \; \omega_1 = \sqrt{g k_1}, \; \omega_2 = \sqrt{g k_2}. \tag{10.1.11}$$

[2] There is a misprint in the original Eq. 8a) of this reference: the factor $4\pi K_0$ should read $4\pi K_0^4$.

Here, we have regrouped under the same kernel the EM and hydrodynamical coupling coefficient:

$$\Gamma = \Gamma_{EM} + \Gamma_H \tag{10.1.12}$$

The hydrodynamical kernel given by (2.7.38) can be simplified by taking into account the Bragg resonance condition (10.1.11) and the Dirac constraint under the integral (10.1.10):

$$\Gamma_H(\mathbf{k_1}, \mathbf{k_2}, n_1\omega_1, n_2\omega_2) = -\frac{\mathrm{i}}{2}\left(k_1 + k_2 - n_1 n_2 \sqrt{k_1 k_2}(1 - \widehat{\mathbf{k_1}} \cdot \widehat{\mathbf{k_2}})\frac{gk_B + \omega^2}{gk_B - \omega^2}\right). \tag{10.1.13}$$

In this expression we recall that $g = 9.81\ m.s^{-2}$ is the Earth gravity and $\widehat{\mathbf{k_j}}$ is the unit vector along $\mathbf{k_j}$. For strong conductivity, the denominator of the EM coupling coefficient (10.1.9) is very small whenever $\mathbf{k_1}$ and $\mathbf{k_2}$ are mutually perpendicular. This induces a sharp amplitude peak for the coupling coefficient, known as the "corner reflector" effect. It can be interpreted as a double bouncing at right angles on a reflector (Fig. 10.1). Due to the finite impedance term, it is, however, not a true mathematical singularity.

Note that we obtain a "+" sign for the impedance term $+K_0 Z_r$ in the denominator of (10.1.12). This is consistent with the first version of the EM kernel given by Barrick in 1972 [10] but not with the subsequent versions starting with Lipa and Barrick [173], who claimed to have corrected a sign error in restoring a "−" sign. We maintain the "+" sign, which comes out of the general second-order kernel factor $1/(\epsilon_r q_1 + q_1')$ (a justification of which is the presence of various Fresnel coefficients in the second-order kernel, see Appendix G.3). Because of the technical difficulty it involves, it seems that no one really bothered to verify this statement, so that the sign error in the EM coupling coefficient has survived for at least 40 years. However, as was already acknowledged by Lipa and Barrick [173], the change of sign makes a negligible numerical difference in the integral. Figure 10.2 shows the magnitude of the EM and hydrodynamical coupling coefficients in the (k_{1x}, k_{1y}) plane and Fig. 10.3 shows their relative importance in the second-order Doppler spectrum.

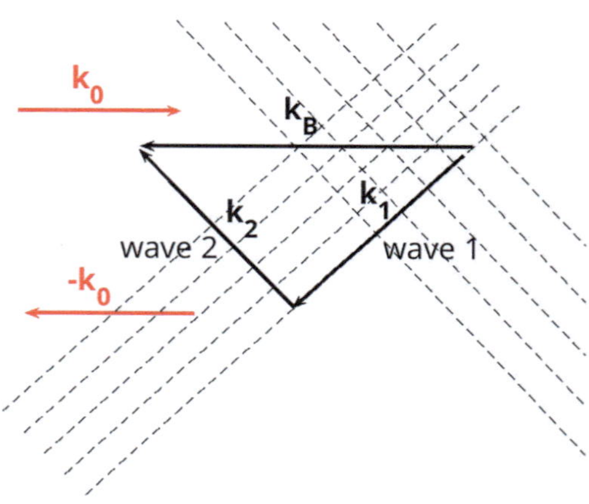

Fig. 10.1 Second-order scattering coupling two waves which propagates at right angle enhance the backscattering direction and produce the so-called "corner reflector" effect

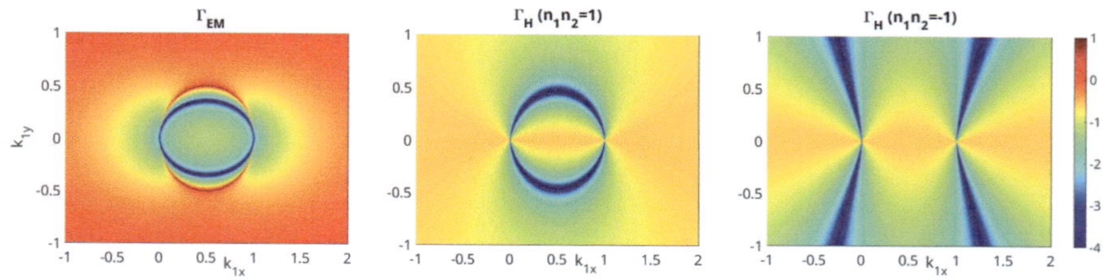

Fig. 10.2 Left panel: squared modulus of the EM kernel in logarithmic scale ($20 \log_{10} |\Gamma_{EM}|$) in the (k_{1x}, k_{1y}) wavenumber plane. All units have been normalized by the Bragg wavenumber (which amounts to take $k_B = 1$). The Bragg wave vector is assumed to be directed along the x axis. The "corner reflector" singularity is visible on the circle of center $(1/2, 0)$ and radius $1/2$, corresponding to the locus $\mathbf{k_1} \cdot (\mathbf{k_B} - \mathbf{k_1}) = 0$. Middle and right panel: Same for the hydrodynamical kernel ($20 \log_{10} |\Gamma_H|^2$) for the 2 cases $n_1 n_2 = \pm 1$. The colorbar is common to the 3 plots

The expression (10.1.10) of the second-order Doppler cross section was given for the first time in [16] and can be found commented in greater details in [173]. In the derivation of the Doppler cross section, we obtained a common normalization factor $\mathcal{N} = 2^5 \pi K_0^4$ for the first- and second-order component (10.1.4) and (10.1.10), respectively. It is worthy to mention that the normalization factors $\mathcal{N}_1$ for $\sigma_1(\omega)$ and $\mathcal{N}_2$ for $\sigma_2(\omega)$ vary in the literature depending on the conventions that are used for the wave spectrum and the Doppler cross section, which might be misleading for the new practitioner of HFR. In his 1972 papers [10,15,16], Barrick applied Rice perturbation theory [222] defining the wavenumber frequency spectrum as the Fourier transform of the spatio-temporal correlation function with a cumbersome factor $1/\pi^3$ instead of $1/(2\pi)^3$ as in (A.1.4); another source of confusion is the original choice to define the Doppler cross section by $\sigma^0 = \frac{1}{2} \int_{-\infty}^{\infty} \sigma(\omega) d\omega$, making it twice as large as the standard Doppler cross section, which is defined through (9.4.32). This results in $\mathcal{N}_1 = \mathcal{N}_2 = 2^4 \pi K_0^4$ in these early papers. In later works [17,18], the standard $1/(2\pi)^3$ normalization for the wavenumber frequency spectrum was adopted but the directional ocean wavenumber spectrum was defined as the sum of two separate symmetric spectra, differing by a factor 2 from the directional ocean wavenumber $S(\mathbf{k})$ used by oceanographers, see (9.4.40). This results in $\mathcal{N}_1 = 2^7 \pi K_0^4$ and $\mathcal{N}_2 = 2^8 \pi K_0^4$ in these papers. The convention was stabilized in the papers with Lipa [170,171,173] by restoring the missing 1/2 factor in the definition of the directional ocean wavenumber, leading to equal normalization factors $\mathcal{N}_1 = \mathcal{N}_2 = 2^6 \pi K_0^4$. However, they still differ from ours by a factor 2, which can be attributed to the implicit definition of the Doppler cross section through (9.4.32).

The second-order theory for the Doppler spectrum has been developed in the absence of surface current. If one assumes a uniform surface current $\mathbf{U}$, with no variation in depth, then the gravity wave dispersion relation is changed to:

$$\omega(\mathbf{k}) = \sqrt{gk} + \mathbf{k} \cdot \mathbf{U} \tag{10.1.14}$$

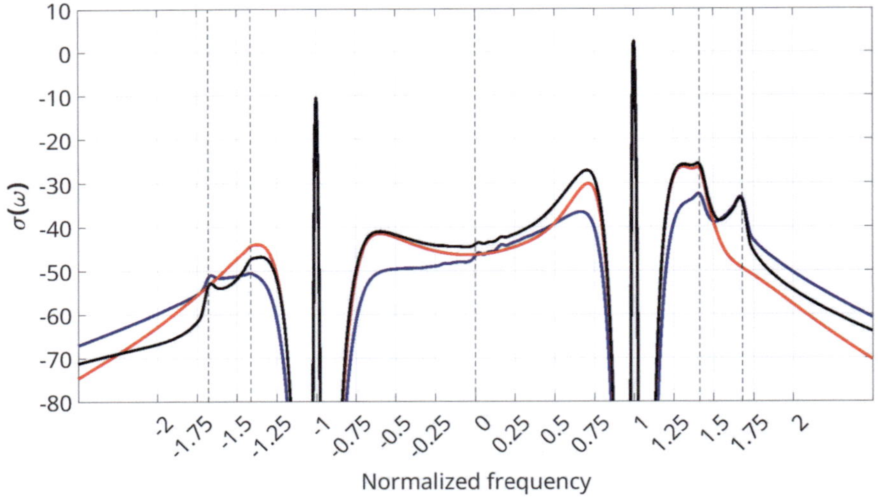

Fig. 10.3 Second-order Doppler spectrum at radar frequency $f_0 = 16$ MHz as a function of the normalized Doppler frequency ω/ω_B (black line). The ocean wave spectrum is a directional ocean Pierson-Moskowitz spectrum at 10 m/s wind speed with the radar looking upwind (black solid lines). The Doppler spectrum has been smoothed over 0.01 Hz to take into account the finite integration time. The red line shows the effect of the hydrodynamical kernel alone ($\Gamma = \Gamma_H$) and the blue line shows the effect of the EM kernel alone ($\Gamma = \Gamma_{EM}$). The black dashed vertical lines mark the $\pm 2^{1/2}$ and $\pm 2^{3/4}$ singularities

Hence, the Dirac condition in the second-order integral (10.1.10) is modified in the following way:

$$\delta(\omega - n_1\omega_1 - n_2\omega_2) \rightarrow \delta(\omega - n_1\omega_1 - n_2\omega_2 - (\mathbf{k_1} + \mathbf{k_2}) \cdot \mathbf{U}), \tag{10.1.15}$$

where as usual $\omega_1 = \sqrt{gk_1}$ and $\omega_2 = \sqrt{gk_2}$. In view of the Bragg resonance condition $\mathbf{k_1} + \mathbf{k_2} = \mathbf{k_B}$, this entails:

$$\omega = n_1\omega_1 + n_2\omega_2 + \mathbf{k_B} \cdot \mathbf{U}, \tag{10.1.16}$$

which implies that the second-order Doppler spectrum undergoes the same overall translation $\mathbf{k_B} \cdot \mathbf{U}$ as the first-order Doppler spectrum, a result that was first shown by Robson [227]. Things happen to be more complicated when the current is varying with depth because the effective current $\mathbf{U}_e$ that enters in the dispersion relation depends on the wave vector [74,151,251]. In that case, the second-order spectrum is not uniformly translated, hence deformed. It seems, however, that little research has been made on this topic.

10.2 Bistatic Geometry

In the bistatic geometry, the source and the receiver are situated at distant locations. Even though the concepts are identical to those of the monostatic case, the technical difficulty of the derivation of the radar and Doppler cross section is much greater.

For this reason, the bistatic cross section was only given at first-order in the beginning [15] and it took three more decades before the first bistatic expressions of the second-order Doppler spectrum appeared [89,92]. However, these expressions rely on an alternative mathematical approach with different hypothesis and are not fully equivalent to Barrick's theory when specialized to the monostatic case (this will be discussed in details in Sect. 10.3). A consistent extension of second-order Barrick's theory to the bistatic case was developed only half a century later in the works by Hardman and Wyatt [114,115]. We provide here the bistatic results as a natural extension of the monostatic theory in the framework of the perturbation approach of Chap. 9.

10.2.1 Bistatic Doppler Effect

To establish the Doppler shift formula in the bistatic case it suffices to mimic the demonstration of the Doppler effect in the monostatic case (Sect. 2.2). Consider a moving target at position $\mathbf{R}(t)$ travelling at speed $\mathbf{V}$ and illuminated by a monochromatic source at position $\mathbf{R}_t$, with harmonic time dependence $e^{-i\omega_0 t}$. According to Huyghens principle, the target radiates a secondary wave which is in phase with the excitation field. This scattered field is measured at some other position $\mathbf{R}_r$ (Fig. 10.4). Accounting for the change of phase along a path connecting the source to the receiver via the target, we may write the phase of the complex received field as:

$$\Phi(t) = -\omega_0 t + \mathbf{K_0} \cdot \mathbf{R_1} - \mathbf{K} \cdot \mathbf{R_2}, \tag{10.2.17}$$

where $\mathbf{R}_1 = \mathbf{R}(t) - \mathbf{R}_t$ and $\mathbf{R}_2 = \mathbf{R}(t) - \mathbf{R}_r$ are the instantaneous position vector of the target seen from the transmitter and receiver, respectively. Differentiating the phase in time we therefore have:

$$\frac{d\Phi}{dt} = -\omega_0 + (\mathbf{K_0} - \mathbf{K}) \cdot \mathbf{V}(t) = -\omega_0 + \|\mathbf{K} - \mathbf{K_0}\| \, V_n(t), \tag{10.2.18}$$

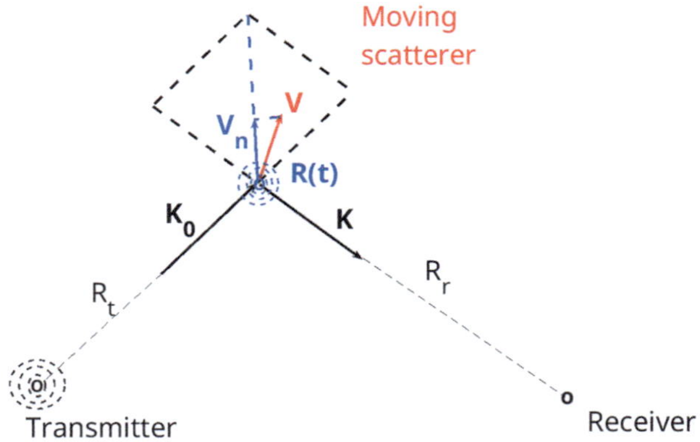

Fig. 10.4 The Doppler frequency shift caused by a moving target seen from a bistatic receiver is proportional to its elliptical velocity component V_n

where $V_n = \mathbf{V} \cdot \mathbf{n}$ is the component of the speed vector $\mathbf{V}$ along the direction:

$$\mathbf{n} = \frac{\mathbf{K} - \mathbf{K_0}}{\|\mathbf{K} - \mathbf{K_0}\|} \tag{10.2.19}$$

It follows that the received signal undergoes an instantaneous Doppler frequency shift:

$$f = -\frac{1}{2\pi} \|\mathbf{K} - \mathbf{K_0}\| \, V_n(t) \tag{10.2.20}$$

When the receiver and the source are colocated, we have $\|\mathbf{K} - \mathbf{K_0}\| = 2K_0 = 2\omega_0/c_0$, where c_0 is the speed of light in vacuum, and V_n is the radial speed V_r, so that the classical monostatic Doppler frequency shift (2.2.10) is recovered.

10.2.2 The Bistatic Bragg Frequency

A bistatic Bragg wave can be easily determined using the same simple physical argument as was employed in Sect. 2.3. Consider a sinusoidal surface whose spatial period is given by the horizontal vector $\boldsymbol{\lambda}$, illuminated by a plane wave with wave vector $\mathbf{K_0}$. The scattered field in the direction $\mathbf{K}$ interfers constructively with the incident field whenever $(\mathbf{K} - \mathbf{K_0}) \cdot \boldsymbol{\lambda} = (\mathbf{k} - \mathbf{k_0}) \cdot \boldsymbol{\lambda}$ is a multiple of 2π (Fig. 10.5). This defines a bistatic Bragg wavenumber:

$$\mathbf{k_B} = \mathbf{k} - \mathbf{k_0} \tag{10.2.21}$$

and a Bragg wavelength:

$$\lambda_B = \frac{2\pi}{k_B} \tag{10.2.22}$$

These last quantities can be advantageously described with help of the so-called *bistatic angle* φ , which is defined by:

$$\mathbf{k_0} \cdot \mathbf{k} = -\cos(2\varphi), \tag{10.2.23}$$

yielding the following simple expressions

$$k_B = 2K_0 \cos\varphi, \; \lambda_B = \frac{2\pi}{K_0} \cos\varphi \tag{10.2.24}$$

Using the dispersion relation of gravity waves $c_B = \sqrt{g/k_B}$ combined with the bistatic Doppler formula (10.2.20), we obtain the *bistatic Bragg frequency*:

$$f_B = \sqrt{\frac{g \cos\varphi}{\pi \lambda_0}} \tag{10.2.25}$$

Monostatic expressions for the Bragg wavenumber, wavelength and frequency are recovered with $\varphi = 0$.

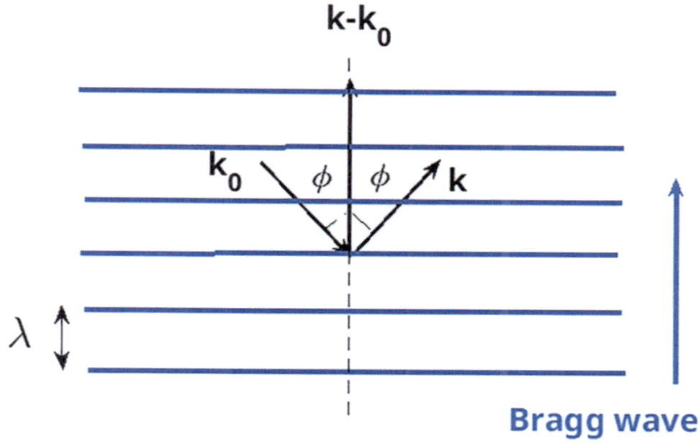

Fig. 10.5 The Bragg wave is oriented along $\mathbf{k} - \mathbf{k_0}$ and depends on the bistatic angle φ

10.2.3 Bistatic Surface Current Mapping

Bistatic HFR can be used to map surface currents using the same principles as in the monostatic case with a few technical adaptations:

- The Doppler shift of the main Bragg lines must be measured from the bistatic Bragg frequency, which varies over the radar coverage.
- When sorting the bistatic returns in time, the monostatic range R_0 should be replaced by the bistatic range $(R_1 + R_2)/2$.

The range rings at a given bistatic distance describe ellipses encircling the transmitter and the receiver that represent its focal points (Fig. 10.6). Note that the direction **n** of the Bragg wave vector is locally perpendicular to these ellipses. For this reason, the speed component V_n is referred to as the *elliptical velocity component*. Bistatic surface current mapping has been applied only by a few groups in the HFR community [63, 102, 179, 263, 264]. It is slightly more complicated to install as it requires an extra radar site and a perfect synchronization of the remote transmitters and receivers. However, it can be advantageous in certain situations where installing a monostatic system is difficult for various reasons: lack of physical space, strong direct coupling between TX and RX, propagation losses due to natural obstacles on the round-trip [100]. Nevertheless, the bistatic geometry has also some limitations such as the increased spread of Bragg lines [124] or spurious peaks in the Doppler spectra caused by the variation of the bistatic Bragg frequency across the radar cells, which can are enhanced by the secondary lobes of the range or azimuthal processing windows [100].

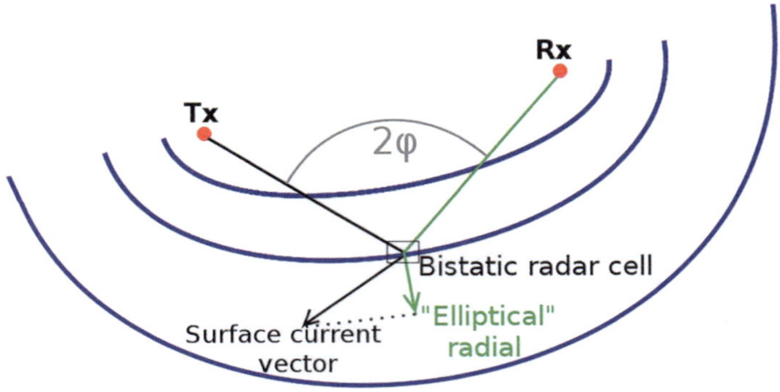

Fig. 10.6 The range cells in bistatic geometry are located along ellipses with focii at the transmitter and receiver locations. The half-angle φ between the transmitter and receiver direction is called the bistatic angle

10.2.4 Bistatic Radar and Doppler Cross Sections

An explicit formulation of the second-order bistatic radar and Doppler cross sections in the context of HFR can be obtained, again, by specializing the general expression of Sects. 9.3 and 9.4 to the case of a vertically polarized electric field at grazing angle over a strongly conducting medium. Taking $q = q_0 \to 0$ in expression (G.3.21) gives:

$$q\mathsf{B}_1 \simeq -4K_0^2 \cos^2 \varphi \tag{10.2.26}$$

and from Eq. (9.3.20) we obtain the first-order bistatic radar cross section:

$$\sigma_1^0 = 2\mathcal{N} \cos^2 \varphi \; S_0(\mathbf{k_B}) \tag{10.2.27}$$

and the first-order bistatic Doppler cross section:

$$\sigma_1(\omega) = \mathcal{N} \cos^2 \varphi \left(S_0(\mathbf{k_B})\delta(\omega - \omega_B) + S_0(-\mathbf{k_B})\delta(\omega + \omega_B)\right), \tag{10.2.28}$$

where as usual $\mathcal{N} = 2^5 \pi K_0^4$ and $\omega_B = 2\pi f_B$ is the radian bistatic Bragg frequency corresponding to (10.2.25). The second-order bistatic cross section can be obtained by mimicking the derivation of the monostatic case with the appropriate kernels. For the bistatic second-order kernel we obtain from (G.3.22) after some simplifications:

$$q\mathsf{B}_2(\mathbf{k}, \mathbf{k_0}, \mathbf{k_0} + \mathbf{k_1}) = -\frac{2(\mathbf{k} \cdot \mathbf{k_1})(\mathbf{k_0} \cdot \mathbf{k_1}) + 4K_0^2 \cos^2 \varphi (\mathbf{k_0} \cdot \mathbf{k_1}) + 4K_0^2 k_1^2 \cos^2 \varphi}{\sqrt{K_0^2 - \|\mathbf{k_0} + \mathbf{k_1}\|^2} + K_0 Z_r} \tag{10.2.29}$$

After some algebra this can be written as:

$$q\mathsf{B}_2 = q\mathsf{B}_1 \Gamma_{EM}^{bi}, \tag{10.2.30}$$

with

$$\Gamma^{bi}_{EM} = \frac{1}{2}\frac{k_B^{-2}(\mathbf{2k}\cdot\mathbf{k_1})(-\mathbf{2k_0}\cdot\mathbf{k_2}) - \mathbf{k_1}\cdot\mathbf{k_2}}{\sqrt{\mathbf{k_1}\cdot[\mathbf{k_2}-(\mathbf{k}+\mathbf{k_0})]}+K_0 Z_r}, \tag{10.2.31}$$

where $k_B = 2K_0\cos\varphi$ and $\mathbf{k_2} = \mathbf{k_B} - \mathbf{k_1}$. From (9.4.49) it follows that:

$$\sigma_2(\omega) = \mathcal{N}\cos^4\varphi \sum_{n_1=\pm1,n_2=\pm1} \int S(n_1\mathbf{k_1})S(n_2\mathbf{k_2})\,|\Gamma|^2\,\delta(\omega - n_1\omega_1 - n_2\omega_2)d\mathbf{k_1} \tag{10.2.32}$$

with $\Gamma = \Gamma^{bi}_{EM} + \Gamma_H$, where the hydrodynamical kernel (10.1.13) is unchanged. Note that the monostatic expression (10.1.9) is recovered for $\mathbf{2k} = -\mathbf{2k_0} = \mathbf{k_B}$. By symmetrizing the kernel B_2 as in (G.2.2), it is seen to coincide with the expression derived in [114]. The second-order bistatic Doppler cross section has essentially the same features as the monostatic ones, with four side bands surrounding the Bragg peaks. It presents the same "hydrodynamical" singularity at $\omega = \sqrt{2}\,\omega_B$ and a "corner reflector" singularity at a frequency that now depends on the bistatic angle [92,114]:

$$\omega = \pm 2^{3/4}\omega_B\left(\frac{\sqrt{1\pm\sin\varphi}}{\cos\varphi}\right)^{1/2} \tag{10.2.33}$$

10.3 Walsh and Gill Theory

The radar and Doppler cross section derived within the framework of Barrick's theory, and which we have re-derived using the general concept of scattering amplitude at arbitrary incidence, are based on the simplifying assumption of an infinite sea surface illuminated by a perfectly monochromatic plane wave. Consideration of the finite sea surface patch only comes into play at the level of the radar equation, which uses the radar cell cross section as if it were from an infinite surface. Since the radar cell is generally much larger than the Bragg wave, this is a very good approximation in most cases and has the advantage of simplifying the formalism considerably. However, there may be some extreme cases where the Doppler spectrum becomes sensitive to the waveform of the transmitted field; this occurs at very high spatial resolution when the Bragg wavelength is no longer negligible compared to the range resolution. An alternative to Barrick's theory has been developed and pursued for 25 years by the group of Walsh and Gill at the Memorial University of Newfoundland (Canada). It is more complex mathematically but can be applied to any type of illumination with no limitation on size of the sea surface patch. We will refer to this approach as the "Walsh and Gill" theory.

In a series of papers in the 1980s, Walsh and co-workers [273,274,276] developed a rigorous theory for modeling EM scattering from rough surfaces using the mathematical framework of generalized functions. They make use of Heaviside functions to define characteristic functions of the different geometrical domains (above or below the surface) and the associated fields. For an arbitrary current source, they obtained a general expression for the electric field at any point in space as an integral equation involving only the free space Green's function and the gradients of these characteristic functions. This formalism was extended in [275] to give rigorous expressions for the tangential and normal electric fields at the surface, as well as the scattered field above the surface, in the form of integral operators involving the roughness function and its normal vector (Eqs. (27)–(28) of that reference). These operators can be simplified and inverted under the assumption of good conducting surface and small surface roughness and slopes. In the end, general integral expressions are obtained for the scattered field, the calculation of which is tractable if a specific form is assumed for the current source and if a perturbation expansion in scattering orders is performed for the scattered field. Once the time-dependent scattered field is obtained, its auto-correlation function can be calculated to obtain the bistatic Doppler cross section at different orders.

A very simple (though not the most realistic) model for the current source is the pulsed sinusoidal dipole:

$$\mathbf{j}(\mathbf{R}, t) = e^{-i\omega_0 t} W(t/\tau)\delta(\mathbf{R})\mathbf{j_0}, \tag{10.3.34}$$

where W is the rectangular window (3.4.23) and τ is the pulse duration. For this source it has been shown in [92] that the first-order bistatic Doppler cross section is given by:

$$\sigma_1(\omega) = 2^4 \pi K_0^4 \, \Delta R \, K_0^{5/2} g^{-1/2} \cos\varphi \, \text{sinc}^2 \left[\frac{\Delta R}{2} \left(\frac{k}{\cos\varphi} - 2K_0 \right) \right] \sum_{n_1=\pm 1} S(n_1 \mathbf{k_B}), \tag{10.3.35}$$

where φ is the bistatic angle, $\Delta R = \frac{1}{2} c\tau$ is the range resolution and $k = \omega^2/g$, and the second-order bistatic Doppler cross section is given by:

$$\begin{aligned} \sigma_2(\omega) = 2^3 \pi K_0^2 \cos\varphi \, \Delta R \sum_{n_1,n_2} \int_0^\infty \Bigg(\int |\Gamma|^2 k^2 \, S(n_1 \mathbf{k_1}) S(n_2 \mathbf{k_2}) \\ \times \, \text{sinc}^2 \left[\frac{\Delta R}{2} \left(\frac{k}{\cos\varphi} - 2K_0 \right) \right] \delta(\omega - n_1\omega_1 - n_2\omega_2) d\mathbf{k_1} \Bigg) dk \end{aligned} \tag{10.3.36}$$

with the constraint $\mathbf{k_1} + \mathbf{k_2} = k\widehat{\mathbf{k_B}}$ and $\omega_1 = \sqrt{gk_1}$, $\omega_2 = \sqrt{gk_2}$. Here again, the coupling coefficient Γ combines the hydrodynamic and EM second-order interactions. In the limit of an infinitely large sea surface patch ($k_B \Delta R \to \infty$), one can replace the sinc function by a Dirac function:

$$Lx \, \text{sinc}^2(Lx) \to \pi\delta(x), \; L \to \infty \tag{10.3.37}$$

Together with the gravity wave dispersion relation $\omega = \sqrt{gk}$, this leads to:

$$\sigma_1(\omega) = 2^6 \pi K_0^4 \cos^2\varphi \, (S(\mathbf{k_B})\delta(\omega - \omega_B) + S_0(-\mathbf{k_B})\delta(\omega + \omega_B)) \tag{10.3.38}$$

and

$$\sigma_2(\omega) = 2^6 \pi K_0^4 \cos^4 \varphi \sum_{n_1, n_2} \int |\Gamma|^2 \, S(n_1 \mathbf{k_1}) S(n_2 \mathbf{k_2}) \delta(\omega - n_1 \omega_1 - n_2 \omega_2) d\mathbf{k_1} \tag{10.3.39}$$

This gives the first- and second-order bistatic cross sections (10.2.28) and (10.2.32) derived for an infinite sea surface (except for a factor of 2, again due to the different convention that the integral of the Doppler cross section is only half of the radar cross section). For a finite range resolution, this shows that there is a broadening of the first-order Doppler peak due to the finite pulse duration. A simple calculation using the squared sampling function in (10.3.35) shows that the FWHM of the first-order Bragg peak is $\sim g/(f_B \Delta R)$. For a typical HFR installation operating at 16 MHz ($f_B = 0.41$ Hz) with a range resolution $\Delta R = 1500$ m, this gives 0.01 Hz, which is much less than the peak broadening due to the mere influence of the long-wave orbital velocity alone ($\Delta f = 0.02 - 0.06$ Hz, see Sect. 3.5). It would therefore require an exceptionally high range resolution (e.g. 300 m, corresponding to a radar transmission bandwidth $B = 500$ kHz), to be sensitive to the shape of the transmitted waveform in the first-order Bragg peak. Such high resolution are quite rare because of the customary bandwidths allowed. Hence, in most of the case, it is sufficient to use the infinite patch limit (10.3.38) and (10.3.39) of Walsh and Gill theory.

Gill and Walsh [92] were the first to explicitly derive the expression for the second-order Doppler spectrum in a bistatic configuration. However, it should be noted that their coupling coefficient is not fully consistent with the expression (10.2.31) derived with a classical perturbation expansion. While the hydrodynamic part Γ_H is derived from the classical nonlinear expansion by Barrick and Weber [27,283], the EM coupling coefficient is specific to their approach. It is given by Eq. (25) of [89], adapted to our notations):

$$\Gamma_{EM}^{GW} = \frac{1}{2\mathrm{i} K_0^2 \cos \varphi} \frac{\mathbf{k_1} \cdot \mathbf{k_0} (\mathbf{k_2} \cdot (\mathbf{k_1} - \mathbf{k_0}))}{\sqrt{K_0^2 - \|\mathbf{k_0} + \mathbf{k_1}\|^2}}, \tag{10.3.40}$$

where the "GW" subscript stands for "Gill and Walsh". Apart from the missing impedance term in the denominator, which has been heuristically corrected in later work [277], this expression differs substantially from the previously derived kernel (10.2.31). This is easier to see in the monostatic case. After symmetrization of the kernel (10.3.40), it was shown [89] that it can be brought into the form:

$$\Gamma_{EM}^{GW} = -\frac{1}{2k_B^2} \frac{\|\mathbf{k_1} \times \mathbf{k_2}\|^2}{\sqrt{\mathbf{k_1} \cdot \mathbf{k_2}}}, \tag{10.3.41}$$

which can be rearranged, using $\mathbf{k_2} = \mathbf{k_B} - \mathbf{k_1}$, in:

$$\Gamma_{EM}^{GW} = \frac{1}{2k_B^2} \frac{(\mathbf{k_1} \cdot \mathbf{k_B})^2 - k_B^2 k_1^2}{\sqrt{\mathbf{k_1} \cdot \mathbf{k_2}}} \tag{10.3.42}$$

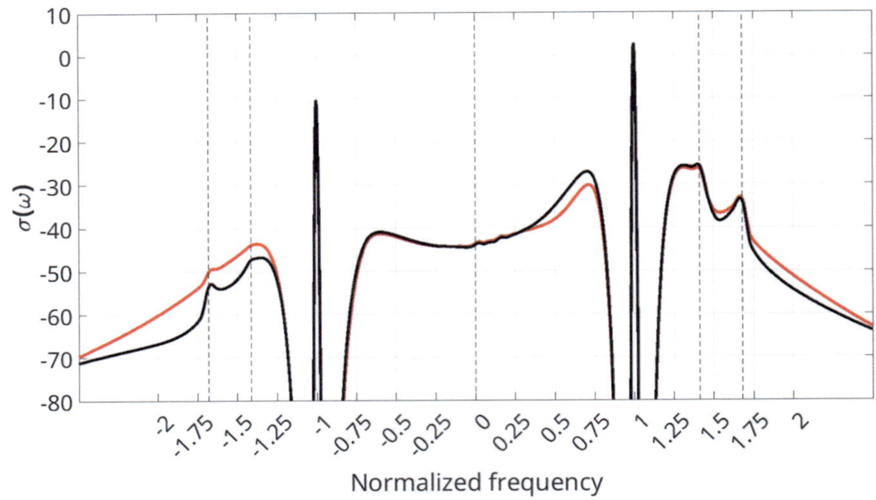

Fig. 10.7 Comparison of the monostatic Doppler cross section using Barrick's (black line) or Gill and Walsh's coupling coefficient (red line), for a Pierson-Moskowitz upwind spectrum by 10 m/s wind speed at the operating frequency $f_0 = 16$ MHz. The maximal difference is of the order of 3 dB

In comparison with the monostatic kernel (10.1.9) from Barrick's theory (in which we ignore the impedance term), there is a difference:

$$\Gamma_{EM}^{GW} - \Gamma_{EM}^{Barrick} = \frac{1}{2}\frac{2(\mathbf{k_1} \cdot \widehat{\mathbf{k_B}})^2 + (\mathbf{k_1} \cdot \mathbf{k_B}) - 3k_1^2}{\sqrt{\mathbf{k_1} \cdot \mathbf{k_2}}}, \tag{10.3.43}$$

which due to the quadratic terms cannot be removed by using one of the allowed transformations of the kernel (see Appendix G.3). Therefore, the two coupling coefficients have a different net contribution to the second-order integral of the Doppler spectrum. The reason for this inconsistency lies in the different approach of Walsh and Gill, which is not really a perturbation expansion in wave height, although it does small-height assumption at some point. Figure 10.7 illustrates the difference between the Barrick and Gill and Walsh formulations of the second-order Doppler spectrum using a numerical simulation with a Pierson-Moskowitz spectrum at 10 m/s wind speed. The discrepancy remains less than 3 dB, which may be insufficient for an experimental evaluation of the best method given the uncertainty of the ocean wave spectra.

Part III

From Principles to Reality

11 The HFR Theory in Practice

It is now time to illustrate the theoretical concepts developed in the course of this book with some specific examples, mainly taken from our daily practice of the HFR networks in the region of Toulon and Nice, France. We briefly present the instruments and sites and illustrate the main properties of the radar signal, both in the time and frequency domain, with some elementary examples. We introduce the range Doppler representation and show the difference between the omnidirectional Doppler spectra, which are not resolved in azimuth, and the directional Doppler spectra resulting from azimuthal BF processing. We also show the difference between monostatic and bistatic systems. We then discuss the choice of integration time and the number of incoherent sample spectra, and give some guidelines on how to obtain radial surface current maps with the different types of azimuthal processing. Finally, we discuss the issue of calibration, which has a major impact on the accuracy and resolution of the current maps.

11.1 How Does a HFR Installation Look Like?

The Mediterranean Institute of Oceanography (MIO)[1] has been operating two HFR networks on the French mediterranean shore for about 15 years.[2] The first network consists of a couple of WERA[3] HFR systems working at 16.15 MHz installed in the

[1] https://www.mio.osupytheas.fr/.

[2] These instruments were purchased with the support of the Centre National de la Recherche Scientifique (CNRS) and the University of Toulon. They have been installed by the former MIO staff of engineers Yves Barbin, Joel Gagelli, Pascal Guterman and Céline Quentin. We acknowledge the outstanding contribution of Yves Barbin (retired in 2014) to the MIO HFR network. He pioneered the bistatic configuration and the DF processing of irregular arrays.

[3] "WEllen RAdar", https://helzel.com/.

C.-A. Guérin, *High-Frequency Radars for the Oceanographer*, Geophysics and Environmental Physics, https://doi.org/10.1007/978-3-031-95642-3_11

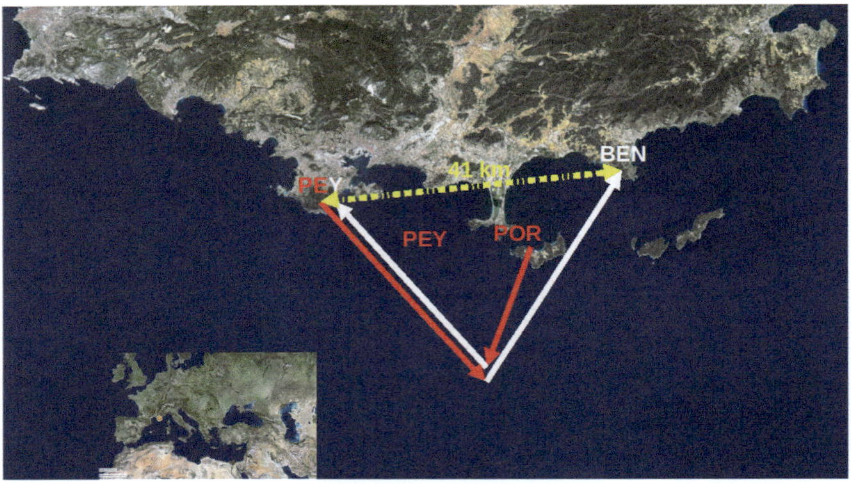

Fig. 11.1 The multistatic HFR network in Toulon is composed of one TX/RX station in Fort Peyras ("PEY"), one RX station in Cap Bénat ("BEN") and one stand-alone TX in Porquerolles ("POR")

region of Toulon and has been operated continuously since 2011. The second network consists of 2 SEASONDE CODAR stations[4] working at 13 MHz and installed in Nice [103] and Menton, which have been operated from 2015 to 2024 and 2020 to 2024, respectively. The WERAs have been deployed on 3 different sites and work in multistatic configuration [102]. The main radar site, which houses both a TX and a RX station is located at Fort Peyras, henceforth referred to as "PEY", about 10 km south-west of Toulon, at an altitude of 160 m; the second site, which is located at the Cap Bénat (henceforth referred to as "BEN") about 40 km East of Toulon at an altitude of about 165 m, hosts a RX station; the third site, located South of Porquerolles island, henceforth referred to as "POR", hosts a stand-alone TX (Fig. 11.1).

The 2 TX and the 2 RX are synchronized, thus providing the measurements of 4 different TX-RX pairs. The pair PEY-PEY is monostatic while the other 3 pairs PEY-BEN, POR-BEN and POR-PEY are bistatic. The RX antenna systems are linear arrays of 12 antennas. This HFR network covers the coastal region of Toulon up to 60–80 km offshore, depending on the days. Figure 11.2 shows the main elements of the HFR installation in Fort Peyras. Without entering into technical details, there are the electronic bay hosting the waveform generator, the mixers, the receiver input filters and channels, the Analogic-to-Digital converters and the computer for storing and processing the data; the generated signal is sent to the power amplifier and then radiated as EM waves by the transmit array. The backscattered EM waves are collected by the receive array and conducted to the receivers, where they are mixed with a copy of the transmitted waveform. Coaxial wires connect the transmit and receive arrays to the electronic bay. To minimize the optical path difference between the various antennas, these cables are have identical indices and length.

[4] https://codar.com/seasonde/.

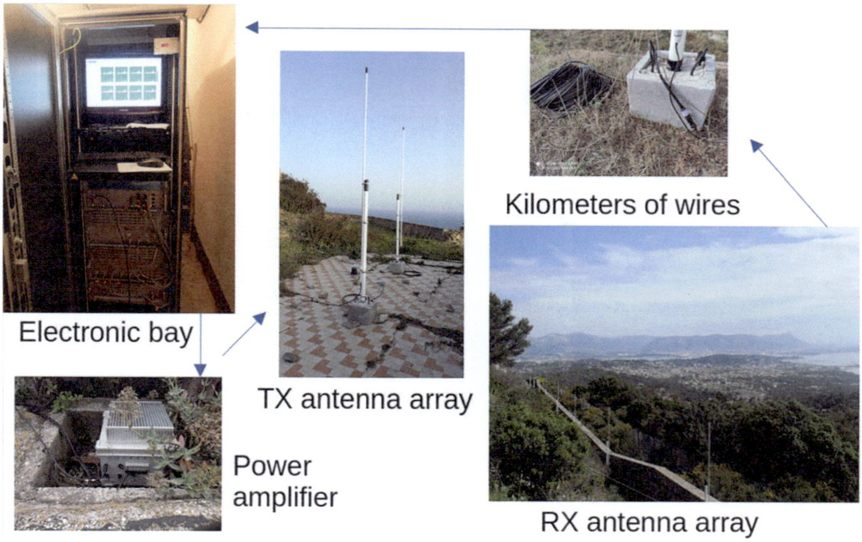

Fig. 11.2 A HFR phase array system is composed of an electronic bay, a power amplifier, a TX and RX antenna arrays and coaxial cables connecting the different parts

The transmitter generates continuous pulses with a chirp waveform of duration 0.26 s around the central frequency 16.15 MHz ($f_B = 0.41$ Hz) within a band of $B =$ 100 kHz allowing for a range resolution $\Delta R = 1.5$ km (see (3.2.3)). The based band I/Q signal is digitized over 16 bits at a rate of 5907 Hz (that is, every chirp is sampled with 1536 points). After FFT, there are therefore 1536 range cells but only the 65 first ones are retained, corresponding to the distance within 96 km from the radar, beyond which the signal is dominated by noise. This operation is performed continuously for every received antenna over N_{chirps} contiguous chirps (to be chosen by the user). In the end the range-sorted complex signal is stored in a ".SORT" file that contains (besides the header) a complex matrix of dimensions $N_{\text{chirps}} \times N_{\text{ranges}} \times N_{\text{antennas}}$. Note that the signal that is contained in a SORT file is omnidirectional, that is prior to any azimuthal processing. More details on the technical design and processing steps of the WERA can be found in [108].

The MIO CODAR station in Nice is shown in Fig. 11.3. It is operating at 13 MHz ($f_B = 0.37$ Hz) with a bandwidth of 100 kHz. The TX/RX antenna system is composed of a unique mast hosting a wip antenna that serve both on transmit and receive and 2 small RX magnetic loop antennas. It is connected to an electronic bay, which contains the similar elements to the WERA system. The radar data are optionally stored under different formats according to the required level of processing. The range-sorted times series as stored in the ".RS" files, that contain a complex matrix of dimensions $N_{\text{chirps}} \times N_{\text{ranges}} \times N_{\text{antennas}}$ with $N_{\text{antennas}} = 3$. The antenna cross-spectra that are used for the DF processing are stored in the ".CS" files. The sweep rate of the transmitted waveforms is 2 Hz, which limits the Doppler frequency band to $\pm 1 Hz$ but is sufficient for most applications.

Fig. 11.3 A compact HFR system (here the MIO CODAR near Nice) consists mainly of a unique TX/RX antenna mast and an electronic bay

11.2 The Radar Signal in the Time Domain

Even though the signal from HFR is rarely used nor visualized in the time domain, it is interesting to start with a look at the time series of the received voltage attached to a given radar cell, that is after a preliminary processing in range and azimuth. Figure 11.4 shows an example of I/Q signal extracted from a SORT file of the monostatic HFR station in Fort Peyras (PEY-PEY) for one fixed direction (South-East) and various ranges. The oscillations at the Bragg frequency are clearly visible in the short- and middle range signal but they are no longer seen in the far range where the signal is dominated by noise. Why is there a modulation of amplitude in the Bragg oscillation? This is a consequence of the slow frequency shift induced by the surface current, which creates a beat phenomenon between the positive and negative Bragg frequencies. To see this, consider the idealized model of a complex backscattered signal produced by two symmetric Bragg waves, discarding noise and second-order effects:

$$s(t) = Ae^{i\varphi}e^{\mathrm{i}\omega_B t} + Ae^{i\varphi}e^{-\mathrm{i}\omega_B t} \tag{11.2.1}$$

for some amplitude $A > 0$. By taking the real and imaginary parts we obtain the I and Q signals in the form:

$$\begin{aligned} I(t) &= 2A\cos\varphi\cos(\omega_B t), \\ Q(t) &= 2A\sin\varphi\cos(\omega_B t). \end{aligned} \tag{11.2.2}$$

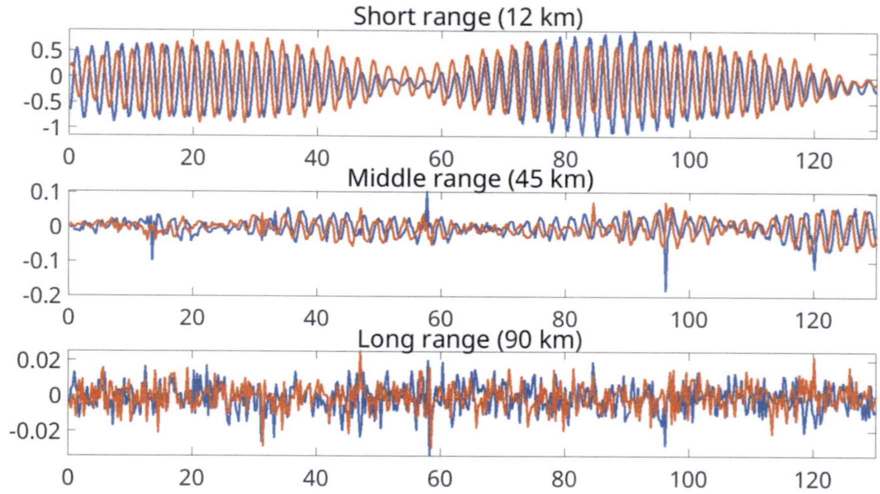

Fig. 11.4 Examples of I/Q signals from a .SORT file of the monostatic HFR station in Fort Peyras (PEY-PEY) at near-, middle- and long-range, respectively. The Bragg oscillation at around 0.4 Hz is plainly seen in the near and middle range with an increasing level of noise. The amplitude modulation is due to the combined effect of surface currents and long waves

In the presence of a surface current, the positive and negative Bragg frequency are Doppler-shifted by the same amount, say ω_U:

$$s(t) = Ae^{i\varphi}e^{\mathrm{i}(\omega_B+\omega_U)t} + Ae^{i\varphi}e^{-\mathrm{i}(\omega_B-\omega_U)t}, \tag{11.2.3}$$

so that the I and Q signal are now given by:

$$\begin{aligned} I(t) &= 2A\cos(\omega_U t + \varphi)\cos(\omega_B t) \\ Q(t) &= 2A\sin(\omega_U t + \varphi)\cos(\omega_B t) \end{aligned} \tag{11.2.4}$$

It can be shown [104] that this modulation formula can be extended to the case where the 2 Bragg peaks have different amplitudes and phases. The factors $\cos(2\pi f_U t + \varphi)$ produces a slow modulation in amplitude of the Bragg frequency (recall that $f_U << f_B$). This beat phenomena between the positive and negative Bragg frequency was identified by Crombie as early as 1966 [57], where he observed these interference patterns in range-time records from an HFR experiment. Note that the identification of this amplitude modulation can serve as a basis to estimate the current, which is a variant of the spectral method [104]. In presence of long waves, the Bragg wave amplitude A is no longer constant and oscillates at the frequency of the dominant wave. This causes another slow modulation of the I and Q signal which mixes up with the current induced modulation.

11.3 The Radar Signal in the Frequency Domain

As explained in Chap. 3, the complex Doppler spectrum can be obtained by performing a simple FFT of the time signal; we recall that it is referred to as *directional* or *omnidirectional* depending on whether the radar signal has been processed in azimuth prior to Doppler processing. Figure 11.5 shows the directional spectra associated to the time series shown in Fig. 11.4.

The spectra have been calculated with 2 min integration time using a Hann tapering window and an incoherent summation over 15 half-overlapping subseries covering

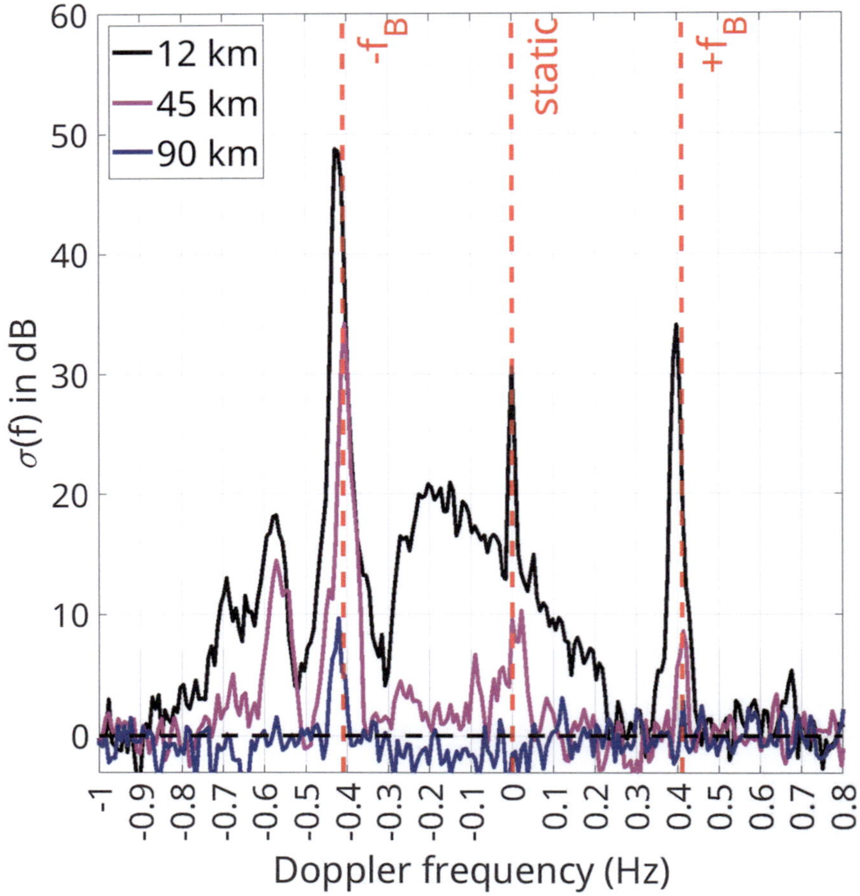

Fig. 11.5 Directional Doppler spectra (measured 40° clockwise from East) associated to the time series shown in Fig. 11.4 for an observation time of about 18 min. The red dashed vertical lines show plus or minus the theoretical Bragg frequency $f_B = 0.41$ Hz and the zero-Doppler peak emanating from static targets. The secondary peaks observed at short (12 km) and middle (45 km) range are the contribution of the second-order Doppler spectrum. At long range (90 km), only the strongest first-order Bragg peak is still visible

18 min observation time (which is the duration of a SORT file). They have been normalized by the noise floor at each range which can be estimated by averaging the tail of the spectra at large Doppler frequencies (say, $f > 1.5$ Hz) where the physical signal is in general too weak to be measured. At short range, the Bragg peaks are well marked with a very high SNR (about 50 dB for the strongest peak). Secondary peaks are also visible in the vicinity of the Bragg frequency; these are the contribution of the second-order components of the backscattered Doppler spectrum. The second-order Doppler spectrum is no longer visible at the largest range (90 km), where only the most energetic peak emerges around the negative Bragg frequency at about 10 dB above the noise level. The sharp peak observed at the zero Doppler in the short and middle range is due to the echo of static targets as for instance metallic objects or structures (antenna tower, buildings, anchored boats, etc.) in the surroundings. This static echoes are not completely eliminated in the directional Doppler spectrum even if the radar cell does not intersect any piece of land; this is because their strong echo can be "caught" by a secondary lobe of the angular window resulting from the beam-forming processing.

Next we investigate the azimuthal dependence of Doppler spectra. Figure 11.6 shows the directional Doppler spectra at a fixed range (45 km) but various bearings for the same case as Fig. 11.5. As before, the spectra have been normalized by their noise level. Superimposed is also shown the omnidirectional Doppler spectrum. Recall that the radial velocity of the surface current varies across the radar coverage for geometrical as well as oceanographical reasons (see Chap. 2). Hence the extra Doppler shifts induced by the surface current depends on the direction of observation so that the position of the Bragg peak changes with the direction. Since the omnidirectional spectrum cumulates the sea echos from all directions, this has the effect of broadening the main Bragg peaks. The directional Doppler spectra, however, only "see" the Doppler shift induced by the surface current in one azimuthal radar call; the corresponding Bragg lines are thus sharper than their omnidirectional counterpart. It is also seen that the omnidirectional Doppler spectrum is more sensitive to the static echoes (zero Doppler frequency) that can arise from certain directions.

11.4 Range-Doppler Maps

A very useful representation of the Doppler spectrum is the so-called *Range-Doppler map*. It plots the Doppler spectrum (expressed in dB) in colorscale in the frequency-range half-plane and thus gives a synoptic view of its structure at various distances from shore. In this map the main two Bragg peaks appear as vertical lines with slight deformations due to the surface current. The second-order Doppler spectrum appears as weaker and broader side-bands. Range-Doppler maps can be formed from either the omnidirectional or directional Doppler spectrum. In the omnidirectional case, the first- and second-order peaks are broader because they cumulate the various shifts caused by the radial surface current at the different radar bearings. Figure 11.7 displays the Range-Doppler map for the omnidirectional and directional Doppler spectra using the same monostatic record as in Fig. 11.5. The maximum

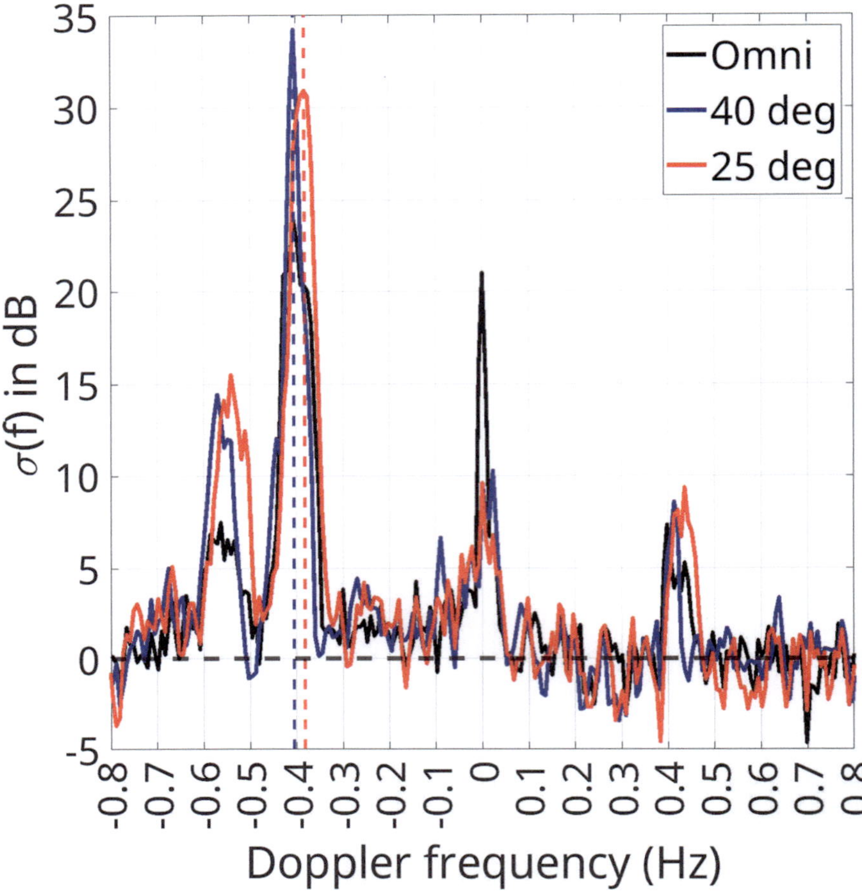

Fig. 11.6 Directional Doppler spectrum obtained in various directions (measured clockwise from East) for the middle range case (45 km) of Fig. 11.5. The omnidirection Doppler spectrum (black line) is shown for comparison. The Bragg peaks are narrower and the SNR is enhanced with the directional Doppler spectrum and the static echoes are more pronounced for the omnidirectional Doppler spectrum

useful range for surface current inversion can be visually seen as the maximal excursion of the strongest Bragg lines (here, about 70–80 km). Again, the directional Doppler spectrum displays more contrasted structures and enhanced SNR compared to its omnidirectional counterpart. In this example, the second-order side bands are clearly visible, especially on the negative Doppler side, and well separated from the first-order Bragg peaks. Ships can also be seen in the range-Doppler map as isolated spots. The bright spots on the zero-Doppler line are not related to the sea echo. They are an artifact of the power grid impurity that generates extra 50 Hz harmonics in the radar signal which, after FMCW processing, are attributed to specific range cells. When the surface currents are very strong and contrasted in azimuth, the First-Order Bragg Region can become very broad and overlap the second-order contributions in

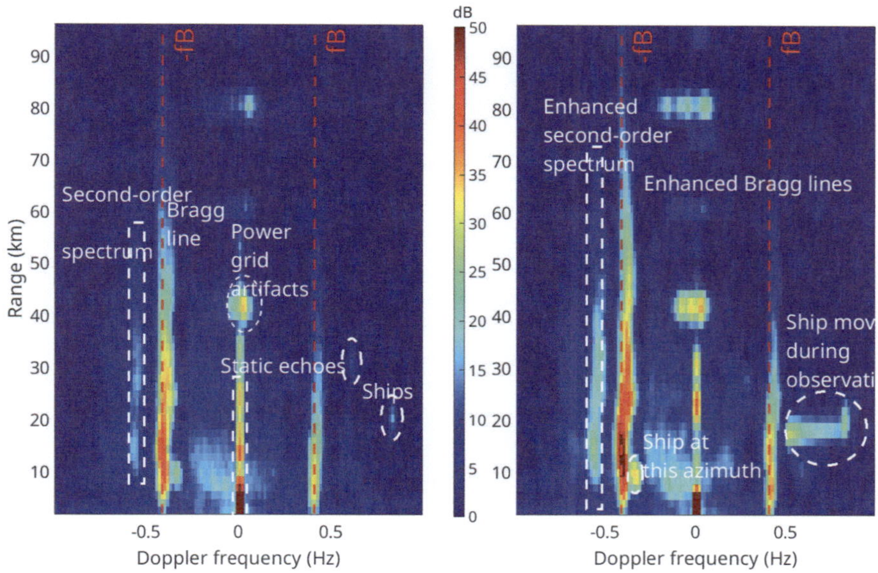

Fig. 11.7 Omnidirectional (left panel) and directional (right panel) monostatic range-Doppler maps obtained over 18 min at the HFR station of Fort Peyras. Directional Doppler spectra have enhanced Bragg peaks and second-order side bands

the omnidirectional spectrum. This is for example observed with the 2 HFR stations (Goury and Jobourg) that monitor the oceanic circulation in the Alderney race, Normandy, France [62, 185]. These phased-array HFR are maintained by the University of Caen (group of Prof. Anne-Claire Bennis) and operate at the dual frequencies 13.5 and 24.5 MHz. The tidal currents in the radar coverage can reach extremely high values in some directions (up to 5 m/s), leading to a strong distorsion of the Bragg peaks. Figure 11.8 shows a range-Doppler map obtained at the Goury station. As seen, the directional Doppler spectra are much thinner than the omnidirectional Doppler spectrum. A range-Doppler map can also be obtained in bistatic configuration, providing the range is replaced by the bistatic range $(R_1 + R_2)/2$ on the vertical axis. In bistatic geometry the Bragg frequency is no longer constant and varies with range and azimuth (see Eq. (10.2.25)). It follows that the Bragg lines in the range-Doppler representation are no longer straight lines but U-shaped lines starting at the range of the transmitter. An example of bistatic range-Doppler map is given in Fig. 11.9 for both the directional and omnidirectional Doppler spectra. They have been obtained with one SORT file (18 min) from the receiver in Cap Bénat coupled with the transmitter in Porquerolles (POR-BEN). At large distances with respect to the TX-RX baseline, the bistatic angle is close to zero and the bistatic Bragg frequency is close to the monostatic one. As a result, the bistatic Bragg lines align to the monostatic Bragg lines at large bistatic distances. Note also the "forking" of the

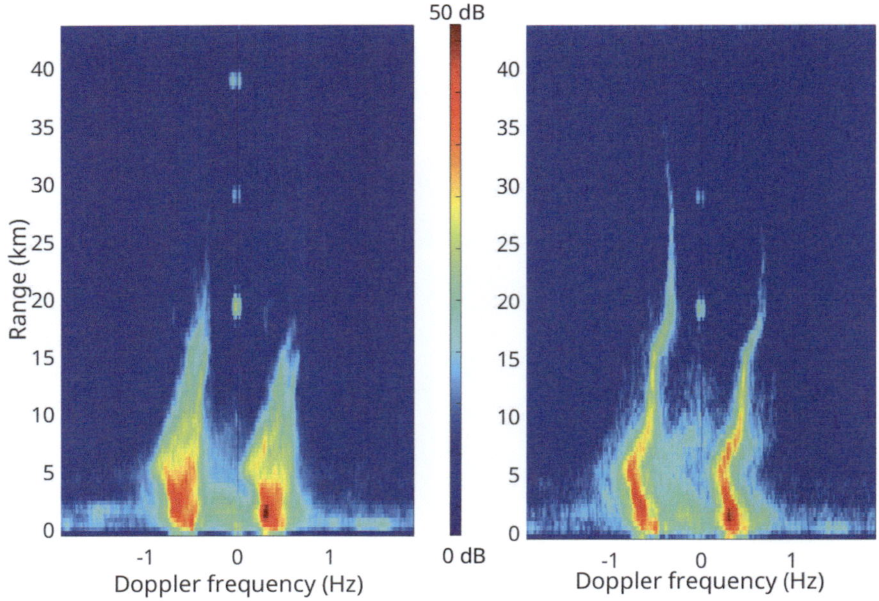

Fig. 11.8 Omnidirectional (left panel) and directional (right panel) monostatic range-Doppler maps obtained over 18 min at the HFR station of Goury. In the presence of strong contrasted surface currents, the First-Order Bragg Region becomes very broad and overlaps the second-order region. Data are courtesy from Anne-Claire Bennis (University of Caen Normandy)

negative first-order Bragg line in the omnidirectional Range-Doppler map, that is the presence of two close but separate Bragg lines. This is the typical effect of strongly contrasted radial surface currents seen in different azimuths that mixed up in the omnidirectional Doppler spectrum. This can be a source of errors when identifying the first-order Bragg lines in a DF processing since the doubling of the Bragg peak can be misinterpreted as the contribution of the second-order spectrum.

11.5 Chosing the Integration Time

As explained in Chap. 3, Doppler spectra are obtained by first performing a Fourier Transform of the complex radar signal over a duration called the integration time, T_i. The frequency resolution ($\Delta f \sim 1/T_i$) as well as the SNR ($\sim T_i$) are improved by increasing the integration time. A further incoherent integration (that is a summation of the power spectra) is performed over a duration $T_0 >> T_i$ called the observation time. This allows to "smooth" the Doppler spectra, that is to operate a variance reduction at the cost of a reduced frequency resolution and SNR. A trade-off must in general be found between these quantities while chosing the integration and observation time. It might be tempting to use the maximal integration time ($T_i = T_0$) to

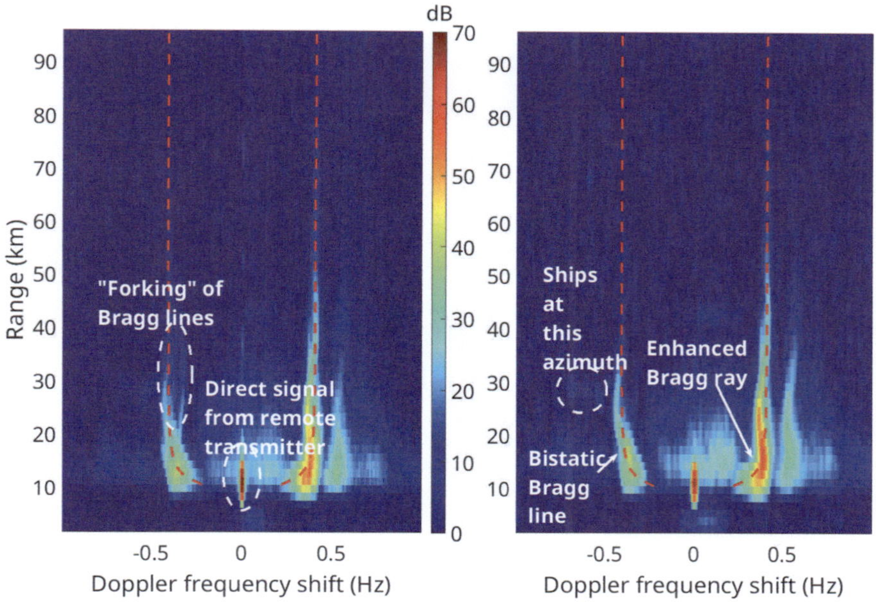

Fig. 11.9 Omnidirectional (left panel) and directional (right panel) bistatic range-Doppler maps obtained over 18 min at HFR station of Cap Bénat (coupled with Porquerolles transmitter). The vertical axis is the bistatic range

increase as much as possible the frequency resolution and the SNR and then to reduce the strong oscillations of the Doppler spectra with a simple moving average. However, from a statistical point of view, smoothing a "long" spectrum is not as good as performing the incoherent average of many "short" spectra as it results in a stronger variance (see the discussion in Sect. 3.3). Another advantage of using short samples is to mitigate transitory parasitic echoes that are captured during a long integration time (typically transcient ship echoes) and jeopardize the Doppler spectrum in some frequency intervals. Taking the incoherent *median* average of short Doppler spectra allows to eliminate those spectra that have been contaminated by a transitory noise. These effects are illustrated in Fig. 11.10 where a range-azimuth revolved Doppler spectrum has been processed with either the full integration time of 18 min (corresponding to the duration of a SORT file) or with a short integration time of 2 min with 15 overlapping subsamples. The long Doppler spectrum has been smoothed with a window that has the same frequency resolution as the short ones and the short spectra have been incoherently averaged using either the mean or the median. As seen, the mean short spectrum displays less oscillations than the smoothed version of the long spectrum, due to a smaller variance. Furthermore, in the frequency interval [0.5–0.8] Hz, the long Doppler spectrum is strongly corrupted by ships crossing the radar cell (as seen in the range-Doppler map of Fig. 11.7 around the range 18 km).

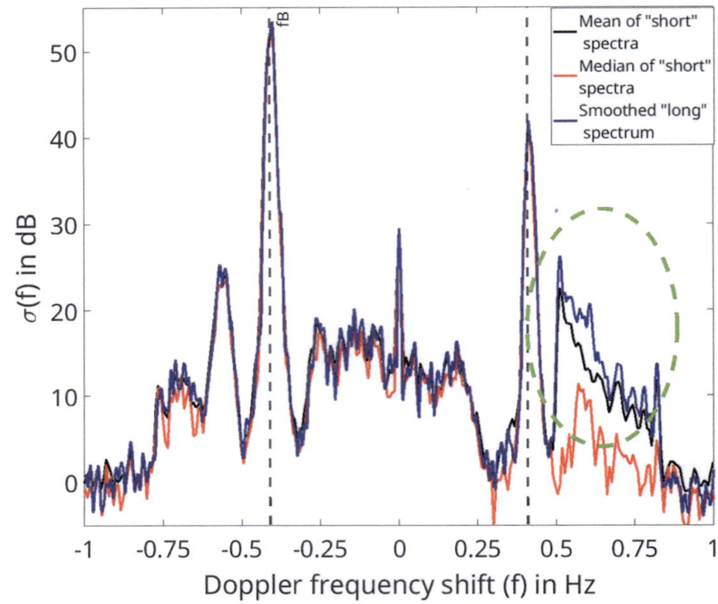

Fig. 11.10 Directional Doppler spectrum at range 18 km for the case of Fig. 11.7. The Doppler spectrum has been processed in 3 different ways: **a** through an incoherent mean of 15 overlapping "short" spectra with 2 min integration time (black solid line); **b** same using the median incoherent average; **c** using a single "long" spectrum of 18 min integration time and smoothing it at the same frequency resolution than the short spectra (red solid line). Random oscillations are weaker in case (a), indicating a stronger variance reduction. Case (c) is the most impacted by the transient echo of a ship in the frequency interval [0.5–0.8] Hz (dashed ellipse), which is clearly seen in the Range-Doppler map of Fig. 11.7. The median average (case b) is almost not affected by this artifact and allows to unveil the second-order peak that is present in this range of frequencies

This effect is reduced with the mean short spectrum and eliminated when using the median short spectrum.

11.6 Radial Maps

Maps of the radial surface current, hereafter simply referred to as *radial maps*, can be obtained from each pair of transmitter and receiver by evaluating the frequency shift of the first-order Bragg lines for each radar cell. We use the sign convention that a radial current is positive if it is flowing toward the radar (negative otherwise), thereby corresponding to a positive frequency shift from the Bragg frequency. Radial maps are obtained in different ways according to whether a BF or DF azimuthal processing is applied.

11.6.1 BF Processing

For phased-array radars, directional Doppler spectra can be obtained after a preliminary BF processing for each range and azimuth and the corresponding radial surface current can be estimated by extracting the peak positions. As seen in Sect. 3.5, the barycenter technique allows to estimate the latter with a better accuracy than that involved by the FFT grid resolution. Some simple control criteria are usually employed to check the validity of an estimated peak position (Fig. 11.11):

1. A FOBR is defined as the frequency domain over which it is legitimate to search a Bragg peak. Given an assumed maximum radial current velocity U_{max}, it follows from the Doppler shift formula (2.4.20) that it is given by two symmetric intervals around the theoretical Bragg frequency:

$$\mathrm{FOBR} = \{f : |f \pm f_B| < 2U_{\max}/\lambda_0\}, \qquad (11.6.5)$$

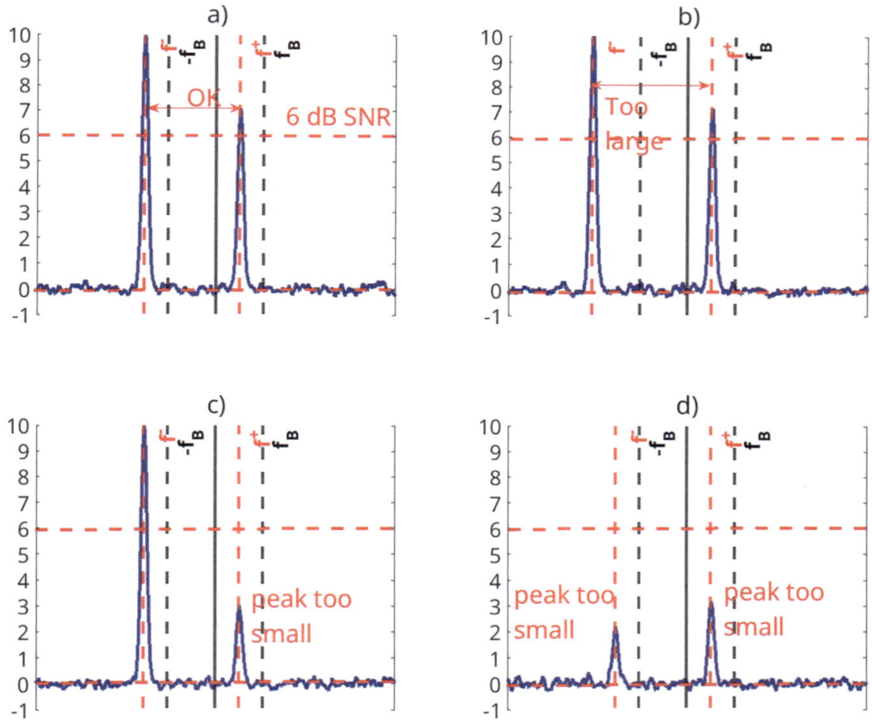

Fig. 11.11 Estimation of the radial current in one radar cell from the Bragg peak position. Case **a**: there are two admissible peaks satisfying both the 6 dB threshold and $2f_B$ spacing condition, providing a reliable estimate $U = (U^- + U^+)/2$ of the radial current. Case **b**: the two peaks satisfy the 6 dB threshold but not the $2f_B$ condition; no value is retained. Case **c** and **d**: one single peak satisfies the 6 dB threshold condition

where we recall that λ_0 is the EM wavelength. The value of $U_{\max}$ depends on the oceanographical features of each geographical site. For example, $U_{\max} = 1$ m/s in Toulon.

2. A peak of the Doppler spectrum found within the FOBR is retained if it exceeds the background level by a certain amount, typically 3 dB or 6 dB.
3. If both the negative and positive Bragg peaks f^- and f^+ are admissible, each providing an estimate U^- and U^+ of the radial current, the mean value $U = (U^- + U^+)/2$ is retained. A further quality control is to ensure that the mutual spacing between these two peaks is close to the theoretical value $2f_B$:

$$\left| f^+ - f^- - 2f_B \right| < \varepsilon \tag{11.6.6}$$

where ε is a small number that can be expressed in terms of current accuracy δU with $\varepsilon = 2\delta U/\lambda_0$.

4. If only the negative (respectively, positive) peak passes the threshold, it is U^- (respectively, U^+) which is retained to estimate the radial current. However, in that case, the previous "$2f_B$ test" cannot be applied. It is the operator's choice to keep this estimate or not, resulting from a trade-off between spatial coverage and reliability of the estimates.
5. If none of the peaks exceeds the requested threshold, the radial current has no reliable estimate and is affected to a NaN ("Not a Number") value.

11.6.2 DF Processing

For compact radars, the directional analysis can only be performed using a DF processing of the omnidirectional complex Doppler spectra. This approach has its own merits and limitations and we will here discuss the main difference with the BF processing. There is no well-defined notion of the angular resolution for DF processing even though it is somehow related to the width of the MUSIC factor (see Chap. 5). As we have seen, the bearing of a source can be determined with an arbitrary large accuracy in case of a high SNR and a perfect match of the actual and ideal antenna gains. In practice, they are many factors that limits the angular resolution obtained with DF, such as an imperfect estimation of the antenna covariance matrix, a wrongly assumed number of sources, an imperfect knowledge of the antenna lobes or an insufficient level of SNR. Therefore, the actual angular resolution depends very much on the given installation, unlike phased-array radars. Moreover, for the purpose of surface current mapping, the possible directions of sources must be binned on a radial grid. Note that, in this case, it is the grid angular step that limits the angular resolution (contrarily to BF, where the actual resolution is coarser than the angular step). The angular bins cannot be chosen too small otherwise they are very likely to remain empty after detection of the source bearings; they cannot be taken too large either, otherwise the angular resolution would be too coarse. Hence a trade-off must be found between the lacunarity of the radial map obtained from a DF processing and the angular resolution which is limited by the angular bin. As an example, Fig. 11.12 shows a hourly radial map obtained with the CODAR Seasonde operated by the MIO

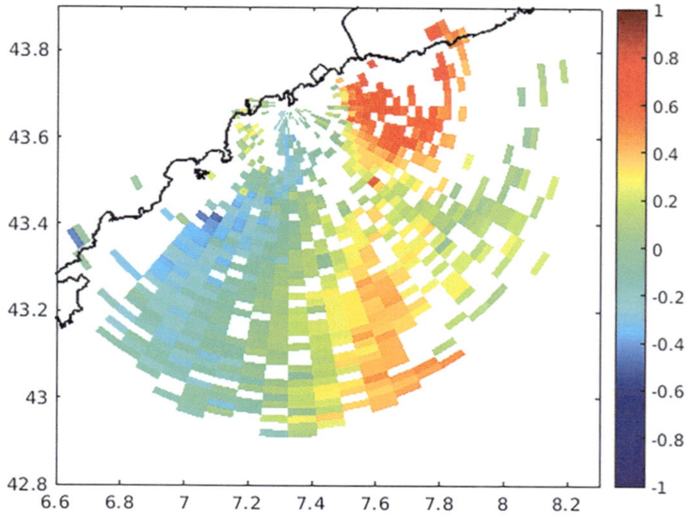

Fig. 11.12 An hourly map of the radial surface current measured with CODAR station of Saint-Jean Cap Ferrat, France. The azimuthal processing has been performed with the "RADAR PROCESSING MIO" software

in the vicinity of Nice, France [103]. The pattern of oceanic circulation is clear, with a main stream flowing westwards alongshore. However, the map is quite lacunary, even with a 5° angular step. This can be partially compensated by "temporal stacking" techniques [103] or gap filling post-processing techniques (e.g. [123]).

Another limitation of DF azimuthal processing is related to the integration time. As explained in Sect. 3.3, the coherent integration time determines the frequency resolution and impacts the SNR. However, it must be kept short enough to allow for an extra incoherent average of the Doppler spectra within a "reasonable" observation time over which ocean processes can be assumed stationary. For phased-array radars, a coherent integration time of a few minutes and an observation period of about 20 min to 1 h is usually employed but depending on the application these durations can be reduced. The extraction of radial currents with a DF processing requires in general a longer observation time (typically, one hour) and a smaller integration time (typically, 1–2 min) because the estimation of the antenna covariance matrix of the complex Doppler spectra requires a large number of samples.

The available values of the radial current are those obtained from a discrete set of Bragg rays at the FFT grid resolution. Contrarily to the BF processing, there is no way to interpolate these discrete set of frequencies by a barycenter mean and thus the estimated values of the radial current are quantified. Since the integration time cannot be long, the related frequency resolution remains coarse, resulting in a limited accuracy of the radial current estimate. This is, however, compensated by the gain in azimuthal resolution. Exactly how much accuracy is gained or losed in either case is difficult to quantify in general.

The choice of the Bragg rays that will be processed with the MUSIC algorithm requires the preliminary identification of the FOBR. As explained in Sect. 5.5, this can be problematic as the first- and second-order contributions may overlap in the omnidirectional Doppler spectrum, especially in the case of strong currents. Several methods have been developed in the recent literature to improve the identification of the first-order Bragg region, based on contour detection [154] or dynamical frequency intervals deduced from the statistics of radial currents [64].

11.6.3 DF with Arrays

Even though BF is the standard azimuthal processing method for antenna arrays, there is also the possibility to resort to DF processing techniques when the array is too short (e.g. [102,156]). The need for an improved angular resolution is obvious in certain situations where the surface current display meandering structures or eddies. A classical BF processing with an insufficient number of antennas (for example 8 or 12) can result in the smoothing or cancellation of these structures and lead to unphysical results. Figure 11.13 shows an hourly radial map obtained with the Bénat RX station (linear array of 12 antennas) using either BF or DF azimuthal processing on a polar grid with one degree step. As seen, the radial current pattern obtained with BF is smeared in azimuth, the effect being more prononced at the edges of the domain. The radial map obtained with DF unveil finer structures but is lacunary, since many cells of the polar grid have not been assigned to a source direction (most often because a threshold criterion with the MUSIC peak was not met). In addition, some directions have been wrongly assigned, resulting in outliers which appears as isolated spots in the radial map and must be eliminated a posteriori.

The lacunarity issue of DF processing when dealing with antenna arrays can be addressed by using the *antenna grouping technique* [64]. The idea is that unlike BF, the DF processing can be performed with subgroups of consecutive antennas in the array with a limited loss in azimuthal resolution if the size of the subgroups is not too small. Hence, there are as many sample radial current maps as possible subgroups. Since the "holes" in the individual lacunary maps occur at various positions of the radar grid, a sample average allows to produce a full map, wherein a stronger weight is given to the larger arrays which have in principle a better azimuthal resolution. For arrays with more than 8 antennas, the number of subgroups is important and antenna grouping brings a substancial improvement of the filling ratio. This method has been implemented in the "RADAR PROCESSING MIO" software,[5] an output of which is shown in Fig. 11.14 with hourly radial maps obtained with the 2 MIO radar stations in Toulon. In addition to filling the maps, the antenna grouping technique is robust to missing or defective antennas in the array, because it can compensate them with adjacent subgroups.

Another interesting feature of DF processing with arrays is that it be combined with a premilinary BF analysis for a better identification of first-order Bragg region [62,279], a crucial step in presence of strong currents. This recent improved DF as been referred to as the *hybrid BF-DF method* [62].

[5] Dylan Dumas and Charles-Antoine Guérin, software registered at the "Agence pour la Protection des Programmes", number 198307662.

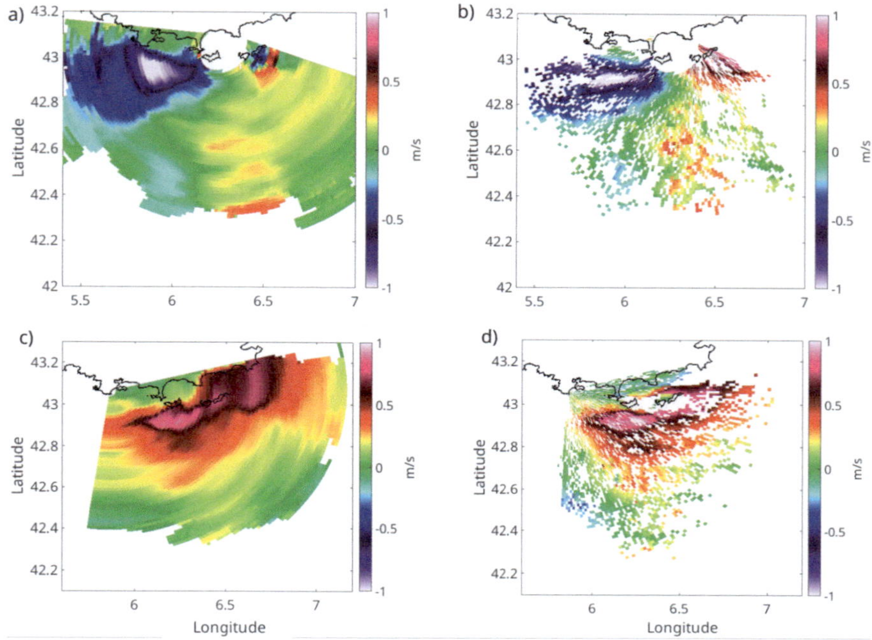

Fig. 11.13 Top panel: hourly maps of radial surface current obtained with the Bénat RX station using **a** BF; or **b** DF azimuthal processing. Bottom panel: same for the Peyras TX/RX station using **c** BF; **d** DF azimuthal processing. As seen, the maps obtained with BF are smeared in azimuth while the one obtained with DF unveils finer structures but are lacunary. In each case, a linear array of 12 antennas has been used

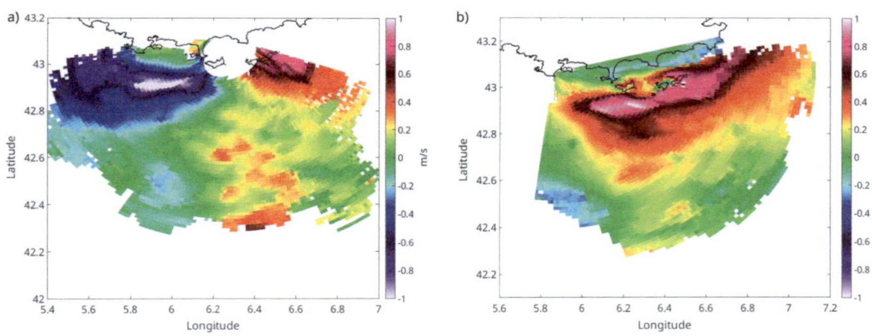

Fig. 11.14 Same as Fig. 11.13 using a DF processing with the antenna grouping technique, in which all subarrays from 8 to 12 antennas have been considered (RADAR PROCESSING MIO software)

11.7 Calibration Issues

One of the key factors that determines the quality of the HFR data is a proper calibration of the receive antenna system since improperly characterized antenna patterns deteriorate the azimuthal processing of surface currents (e.g. [163,212]). Even though the radar manufacturers have well-established procedures to reduce the antenna gain distortion and unbalance, the experimental complex antenna gains often differ from the factory settings and can evolve in time due to several environmental factors (soil moisture, surrounding metallic objetcs, corrosion of cables or connectors, etc.). It is thus important to control them as often as possible. There are several ways to establish a partial or complete calibration of the receive system, depending on the type of antennas. Some systems offer the possibility of an internal calibration to control the balance of the different receiver channels and a calibration box to control the optical length of the cables. However, this is only part of the calibration because it can only correct the various antenna channels with an overall complex factor which does not take into account the possible differences in the antenna patterns. A complete calibration from the antenna to the receiver requires measuring the electric far-field radiated in every direction.

The standard technique, which can be employed with all types of HFR systems, consists in moving a small transmitter (called a *transponder* or *Field Calibration Unit* depending on the manufacturer) around the RX station in such a way to measure the receive signal at every bearing. This can be accomplished by nautical, terrestrian or aerial means. The site configuration (lighthouse, narrow peninsula, etc.) often favors the use of shipborne transmitter to travel around the RX station. Figure 11.15 shows an example of a boat calibration campaign conducted in March 2020 for the SeaSonde CODAR MIO station in Nice [103]. The experimental lobes for the 3 antennas are averaged on the round trip and smoothed in azimuth. As seen on this example, they are quite different from the ideal lobes, with shift of the main bearing, secondary lobes and an overall shape deformation. However, in this case they remain sufficiently close to the ideal settings to provide good quality current maps. The inconvenient of such calibration campaign is the dedicated time and manpower it requires as well as the need to arrange a boat. It can therefore only be done once in a while. An alternative is to use external sources of opportunity such as ship echoes with AIS positioning [44,76,81–83,298] or to deploy other types of light vehicles such as drones [257,281]. For a compact system which has a reduced number of antennas, it is also possible to describe the antenna lobes with a simple parametric form [163] and to optimize the parameters by a mere analysis of the radar data set. The idea is that the optimal lobe parameters should maximize a skill function based on the statistics of the MUSIC DoA functions collected over a long time period [103]. This technique was referred to as a *self-calibration* because it does not require an external source nor specific experimental campaign. It has been tested on several CODAR systems and has been found very efficient in correcting the antenna pattern when those are not too far from the ideal ones. For strongly distorted patterns such as the MIO CODAR installed in the harbor of Menton, however, the method has been found insufficient.

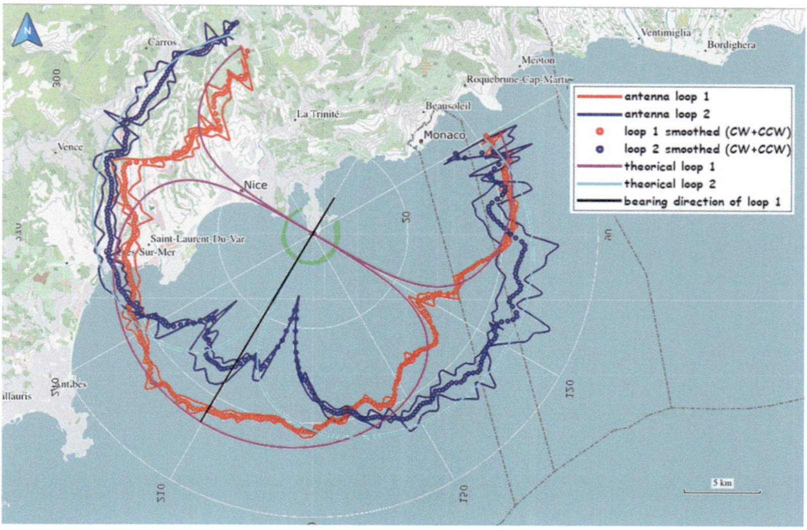

Fig. 11.15 Experimental antenna patterns measured from a ship travelling around the CODAR station of Saint-Jean Cap Ferrat (Nice, France). The ideal lobes are represented with thin magenta and cyan lines. The figure is a courtesy of Céline Quentin (MIO, CNRS). It has been realized with ©OpenStreetMap contributors and ©Seasonde Radial Suite (for loop processing)

Multistatic phased-array systems offer the possibility of another kind of self-calibration, because each separate transmitter allows to calibrate a remote receiver along its bearing. Since the direct electric field incoming from a remote transmitter in direction ϕ_s can be assimilated to a plane wave, the resulting phase shift from one antenna to the other can be predicted with the usual formula $K_0 d \sin \phi_s$ (see Sect. 4.1). To isolate the direct field of the remote transmitter from the sea echo, it suffices to take the signal in a specific range-Doppler cell, namely the one at the zero Doppler shift and the range corresponding to the TX-RX interdistance. This range-Doppler cell can be identified as a bright spot in the range-Doppler map. By considering the phase of this direct signal, one can correct automatically the antenna phase shift in the directions of the various transmitters. This is employed by the phase-array HFR network in Toulon, where 1 or 2 external transmitters are available, depending on the receiver [63, 64, 102], and has shown to be very efficient in correcting the main drift in the antenna gains. The calibration can be automatized and conducted at every cycle of observation (here, 20 min).

In the absence of any remote transmitter or other external source, a "manual" (but still purely numerical) calibration can still be performed in the case of antenna arrays by using the antenna grouping DF technique [64]. The idea is to perform a DF azimuthal processing with each sub-array of 2 consecutive antenna and to compare the $N-1$ resulting radial current maps. Systematic phase errors in the complex antenna gains result in visible rotations of the associated maps. The phase of individual antennas can thus be corrected one after the other by cancelling the rotations between successive radial maps. The reference for adjusting the rotation is

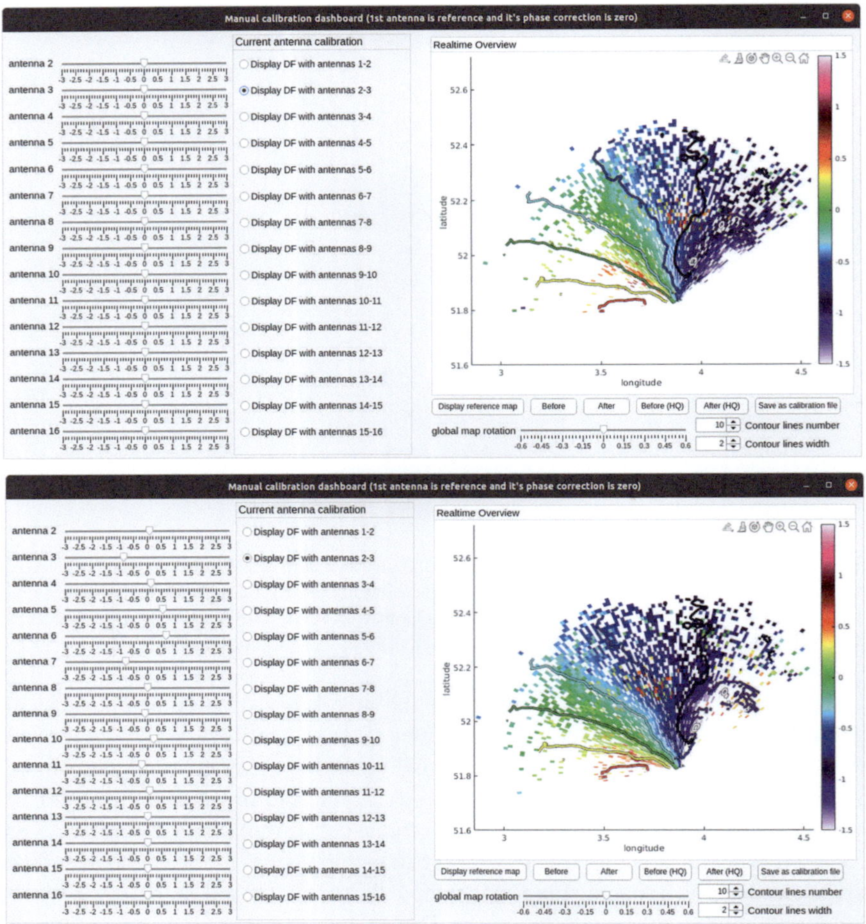

Fig. 11.16 Illustration of the manual calibration technique for phased-arrays using DF processing of 2-antenna groups. The colored solid lines are the contours of the radial map obtained with a DF processing of the full array (here 16 antennas). The map obtained with antennas 2–3 is obviously mispositioned with respect to these contours (upper panel). After phase correction by visual inspection using the phase cursors on the left column, the map is rotated to the correct position (bottom panel)

the radial current map obtained with a DF using the complete antenna array, which is more robust to rotations. If there is a coastline in the field of view of the radar (which is most often the case), it can be used to check and correct the global orientation of the radial map obtained with the full array. The manual calibration technique has been implemented in the RADAR PROCESSING MIO software, where the phase corrections can be done in a very simple way by scrolling various cursors (Fig. 11.16).

The Geophysical Products of HFR 12

This chapter presents the main geophysical applications of HFR, based on the main principles developed throughout this book. The most successful and specific application of HFR is the near real-time measurement of hourly surface currents in coastal areas, based on the measurement of radial currents from the shift of the first-order Bragg peaks. Another standard product today is the estimation of ocean wave spectra and global ocean wave parameters. As this involves the second-order Doppler spectra and complex inversion procedures, it is less accurate and more challenging than surface current measurements, and has less spatial coverage than the latter. The last two by-products of HFR are the surface wind vectors and the possible detection of tsunamis, which we will briefly review.

12.1 Measuring Currents

12.1.1 Vector Field Reconstruction

The primary product of HFR is the spatial field of surface current. Once at least 2 radial velocity maps have been obtained with separate HFR stations, the vector field **U**, referred to as the *total current velocity* can be in principle reconstructed under the common coverage. The total current velocity is usually decomposed into a northward or *meridional* (u) and eastward or *zonal* component (v). Let us first consider the simple case of 2 radial velocities[1] U_{r_1} U_{r_2} measured at the same position and let us denote ϕ_1 and ϕ_2, the associated radar look directions, respectively, where the bearings are measured counterclockwise with respect to East (Fig. 12.1). Then the

[1] These could be as well elliptical velocities obtained with a bistatic radar.

C.-A. Guérin, *High-Frequency Radars for the Oceanographer*, Geophysics and Environmental Physics, https://doi.org/10.1007/978-3-031-95642-3_12

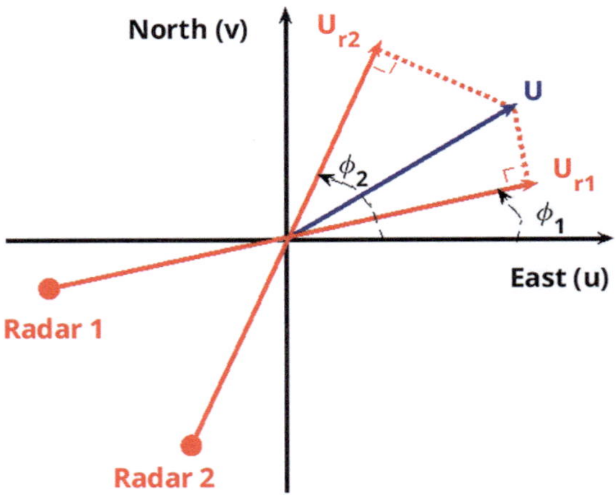

Fig. 12.1 The total velocity current **U** is decomposed into a northward (u) and eastward component (v). The bearing of the radial velocities U_{r_1} and U_{r_2} measured from the 2 radar stations are measured counterclockwise with respect to East

radial and total velocity components are linked with the 2 × 2 linear system:

$$\begin{pmatrix} \cos\phi_1 & \sin\phi_1 \\ \cos\phi_2 & \sin\phi_2 \end{pmatrix} \begin{pmatrix} u \\ v \end{pmatrix} = \begin{pmatrix} U_{r_1} \\ U_{r_2} \end{pmatrix} \tag{12.1.1}$$

Whenever the 2 bearings are different ($\phi_2 \neq \phi_1$), the system can be inverted leading to:

$$\begin{aligned} u &= \frac{\sin\phi_2 U_{r_1} - \sin\phi_1 U_{r2}}{\sin(\phi_2 - \phi_1)} \\ v &= \frac{-\cos\phi_2 U_{r_1} + \cos\phi_1 U_{r2}}{\sin(\phi_2 - \phi_1)} \end{aligned} \tag{12.1.2}$$

From this last formula it is plain that the accuracy of the estimation of the total current depends on the accuracy of the radial velocities but also on the angle between the intersecting beams. More precisely, let us consider the radial velocities U_{r_1} and U_{r_2} as independent random variables with some common variance σ_U^2 that represents the uncertainty in their estimation based on the Doppler spectra and the uncertainty due to the noise. Then the variance of the projected and the total current is given by:

$$\begin{aligned} \mathrm{Var}[u] &= G_e \sigma_{U_r}^2 \\ \mathrm{Var}[v] &= G_n \sigma_{U_r}^2 \\ \mathrm{Var}[U] &= \mathrm{Var}[u] + \mathrm{Var}[v] = G_t \sigma_{U_r}^2 \end{aligned} \tag{12.1.3}$$

with:

$$\begin{aligned} G_e &= \frac{\sin^2(\phi_2) + \sin^2(\phi_1)}{\sin^2(\phi_2 - \phi_1)} \\ G_n &= \frac{\cos^2(\phi_2) + \cos^2(\phi_1)}{\sin^2(\phi_2 - \phi_1)} \\ G_t &= G_e + G_n = \frac{2}{\sin^2(\phi_2 - \phi_1)} \end{aligned} \tag{12.1.4}$$

These geometrical factors are referred to as the east, north and total *Geometrical Dilution of Precision* (GDOP), respectively. This terminology, which has been introduced by [41] in the context of oceanographic HFR,[2] is used for the quantification of positioning errors in satellite navigation systems. The total GDOP is minimal for $\phi_2 = \phi_1 \pm \pi/2$ ($G_e = G_n = 1,\ G_t = 2$) and arbitrarily large when $\phi_2 \to \phi_1$. As expected, the reconstruction is therefore optimal when the 2 radials are perpendicular and extremely poor when the radials are close to parallel. The total GDOP remains small, say $G_t \leq 3$, at a distance on the order of the baseline between the 2 HFR stations. As seen in Fig. 12.3, the addition of an extra HFR station not aligned with the former 2 makes it possible to extend significatively the domain where the GDOP is small.

The total velocity is in general sampled on a regular grid in the longitude-latitude (hereafter abbreviated to "lon-lat") plane, corresponding to a regular sampling of the (u, v) components along their axis. Now, the different radial velocities are obtained on the polar grids associated to the corresponding HFR stations. It follows that the two radials that are combined for the reconstruction are not taken exactly at the same location nor correspond exactly to a grid cell of the total velocity. Therefore, the reconstruction in a given lon-lat cell can only be performed in an approximate way by using the radials that lie within a small region surrounding the center of the grid cell. This is usually done in some circular neighborhood with a given search radius (Fig. 12.2). In this case they may be more than 2 different radials, say $U_{r_1}, \ldots, U_{r_N}$, falling inside the search region, leading to an over-determined linear system:

$$\begin{pmatrix} \cos\phi_1 & \sin\phi_1 \\ \cos\phi_2 & \sin\phi_2 \\ .. & .. \\ \cos\phi_N & \sin\phi_N \end{pmatrix} \begin{pmatrix} u \\ v \end{pmatrix} = \begin{pmatrix} U_{r_1} \\ U_{r_2} \\ .. \\ U_{r_N} \end{pmatrix} \tag{12.1.5}$$

Such a system is no longer invertible but an approximate solution can be found in the least-square sense [166]. Consider the "cost function":

$$\chi = \sum_{j=1}^{N} w_j (u \cos\phi_j + v \sin\phi_j - U_{r_j})^2, \tag{12.1.6}$$

where $w_j > 0$ are some given weighting coefficients and N is the number of available radial in the grid cell. Then the Weighted Least-Square (WLS) can be found by minimizing this function with respect to the 2 variables u and v:

$$\frac{\partial \chi}{\partial u} = 0, \ \frac{\partial \chi}{\partial v} = 0, \tag{12.1.7}$$

[2] We use here the conventions of [148] for the angles and the GDOP, which are different from [41]. In particular, we define the GDOP factors from the variance and not from the standard deviation. However, our Eqs. (12.1.2–12.1.4) are consistent with Eqs. (3) and (4) of this last reference when taking into account these modifications.

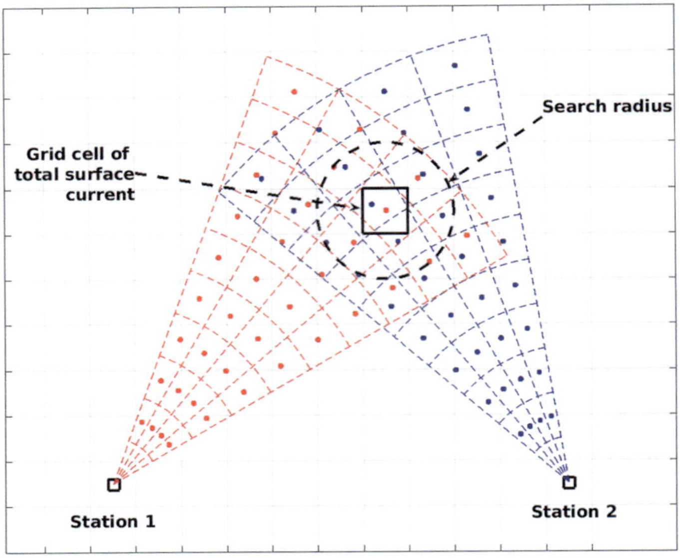

Fig. 12.2 The total current in one cell grid (black square) can be reconstructed from radial velocities found within a search radius around the center of the cell

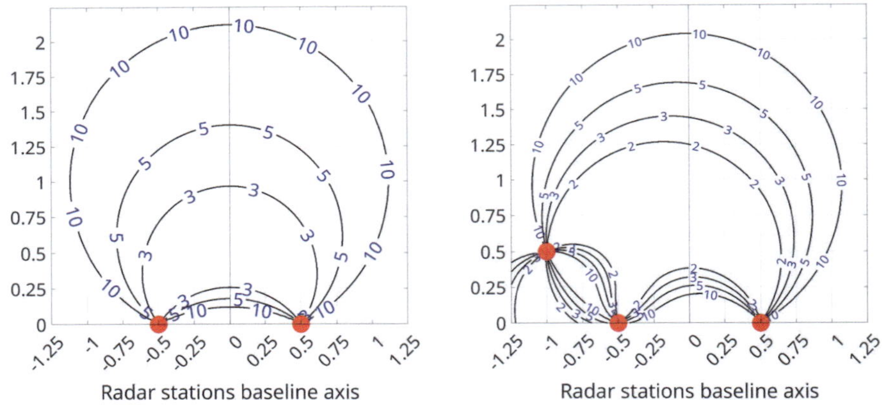

Fig. 12.3 Contour lines of the total GDOP factor G_t in the radar coverage for the reconstruction of the surface current based on 2 (left panel) and 3 (right panel) radials. The red disks show the location of the monostatic HFR stations and the spatial unit is the baseline between the 2 stations on the horizontal axis

from which we obtain easily after solving a 2 × 2 linear system:

$$
\begin{aligned}
u &= \frac{W_{ss}U_{r_1} - W_{cs}U_{r_2}}{W_{cc}W_{ss} - W_{cs}^2} \\
v &= \frac{W_{cs}U_{r_1} + W_{cc}U_{r_2}}{W_{cc}W_{ss} - W_{cs}^2}
\end{aligned}
\tag{12.1.8}
$$

with

$$W_{cc} = \sum_{j=1}^{N} w_j \cos^2 \phi_j, \; W_{ss} = \sum_{j=1}^{N} w_j \sin^2 \phi_j, \; W_{cs} = \sum_{j=1}^{N} w_j \cos \phi_j \sin \phi_j. \tag{12.1.9}$$

As to the uncertainty, we obtain the same formula (12.1.3) with modified GDOP factors:

$$\begin{aligned} G_e &= \frac{W_{ss}^2 + W_{cs}^2}{(W_{cc} W_{ss} - W_{cs}^2)^2} \\ G_n &= \frac{W_{cs}^2 + W_{cc}^2}{(W_{cc} W_{ss} - W_{cs}^2)^2} \\ G_t &= \frac{W_{ss}^2 + 2W_{cs}^2 + W_{cc}^2}{(W_{cc} W_{ss} - W_{cs}^2)^2} \end{aligned} \tag{12.1.10}$$

When all the weights are equal ($w_1 = \cdots = w_N$), the solution (12.1.8) is referred to as the *Unweighted Least-Square Solution* (UWLS). It can easily be checked that the UWLS solution is identical to the exact inverted solution (12.1.2) for $N = 2$. The minimal values of the GDOP factors decrease as the number N of available radial velocities increases. It is therefore useful to combine more than 2 radials whenever possible to reduce the uncertainty. Precisely, it can be shown [148] for the UWLS case that:

$$G_e \geq \frac{1}{N}, \; G_n \geq \frac{1}{N}, \; G_t \geq \frac{4}{N} \tag{12.1.11}$$

There are several meaningful choices for the weights w_j. They can represent the area of one radial grid cell to favor the most representative radial velocities. They can also be chosen to favor the radial velocities U_{r_j} at the smallest distance r_j from the center of the lon-lat grid cell under consideration and operate a smooth cut-off within the search radius; in this case they are often described by an exponential function:

$$w_j = \exp(-r_j / r_{\max}), \tag{12.1.12}$$

where $r_{\max}$ is the search radius. There are no simple formulas for the lower bounds of the GDOP factors in the WLS estimates. However, it is clear that they will decrease with the number of available radials.

The choice of the lon-lat grid for the total current is driven by the necessity to find sufficiently many radials within a given search radius, which must be comparable to the size of the grid cell to avoid an excessive smoothing. When the radial velocity maps or their intersection are lacunary, increasing the search radius helps filling the gaps, at the expense of the spatial resolution. Conversely, increasing the resolution and refining the grid implies diminishing the number of available radials, at the cost of a reduced spatial coverage. Figure 12.4 shows an example of radial velocities obtained over one hour with the HFR station in Fort Peyras and Cap Bénat, respectively, with a common range resolution of 1.5 km and an angular step of 1°. The corresponding total surface current map is obtained on a regular grid with a cell of size 1 × 1 km and a search radius of 1.5 km.

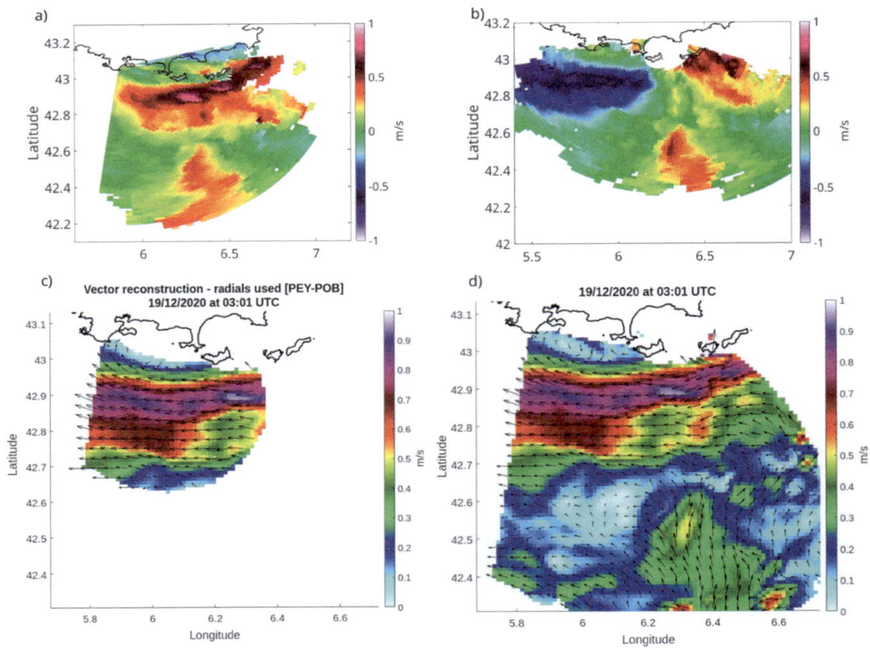

Fig. 12.4 **a** Radial component of the hourly surface current measured on 19 December 03:00 UTC, 2020, with the HFSWR monostatic station of Peyras; **b** Elliptical component of the surface current measured on the same date with the bistatic station of Bénat (the transmitter is in Porquerolles). The surface current vector reconstruction is shown in the bottom panels using a total GDOP **c** $G_t = 5$; **d** $G_t = 25$

12.1.2 What Current Do HFR Actually Measure?

There are various types and various scales of currents in the ocean. What surface currents the HFR measure and how do they compare with those measured with in situ instruments is a recurring question among the oceanographic community. It raises the question of both the interpretation of the radar-based surface currents and their validation by direct comparisons with drifters, buoys or Acoustic Doppler Current Profilers (ADCP). The most obvious difference is that HFR measurement is averaged over a radar cell of a few km2 while in situ instruments make a (quasi) point-like measurement. Another important difference is related to various observation times. While HFR-based surface currents are typically processed at an hourly rate, other instrument can operate on shorter periods. In addition, the effective temporal resolution of HFR measurements deteriorates with the level of noise, as was shown by Forget by analyzing the power density spectra of radial currents records [84]. At last, there is a difference in the vertical layer which is measured by the various instruments. Stewart and Joy [251] showed that, in presence of a horizontally uniform

Eulerian current $\mathbf{U}(z)$, the celerity of the Bragg wave is modified by an amount[3]:

$$U_e = 2k_B \int_{-\infty}^{0} e^{2k_B z} \mathbf{U}(z) \cdot \widehat{\mathbf{k_B}} \, dz, \qquad (12.1.13)$$

which is equal to the radial component of the current $\mathbf{U}$ averaged over a surface layer whose width is controlled by the Bragg wavelength. For a vertical linear profile, this effective depth is $\lambda_0/(8\pi)$, where λ_0 is the radar wavelength, which has the value of 92 cm or 75 cm at the typical operating frequencies 13 MHz or 16 MHz. Now, drifters are usually driven by the upper layer of currents and are also sensitive to direct wind effect, which makes them incompatible with the HFR measurements. However, by the addition of a drogue and the reduction of the emerging part of the body, they can be brought to approximately match the HFR integration depth. This is the case, for example, for CARTHE drifters, which have been shown relevant for HFR comparisons at 16 MHz [63], with a root-mean-square error of the hourly currents limited to a few cm/s for each radial component. ADCP scan the water column over a certain number of vertical bins but the closest bin to the interface is often much deeper than the HFR effective depth, resulting in significant quantitative differences between the currents measured by the two types of instruments, even through the trends are similar. There have been many comparisons in the literature between in situ and HFR measurements (e.g. [31,39–41,49,78,94,95,147,162,180, 187,189,196,210,212,226,234]). There is a wide range of results depending on the quality of the radar data (measurement range, integration time, signal-to-noise ratio, spatio-temporal resolution, etc.), the type of in situ instruments (drifters, ADCP, ship-based sensors, etc.) and the oceanographic conditions (strong or weak currents, sea state, etc.). However, most of the observed root-mean-square error for the total surface current are in the order of 10–20 cm/s.

Stewart and Joy's formula (12.1.13) is based on a first-order perturbation expansion of the velocity field in terms of a small dimensionless parameter $\alpha = c_B/U(0)$, given by the ratio of the Bragg wave velocity to the surface current, an expansion pushed at the second-order $\sim \alpha^2$ by Kirby and Chen [151]. It is thus limited to small values of the current and might break in some extreme cases with a strong tidal current and a reduced Bragg wave celerity in the upper part of the HF frequency band. Another effect not considered in the Stewart and Joy approach is the nonlinear interaction between the waves and the induced Stokes drift, which is not to be neglected at high sea states. We recall that the Stokes drift is a net horizontal mean velocity of fluid particles in the direction of propagation of waves, resulting from their orbital motion (e.g. [217]). It is of Lagrangian nature, as it characterizes the motion of individual particles, and is an effect of the nonlinear terms in the Lagrangian equation of motion. There has been a long series of papers in the literature questioning the Lagrangian or Eulerian nature of the HFR measurements and trying to determine whether or not the Stokes drift is included [6,37,42,138,228]. As is well-known,

[3] We recall the notation $\widehat{\mathbf{k}} = \mathbf{k}/k$ for a unit vector along $\mathbf{k}$.

the dispersion relation of gravity waves is modified by nonlinear hydrodynamical interactions and thus also the speed of the Bragg wave. Barrick and Weber [283] showed that, taking into account the third-order interactions in the wave amplitudes, the phase velocity c of the Bragg wave undergoes a relative change:

$$\frac{\Delta c_B}{c_B} = \frac{1}{2} \int k^2 F(\mathbf{k_B}, \mathbf{k}) S(\mathbf{k}) \mathbf{dk}, \tag{12.1.14}$$

where F is some kernel that does not depend on the wave field and $S(\mathbf{k})$ is the directional ocean wave spectrum. The expression for the latter (given in Eq. (30) of [283]) is very complex but Broche [37] observed that it can be simplified to a very good approximation to:

$$F(\mathbf{k_B}, \mathbf{k}) \simeq 4(\widehat{\mathbf{k_B}} \cdot \widehat{\mathbf{k}}) \begin{cases} \left(\frac{k}{k_B}\right)^{3/2} & \text{if } k < k_B \\ \left(\frac{k}{k_B}\right)^{1/2} & \text{if } k > k_B \end{cases} . \tag{12.1.15}$$

In this form, the correction term (12.1.14) appears as a filtered Stokes drift around the Bragg wave vector. A coarse upper estimate can be obtained by noting that:

$$\frac{|\Delta c_B|}{c_B} < 2\text{mss}, \tag{12.1.16}$$

where

$$\text{mss} = \int k^2 S(\mathbf{k}) d\mathbf{k} \tag{12.1.17}$$

is the *mean square slope* of the sea surface, a quantity which has been well characterized by Cox and Munk [50] as a function of wind speed U_{10} and is on the order mss $\simeq 5.10^{-3} U_{10}$. Hence, the relative correction is at most of a few percents; it is maximal when the directional ocean spectrum is peaked along the radar look direction ($\widehat{\mathbf{k_B}} \cdot \widehat{\mathbf{k}} \simeq 1$) and null for isotropic ocean wave spectra (because the integral (12.1.14) cancels by symmetry). This result indicates that when measuring the surface current from the shift of the Bragg peak, a wave-induced Stokes-like current must be added to the Eulerian surface current. However, to date there is no conclusive experimental evidence for the inclusion of this Lagrangian contribution to the HFR surface current, because its value is on the order of the uncertainty margin of the measurement [228].

Another challenging aspect in remote sensing of surface current is the ability to measure its vertical shear, i.e. its variation with depth. The frequency dependence of the integration depth of the HFR measurements can be exploited in a very simple way, by using different radar frequencies to estimate the current at different depths below the surface [80,254,255,262]. However, some information on the current shear can also be obtained with a single frequency radar by exploiting the non-uniform shift of the second-order Doppler spectrum. The idea, that was proposed and experimentally tested in [141,245], is to measure the shift of the hydrodynamical second-order peak close to $\omega = \sqrt{2}\omega_B$. This peak is given by 2 equal wavenumbers at half the Bragg wavenumber ($k_1 = k_2 = k_B/2$). Now, at this wavenumber, the effective current predicted by Stewart and Joy theory [251] is integrated over twice the depth that is used to define the effective current which is responsible for the shift of the first-order Bragg peaks. Therefore measurement of the shift of this secondary peak gives access to the current at 2 different depths. Assuming a simple parametric form, this is sufficient to derive the current profile.

12.2 Measuring Waves

The measurement of ocean waves with HFR shares the same long history as the surface currents. The idea of using the second-order ocean Doppler spectra for the inversion of ocean waves dates back to Hasselmann [121] and has been the subject of much research since then. There are two types of wave field characterization. The first is limited to the bulk wave parameters such as the mean wave height and period; the second provides an estimate of the directional ocean wave spectrum, at least in a certain range of frequencies. In most cases, the effect of the surface current is ignored or is assumed to be an overall uniform translation of the Doppler spectrum. We will review the different approaches without going into too much technical detail.

12.2.1 Estimation of the Main Wave Parameters

We here address the problem of estimating the main wave parameters from the directional Doppler spectrum obtained with a single monostatic HFR station after resolution in range and azimuth. It is assumed that Barrick's formulas (10.1.4) and (10.1.10), which have been derived for an infinite surface, hold also on the considered sea patch. Barrick first derived analytical estimates of the significant wave height H_s and the *mean wave period* T_p from approximations to the second-order Doppler spectrum [17,18]. We recall that H_s is the average wave height (through to crest) of the highest third of the waves. It can be obtained as:

$$H_s^2 = 16 \int S(\mathbf{k}) d\mathbf{k}, \tag{12.2.18}$$

where $S(\mathbf{k})$ is the directional ocean wavenumber spectrum. The mean wave period is obtained as the inverse of the centroïd frequency of the ocean wave frequency spectrum $S(\omega)$, which is related to the omni-directional ocean wavenumber spectrum by (A.2.22):

$$T_p = 2\pi \frac{\int_0^{+\infty} S(\omega) d\omega}{\int_0^{+\infty} \omega S(\omega) d\omega} \tag{12.2.19}$$

The main idea to extract H_s from the second-order directional Doppler spectrum (10.1.10) is to proceed as if the coupling coefficient were constant. For this, let us consider the "Zeta function" [101]:

$$\zeta(\omega) = \mathcal{N} k_B^2 \sum_{n_1,n_2=\pm 1} \int S(n_1 \mathbf{k_1}) S(n_2 \mathbf{k_2}) \delta(\omega - n_1\omega_1 - n_2\omega_2) d\mathbf{k_1}, \tag{12.2.20}$$

where as before $S(\mathbf{k})$ is the directional ocean wave spectrum, $\mathcal{N} = 2^5 \pi K_0^2$, $k_B = 2K_0$ is the Bragg wavenumber and the integration variables are linked by the conditions $\mathbf{k_1} + \mathbf{k_2} = \mathbf{k_B}$, $\omega_1 = \sqrt{gk_1}$ and $\omega_2 = \sqrt{gk_2}$. The Zeta function has the same dimension as the Doppler spectrum but is free of its coupling coefficient.[4] The delta function constraint can be removed by integration, yielding:

$$\int_{-\infty}^{+\infty} \zeta(\omega) d\omega \simeq \mathcal{N} k_B^2 \sum_{n_1,n_2=\pm 1} \int S(n_1 \mathbf{k_1}) S(n_2 \mathbf{k_2}) d\mathbf{k_1}, \tag{12.2.21}$$

Most of the wave energy in the directional ocean spectrum is concentrated near the peak wavenumber k_p. Now, except in very calm seas and low radar frequencies, the peak wavenumber is generally much smaller than the Bragg wavenumber ($k_p << k_B$). The main contribution of the integrand to the second-order integral is therefore if one of the variables $\mathbf{k_1}$ or $\mathbf{k_2} = \mathbf{k_B} - \mathbf{k_1}$ is close to $\mathbf{k_B}$. One can therefore approximate $\mathbf{k}_1 \simeq k_B$ or $\mathbf{k}_2 \simeq k_B$ in the integral (12.2.21), which leads to:

$$\sum_{n_1,n_2=\pm 1} \int S(n_1 \mathbf{k_1}) S(n_2 \mathbf{k_2}) d\mathbf{k_1} \simeq 4 \int S(\mathbf{k}) d\mathbf{k} \sum_{n_1=\pm 1} S(n_1 \mathbf{k_B}) \tag{12.2.22}$$

The summation on the right-hand side can be obtained by a simple integration of the first-order Doppler spectrum (10.1.4):

$$\int_{-\infty}^{+\infty} \sigma_1(\omega) d\omega = \mathcal{N} \sum_{n_1=\pm 1} S(n_1 \mathbf{k_B}) \tag{12.2.23}$$

Combining (12.2.18)–(12.2.23) we arrive at a simple expression of H_s in terms of the Zeta function:

$$H_s^2 = \frac{4 \int_{-\infty}^{\infty} \zeta(\omega_B) d\omega}{k_B^2 \int_{-\infty}^{\infty} \sigma_1(\omega) d\omega}. \tag{12.2.24}$$

With a similar technique, the dominant wave period (12.2.19) can be determined from another ratio of second-order integrals:

$$T_p = 2\pi \frac{\int_I \zeta(\omega) d\omega}{\int_I |\omega - \omega_B| \zeta(\omega) d\omega}, \tag{12.2.25}$$

---

[4] Note that the Zeta function $\zeta(\omega)$ and the Doppler spectrum $\sigma(\omega)$ depend on the direction of the Bragg wave vector $\mathbf{k_B}$. However, for simplicity, this dependence does not explicitly appear in the argument of the functions.

where I can be one of the 4 following intervals: $I = (-\infty, -1]$, $I = [-1, 0]$, $I = [0, 1]$ or $I = [0, +\infty)$. To relate the Zeta function to the second-order Doppler spectrum, Barrick introduced a *weighting function* $\mathcal{W}$ such that:

$$\zeta(\omega) \simeq \frac{2^5 \sigma_2(\omega)}{\mathcal{W}(\omega/\omega_B)}, \tag{12.2.26}$$

assuming that the Zeta function and the second-order Doppler have the same directional dependence to $\mathbf{k}_B$. To do this, Barrick chose to average the coupling coefficient $|\Gamma|$ in (10.1.10) over the angular integration variable, after changing to polar coordinates. There are other ways to define such a weighting function depending on the chosen integration variables to express the second-order integral (see Appendix J). With a weighting function approximation, it follows from (12.2.24) that H_s can be obtained as a mere ratio of the second-order to the first-order Doppler spectrum integral:

$$H_s^2 = \frac{32 \int_{-\infty}^{\infty} \sigma_2(\omega_B)/\mathcal{W}(\omega/\omega_B) d\omega}{K_0^2 \int_{-\infty}^{\infty} \sigma_1(\omega) d\omega}, \tag{12.2.27}$$

and the dominant wave period T_p can be determined from another ratio of the second-order integrals:

$$T_p = 2\pi \frac{\int_I \sigma_2(\omega)/\mathcal{W}(\omega/\omega_B) d\omega}{\int_I |\omega - \omega_B| \sigma_2(\omega)/\mathcal{W}(\omega/\omega_B) d\omega}, \tag{12.2.28}$$

where the interval I is as described above. Barrick's approach has been used by many authors to estimate the main wave parameters (e.g. [5,43,96]). The advantage of such a formulation is that it does not require a calibrated Doppler spectrum, because it relies on a ratio of integrated Doppler spectra. However, there are some inherent limitations to the method. First, Barrick's formulae (12.2.27) and (12.2.28) are based on the approximation (12.2.22), which holds true only in the limit $k_p/k_B \to 0$. Furthermore, as first shown by Wyatt [286], the accuracy of the weighting function approximation is strongly dependent on the wave direction and is expected to be better for waves travelling perpendicular to the radar beam. Averaging the coupling coefficient over directions generally results in an overestimation of both H_s and T_p. This is illustrated in Fig. 12.5, where the different estimates for H_s and T_p are calculated from Barrick's equations (12.2.27) and (12.2.28) using a numerically simulated second-order Doppler spectrum (see Appendix I). Here the directional ocean wave spectrum was calculated with a classical Pierson-Moskowitz model and a cardioïd-shaped angular spreading function (A.4.31) with parameters $s = 2$ and $\alpha = 0.05$ (see Appendix A). The dominant wave period is obtained by averaging the four possible expressions in (12.2.28), which are integrated over different frequency intervals I. The estimates of both H_s and T_p depend on the wave direction (here, coinciding with the wind direction) with respect to the radar look direction. The observed error is maximal when the radar is looking upwind or downwind and minimal when the radar is looking crosswind. The overestimation of the radar estimate and the discrepancy between various directions can be reduced but not completely eliminated using an

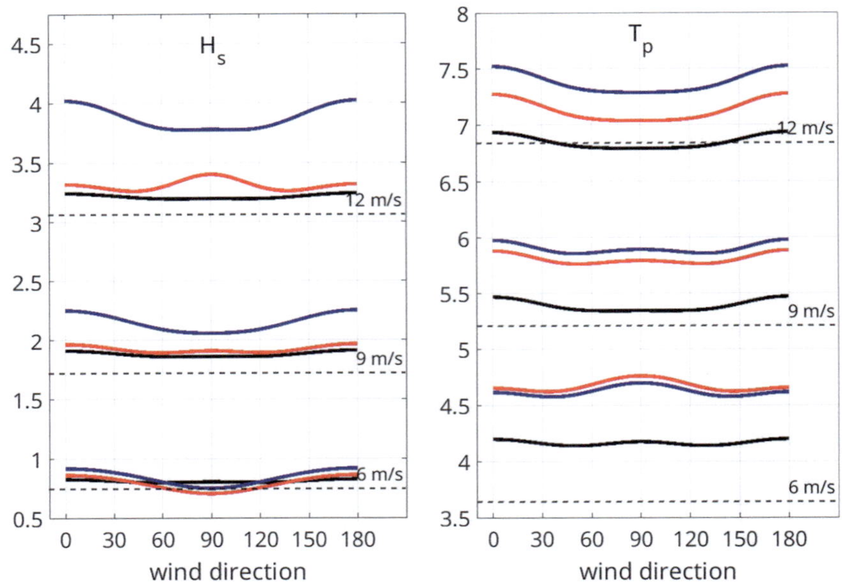

Fig. 12.5 Numerical estimation of H_s and T_p using a simulated second-order Doppler spectrum. A directional Pierson-Moskowitz ocean wave spectrum model has been employed for various wind speed (6, 9 and 12 m/s) and various wind direction measured from the radar look direction (0° is upwind, 90 is crosswind and 180 is downwind). The dashed horizontal lines indicate the actual values of H_s and T_p. The calculation has been performed with the Zeta function (black solid lines) or the weighting function approach using either Barrick's original formulation (blue solid lines) or an alternative weighting function [101] as described in Appendix J (red solid lines)

alternative weighting function. For this reason, some authors chose to correct the formula (12.2.27) by an empirical calibration factor [2,3,129,146,197,219].

Notwithstanding the problem with the weighting function approach, the validity of Barrick's approach for estimating wave parameters is limited by the validity of the second-order theory itself. The perturbation approach requires that the wave height remain small compared to the radar wavelength ($K_0 H_s < 1$). This is generally true for common HFR frequencies and sea states but may not be the case in some extreme cases [246]. Figure 12.6 shows schematically, using a Pierson-Moskowitz ocean wave spectrum, the domain for which the method is applicable, as a function of wind speed and radar frequency. Another practical limitation of Barrick's method is that it requires a clear identification of the second-order contribution. The frequency threshold between the first- and second-order Bragg region is not always well defined,

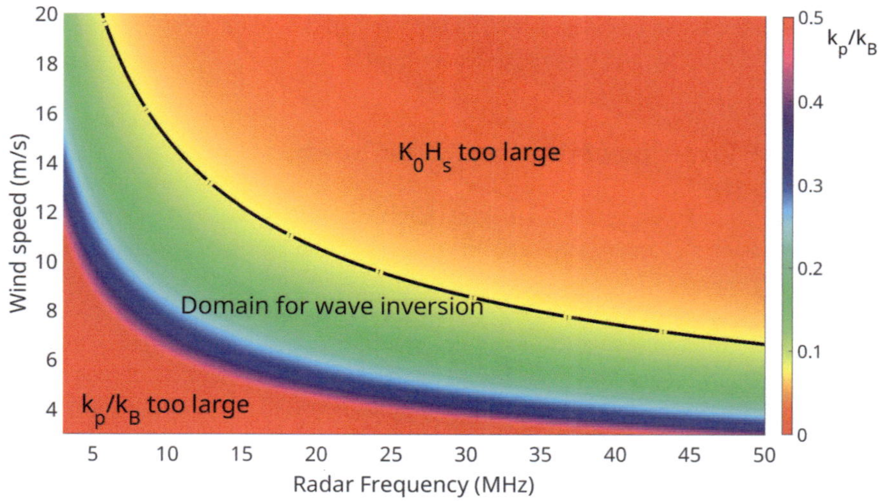

Fig. 12.6 Variation of the peak wavenumber to Bragg wavenumber ratio (k_p/k_B, in color scale) as a function of the radar frequency and wind speed for a Pierson-Moskowitz ocean frequency spectrum. The black line indicates a threshold value $K_0 H_s = 1$. The admissible domain to apply Barrick's method is the intermediate region where $K_0 H_s < 1$ and $k_p/K_B << 1$

especially in case of strong currents or in the absence of azimuthal resolution. For these reasons, many authors have only considered the estimation of the bulk wave parameters using empirical methods based on either first- [77,256,257,259,300] or second-order Doppler spectra (e.g. [77,111,186,260]). Most of these methods require the calculation of Doppler spectra that are resolved in both range and azimuth and can thus be associated with a reduced sea surface patch with constant sea state parameters. In the case of compact antenna arrays, the Doppler spectra are omnidirectional, that is, they cannot be resolved in azimuth. This does not allow for the measurement of the wave frequency spectrum from a single station. Nevertheless, the H_s can be estimated on range arcs by adapting Barrick's original methods. The long-term performance of such measurements has been assessed with buoys and models and is quite variable depending on sea state, range, and wave directions (see e.g. [208,224,231] or [188] for recent reviews).

Recently, the idea of simply using the time variation of the radar signal instead of the Doppler spectrum to estimate the main sea state parameters have been proposed by some authors [197,240]. Within the framework of the Walsh and Gill formalism (see Sect. 10.3), Shahidi and Gill [239,240] showed that, after some mathematical manipulations on the first-order backscattered electric field originating from a patch of the sea surface, H_s is proportional to the variance of the received voltage and that the peak frequency of the ocean spectrum could be obtained from the same time series after integration over range. The idea was taken up in a different form in [197], where a method was proposed to estimate H_s and T_p from the time modulation of the Bragg peak amplitude. At this stage, there is insufficient experimental evidence that these methods can compete with the Doppler-based method. However, they have the merit of simplicity and deserve to be continued.

12.2.2 Full Inversion of the Ocean Wave Spectrum

Better than just estimating the main wave parameters, it is possible to proceed to a full inversion of the ocean wave spectrum from the HFR measurements using the integral expression (10.1.10) of the second-order Doppler spectrum. This reduces the mathematical problem to the inversion of a nonlinear first-kind Fredholm integral, with the difficulty that the unknown ocean wave spectrum appears through the quadratic combinations $S(\mathbf{k_1})S(\mathbf{k_2})$. Lipa [170,171] was the first to propose an inversion method based on this integral equation. For this, she introduced the concept of the so-called "linearization" of the second-order integral, which is valid when the Doppler frequency ω is not too far from the Bragg frequency. The main idea is that, given the double condition $\pm\omega_1 \pm \omega_2 = \omega_B$ and $\mathbf{k_1} + \mathbf{k_2} = \mathbf{k_B}$, at least one of the wavevector, say $\mathbf{k_2}$, is close to $\mathbf{k_B}$. At HFR frequencies, the Bragg wavenumber can be assumed much larger than the peak wavenumber of the ocean wave spectrum ($k_B >> k_p$), except at very calm sea states. Hence, $\mathbf{k_2}$ is in the "saturation" region of the wavenumber ocean spectrum (Fig. 12.7). This short-wave part of the spectrum, which follows essentially a k^{-4} fall-off and is aligned with the wind direction, can be reasonably well described using a simple parametric model (such as typically the Pierson-Moskowitz model for the omnidirectional part and a cardioïd-shaped angular spreading function for the directional part, see Appendix A) and inverted separately [170] or adjusted using the first-order spectrum [172,291]. This reduces the problem to a linear integral equation that can be handled by model fitting techniques [146,170,171,287] or by non-parametric iterative techniques such as the Chahine-Twomey algorithm, henceforth referred to as *Wyatt's method* [288]. In this way, the long wave ocean spectrum can be recovered on a discrete set of frequencies. The directional dependence is obtained either by using a Fourier series [137,177,292] with a few azimuthal harmonics or by sampling the wavenumbers on a grid with no further constraint [288]. The linearized inversion method can be applied to wave frequencies in a frequency interval surrounding the Bragg frequency, typically $0.4 f_B - 1.6 f_B$. We refer to the reviews in [7,289,293] for more details on these methods and assessment of their performances.

A full non-linear inversion scheme was proposed by Hisaki [130,132,133], which allows to extend the range of available Doppler frequencies. However, the convergence of iterative solutions requires the introduction of additional physical regularization constraints such as the energy balance equation and the continuity equation for the wind input vector. Regardless of the chosen method, the quality of the inversion is subject to the same practical limitations: the SNR level, the number of available HFR sites, the sea state, the antenna side-lobes, etc. [289,295]. In particular, estimates from a single radar station are not able to resolve a directional ambiguity that was already encountered using the weighting function approach [290] and have a much lower accuracy than dual-site estimates (e.g. [131]). The ocean wave spectrum

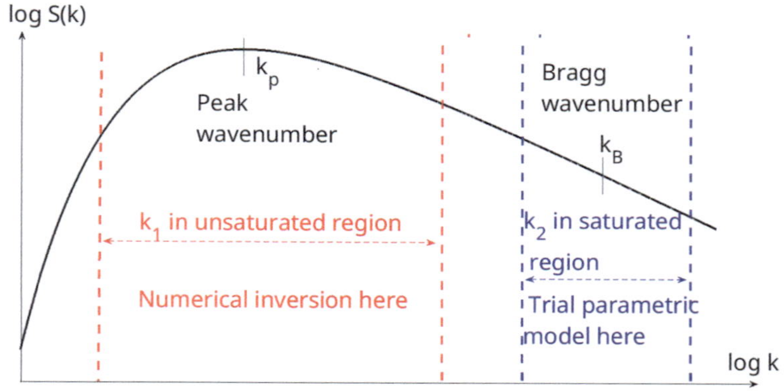

Fig. 12.7 Schematic view of the different regions of the directional ocean spectrum used for the inversion. In the linearized inversion problem, at least one the wave vectors (say $\mathbf{k_2}$) involved in the resonance relation $\mathbf{k_1} + \mathbf{k_2} = \mathbf{k_B}$ is assumed to be in the "saturated" part of the ocean spectrum (see Sect. 2.5), in which the directional spectrum $S(\mathbf{k_2})$ can be represented with a simple parametric model such as the Pierson-Moskowitz spectrum. The directional ocean spectrum $S(\mathbf{k_1})$ corresponding to the other wave vector $\mathbf{k_1}$ is inverted numerically

inversion can also be conducted in the bistatic case, as was shown in [247,298] on the basis of numerical simulation using Walsh and Gill formalism or with experimental data when adapting Wyatt's method to the bistatic case [114].

12.2.3 Swell Inversion

Swell are long quasi-monochromatic ocean waves whose spectral components are often well separated from the wind-wave ocean spectrum. They can be as long as a few hundred meter wavelength and are usually much larger than the Bragg wavelength. Nevertheless, they can measured with a HFR through their nonlinear interaction with the Bragg wave. Assuming a very directional swell with a narrow spectral peak, one can complement the directional wind-wave spectrum $S(\mathbf{k})$ with an additional delta-like component [172]:

$$S(\mathbf{k}) \rightarrow S(\mathbf{k}) + H_{\mathrm{sw}}^2 \delta(\mathbf{k} - \mathbf{k}_{\mathrm{sw}}), \tag{12.2.29}$$

where H_{sw} and $\mathbf{k}_{\mathrm{sw}}$ are the root-mean-square elevation and wave vector of the swell, respectively. We assume that the swell wavelength is at least twice the Bragg wavelength ($k_{\mathrm{sw}} < k_B/2$). This produces an additional term in the second-order integral of the Doppler spectrum (10.1.10), which appears as the combination of the wind-wave spectrum and the swell spectrum:

$$2NH_{\mathrm{sw}}^2 \sum_{n_1,n_2} \int \left\{ |\Gamma|^2 \left[S(n_1\mathbf{k_1})\delta(n_2\mathbf{k_2} - \mathbf{k_s})\delta(\omega - n_1\omega_1 - n_2\omega_2) \right] \right\} d\mathbf{k_1} \tag{12.2.30}$$

with the usual constraints $\mathbf{k_1} + \mathbf{k_2} = \mathbf{k_B}$ and $\omega_1 = \sqrt{gk_1},\ \omega_2 = \sqrt{gk_2}$. Note that a product of swell spectra in (10.1.10) can only arise if $\mathbf{k}_{\mathrm{sw}} = \mathbf{k_B}/2$, a possibility that has been ruled out. The delta functions prescribe the values $\mathbf{k_2} = \pm\mathbf{k}_{\mathrm{sw}}$ and $\omega_2 = \omega_{\mathrm{sw}}$ in the above integral. If the swell wavelength is much longer than the Bragg wavelength ($k_{\mathrm{sw}} << k_B$), one can further approximate $S(\pm\mathbf{k_1}) \simeq S(\pm\mathbf{k_B})$ and

$$\omega_1 \simeq \omega_B \left(1 - \frac{1}{2}\frac{\omega_{\mathrm{sw}}^2}{\omega_B^2}\cos\phi_{\mathrm{sw}}\right), \tag{12.2.31}$$

where ϕ_{sw} is the angle between the swell propagation and the radar look direction. It follows that the Doppler spectrum in presence of a long monochromatic swell has 4 additional delta-like peaks surrounding the 2 Bragg peaks (Fig. 12.8):

$$\omega_{n_1,n_2} = n_1\omega_B \left(1 - \frac{1}{2}\frac{\omega_{\mathrm{sw}}^2}{\omega_B^2}\cos\phi_{\mathrm{sw}}\right) + n_2\omega_{\mathrm{sw}} \tag{12.2.32}$$

Integrating the second-order Doppler spectrum over a very narrow interval $I_{n_1,n_2} = [\omega_{n_1,n_2} - \delta\omega, \omega_{n_1,n_2} + \delta\omega]$ around each swell peak and comparing with the same integral around the Bragg peaks, $J_{n_1} = [n_1\omega_B - \delta\omega, n_1\omega_B + \delta\omega]$, yields:

$$\frac{\int_{I_{n_1,n_2}} \sigma_2(\omega)d\omega}{\int_{J_{n_1}} \sigma_1(\omega)d\omega} \simeq 2H_{\mathrm{sw}}^2\, |\Gamma(\mathbf{k_B}, \mathbf{k}_{\mathrm{sw}}, n_1\omega_B, n_2\omega_{\mathrm{sw}})|^2 \tag{12.2.33}$$

The measurement of the location of the 4 swell peaks therefore allows to invert the swell frequency ω_{sw} and direction ϕ_{sw}, which in turns allows to find the swell wave vector $\mathbf{k}_{\mathrm{sw}}$ and wave height H_{sw} from (12.2.33). These properties were established in the early 80 s [87,172] for the measurement of swell and have recently been used (with some technical refinements) for a systematic characterization of swell over an annual data base [278,279].

12.3 Measuring Winds

Because winds and waves are related, HFR can also be employed to measure nearshore winds. We limit ourselves here to the main principles and refer to e.g. [75,152] for recent surveys. The main idea is that the Bragg wave, which is on the order of 5–15 m in the usual HFR operating frequencies, is quite responsive to the wind actions and adjusts rapidly in both direction and amplitude [128]. Therefore, a change in the local wind vector will result in a change in the absolute power of the first-order Bragg peaks (stronger winds induce stronger peaks) as well as a change in their relative importance (the positive Bragg peak will be stronger than the negative one when the radar is looking upwind and vice-versa downwind).

The simple concept of comparing the first-order Bragg lines was used by Long and Trizna as early as 1973 [181], who laid the basis of wind direction estimation

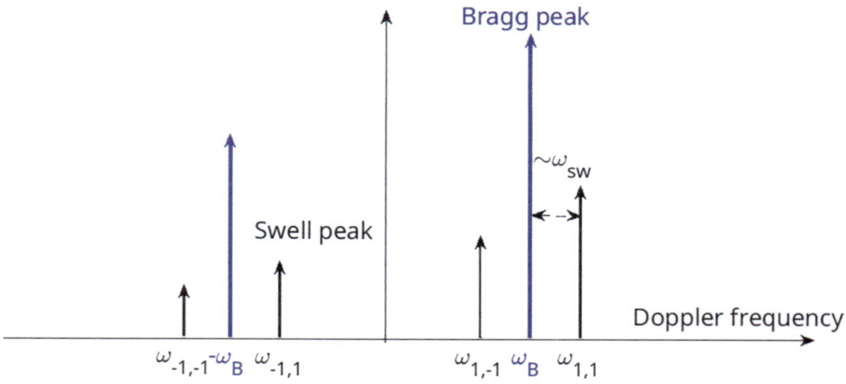

Fig. 12.8 There are 4 swell peaks surrounding the Bragg peaks, approximatively located at $\pm\omega_B \pm \omega_{sw}$ (see (12.2.32))

with HFR. They proposed to use the Bragg Line Ratio (BLR), defined as the ratio of the power of the positive to the negative Bragg peak, expressed in dB:

$$BLR = 10\log_{10}\left(\frac{P^+}{P^-}\right) \tag{12.3.34}$$

Let $D(\phi)$ be the angular spreading function of the ocean wave spectrum (see Appendix A), with the origin of angle taken in the direction of the radar (i.e. $\phi = 0$ when waves propagate upbeam). Then one has from the first-order Doppler spectrum (Eq. (10.1.4)):

$$BLR = 10\log_{10}\left(\frac{D(\phi)}{D(\phi+\pi)}\right) \tag{12.3.35}$$

Assuming a simple cardioid model $D(\phi) \sim \cos^{2s}(\phi/2)$ we can invert the wind direction from the BLR:

$$\phi = 2\ \text{atan}\left(\frac{P^-}{P^+}\right) = 2\ \text{atan}\left(10^{-\frac{BLR}{20s}}\right) \tag{12.3.36}$$

The cardioid parameter is most often assumed to be constant but some physical arguments combined with experimental observations [252] suggest that it depends on wind speed, with $s = 2$ at low wind speed and $s < 2$ at high wind speeds. However, due to the atan dependence, the BLR has a weak dependence on the wind angle ϕ when the radar is looking around the upwind or downwind direction (large or small BLR), making its estimation very inaccurate. On the contrary, the wind angle has a maximum of sensitivity to small BLR, corresponding to a situation where the radar is looking perpendicular to the wind direction (Fig. 12.9). It should be noted that the relationship between the BLR and the wind direction is well approximated by a linear function over a wide angular sector around the crosswind direction:

$$\phi = -\frac{6.6}{s}BLR + 90 \tag{12.3.37}$$

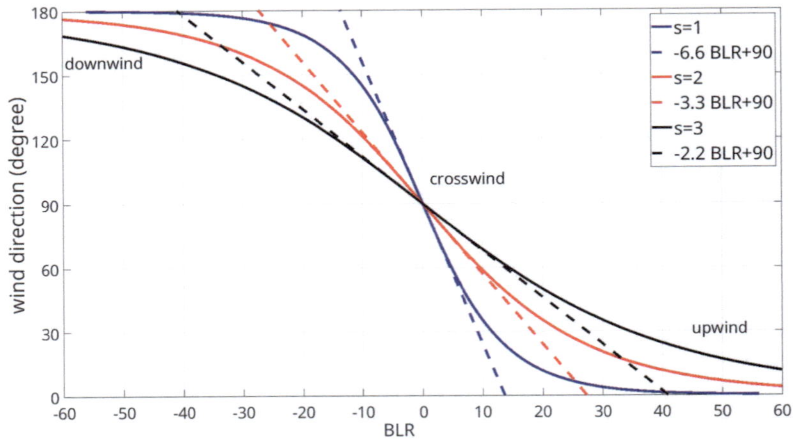

Fig. 12.9 The wind direction can be inverted from the BLR assuming a cardioid $\cos^{2s}(\phi/2)$ model for the angular spreading function. The maximum of sensitivity of the wind direction to the BLR is obtained when the radar is looking in the crosswind direction

Such a linear relation has been found empirically in [116], with $\phi = -3.75\ BLR + 90$, suggesting in this case a cardioid exponent $s = 1.75$.

The validity of the cardioid model for the spreading function has sometimes been questioned because it produces infinite BLR in the upwind/downwind directions. This can be corrected for by introducing an additional backpropagation parameter e as in Eq. (A.4.31) or by choosing another form of the spreading function such as the $\text{sech}(\beta\phi)$ function [111,291,296]. Regardless of the model chosen for the spreading function, the estimation of the wind direction from the BLR suffers from another shortcoming, namely a sign ambiguity of the angle, since waves propagating at opposite angles with respect to the radar look direction share the same radial velocity (see Fig. 12.10). This ambiguity cannot be resolved without additional information, such as a coarse initial estimate of the wind direction or the use of a second radar station as in [111].

Because it requires an azimuthally resolved Doppler spectrum, the BLR technique was originally developed for phased-array radars and can be used in both monostatic (e.g. [128,211,243]) and bistatic [100,139] configuration. From a practical point of view, the negative and positive Bragg peak power in each radar cell can be estimated by taking the maximum of the beamformed Doppler spectrum for negative and positive Doppler frequencies, respectively. However, due to the narrowness of this peak and possible fluctuations if the Doppler spectrum is insufficiently averaged, it is preferable to use the integral of the Doppler spectrum in the first-order region, which does not change the BLR but produces more stable estimates.

In principle, the BLR technique could be adapted to compact radars using DF azimuthal processing. For this, a discrete set of BLR can be obtained from the positive and negative rays of the omnidirectional Doppler spectrum in the first-order Bragg region and each BLR can be assigned to a given bearing through the MUSIC analysis. The problem is that corresponding positive and negative rays can be

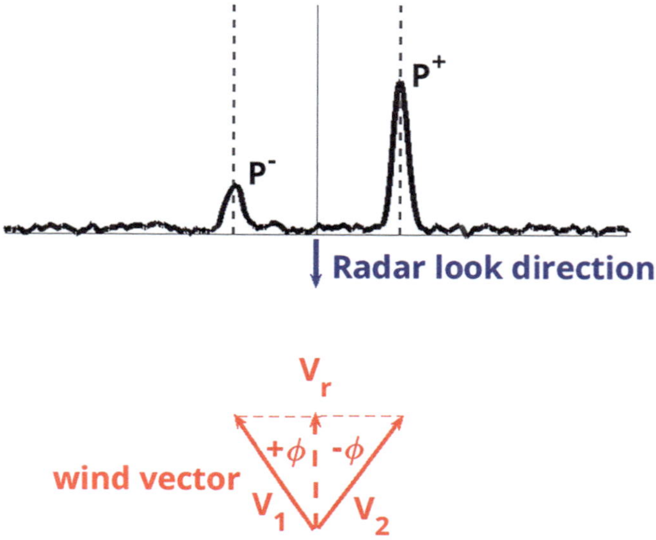

Fig. 12.10 The Bragg Line Ratio P^+/P^- depends on the wind direction with respect to the radar look direction. However, there are two directions with the same radial velocity, so that the estimation of the wind direction is ambiguous

assigned to different bearings in this DF process, either due to inaccurate estimation or to multiple sources. In practice, this means that DF processing is not suited to a direct approach such as the BLR method. Instead, such systems estimate both wind speed and direction by fitting a parametric ocean wave spectrum model (typically a Pierson-Moskowitz ocean spectrum) to the omnidirectional second-order Doppler spectrum (e.g. [177]).

A direct estimate of wind speed, i.e. without the need for wave spectrum inversion, can also be obtained with phased-array radars. It is mostly based on empirical relationships that must be calibrated with external observations [153] or neural networks that must be trained with data base derived from model [244] or in situ observations [65].

12.4 Detecting Tsunamis

A relatively recent application of HFR is tsunami detection and warning. We provide here only a brief overview of the main principles of tsunami detection. More comprehensive and up-to-date reviews can be found in e.g. [125, 174].

12.4.1 How Can HFR See Tsunamis?

In a 1979 paper, Barrick [20] first introduced the idea that HFR could be used for tsunami warning and he gave some orders of magnitude of the detection capabilities. Tsunamis are very long waves that have wavelengths on the order of tens of kilometers, hence much larger than any Bragg wave. So how could HFR possibly detect the tsunamis? The main point is that tsunamis are not seen as waves but as currents ! Tsunami waves have a long time period on the order of 10–30 min, a period that is shared by the induced orbital velocity of the water particles. From the radar prospective, this is indistinguishable from a slowly changing surface current. Therefore, radar-based detection of tsunamis relies on the identification of a specific pattern of surface currents that would correspond to this long wave orbital velocity. The main characteristics of a tsunami wave train can be derived from shallow water linear theory (e.g. [51, 150, 249]), according to which the main geometrical and cinematical parameters satisfy simple scaling laws with the local water depth H. Both the wave celerity and the local wavelength L decrease as the inverse square root of the depth:

$$\begin{aligned} c &= c_i (H/H_i)^{1/2} \\ L &= L_i (H/H_i)^{1/2} \end{aligned} \tag{12.4.38}$$

where H_i, $c_i = \sqrt{gH_i}$ and L_i are the initial values of the water depth, wave celerity and wavelength in the deep ocean, respectively. With the mean value $H_i = 4000$ m, this gives $c_i \simeq 200$ m/s or $c_i = 720$ km/h, implying that the tsunamis propagate through the deep ocean at the speed of an airplane. The wave amplitude a, starting from some moderate value a_i offshore, increases towards shore as a power $1/4$ of the inverse water depth, a relationship known as Green's law (e.g. [192]):

$$a = a_i (H_i/H)^{1/4} \tag{12.4.39}$$

The horizontal orbital velocity V_{orb} of a tsunami wave also scales with depth, with a $-3/4$ power:

$$V_{\text{orb}} = a_i \sqrt{\frac{g}{H_i}} \left(\frac{H_i}{H} \right)^{3/4} \tag{12.4.40}$$

Figure 12.11 shows the evolution of these three parameters with depth for typical deep ocean values of $a_i = 0.5$ and $H_i = 4000$ m. The ability of an HFR to detect the tsunami as early as possible is related to the minimum value of a current that it can measure, which defines a detection threshold. Given the rapid decrease of the orbital velocity with depth, a small variation in this threshold results in a large variation in the water depth at which the tsunami can be detected. Depending on the bathymetry, this can have a large effect on the detection range and, correlatively, on the warning period. Figure 12.12 depicts schematically the range-Doppler map that would be observed for the Bragg rays in the presence of an incoming tsunami. It shows oscillations that becomes larger and faster as the distance from shore decreases. Given the spatio-temporal scales of these deformations of the Bragg rays, there are no "ordinary" oceanic currents (such as those generated by wind, tide or inertial oscillations) that could produce such coherent patterns. There is therefore a typical signature of tsunami orbital currents, which serves as the basis for detection and warning.

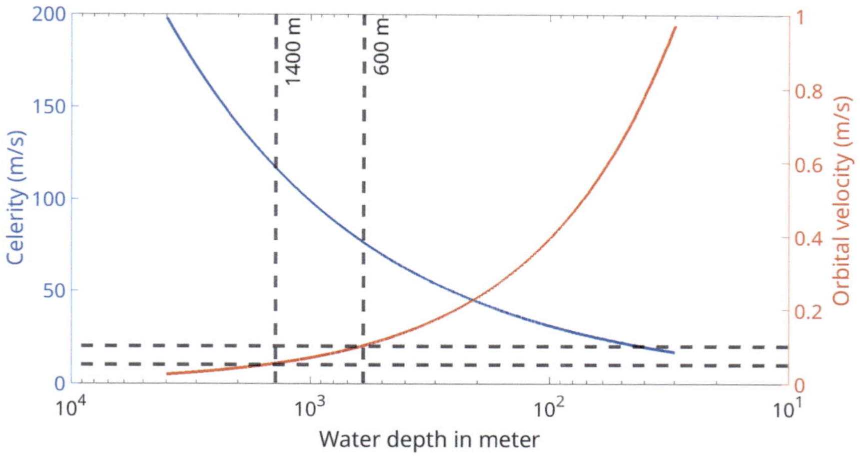

Fig. 12.11 Evolution of the tsunami wave celerity (solid blue line) and horizontal orbital velocity (solid red line) as a function of water depth for an initial amplitude $a_i = 0.5$ m and depth $H_i = 4000$ m in deep ocean. A small variation of the radar detection threshold (set here at the values 5 and 10 cm/s, which are marked by the horizontal dashed black lines) induces a large variation of corresponding the water depth (1400 or 600 m)

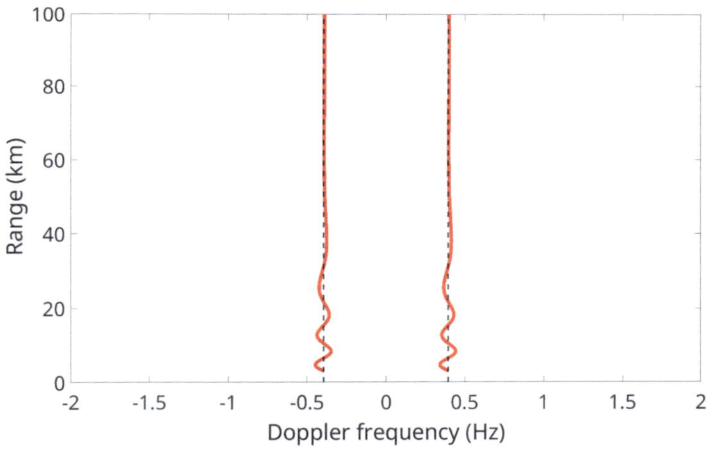

Fig. 12.12 Schematic pattern of the effect of a tsunami on the Bragg rays see in a Range-Doppler map

12.4.2 Detection Algorithms

It was not until the 2004 Indian Ocean tsunami that Barrick's idea was pursued and put in practice by the same author and his co-workers. An algorithm, referred to as the "Q-factor" algorithm, was proposed in [169] to identify a coherent structure that would be characteristic of a tsunami. Since tsunami wave fronts gradually refract to essentially become parallel to the depth contours, the induced currents remain extremely coherent over bands that are parallel to the same contour and extend over a few range cells. The Q-factor at a distance x is defined as the ratio:

$$Q(x) = \frac{\text{Mean}(V_r^*)}{\text{RMS}(V_r^*)}, \tag{12.4.41}$$

where $V_r^* = V_r / \sin\phi$ is the radial current velocity V_r (from which a slowly varying average background current has been subtracted) corrected by the local angle between the direction of the radar beam and the normal to the depth contour, the numerator is its mean value inside the band and the denominator is its standard deviation. An incoming tsunami would produce a peak of $Q(x)$ and a warning could be triggered by selecting the appropriate threshold. This clever algorithm is well suited to compact radars such as the SeaSonde CODAR, as it does not require a fine azimuthal gridding of the surface currents (Fig. 12.13).

Another detection algorithm, suitable for phased array radars, is the so-called "Time-Correlation Algorithm" (TCA) [97,98,105]. It is based on the idea that the main directions of tsunami propagation can be predicted by geometrical optics theory from the bathymetry alone, independent of the source. This allows a preliminary computation of the tsunami propagation time from one radar cell to another (Fig. 12.14). Unlike standard algorithms that detect surface current patterns, the TCA is based on the analysis of the space-time correlations of the radar signal time series between pairs of radar cells, which does not require inverting radial surface currents. Denoting $V_p(t) = s(t)/|s(t)|$ the normalized received signal formed in one radar cell p, we define the inter-cell correlation:

$$C_{pq}(\tau) = \langle V_p(t+\tau)V_q^*(t-\tau_{pq})\rangle, \tag{12.4.42}$$

where τ_{pq} is the pre-computed propagation time from cell p to cell q. Such correlations, which are updated periodically, are maximal at $\tau = 0$ in presence of a

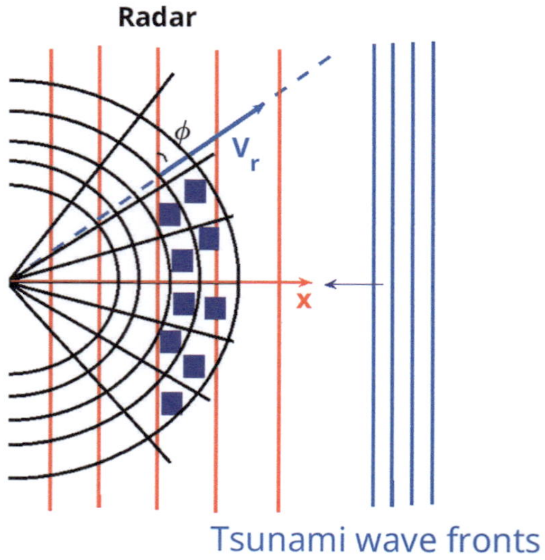

Fig. 12.13 The Q-factor is calculated over bands (red lines) parallel to the tsunami wave front (blue lines) by averaging the projected current $V_r / \sin\phi$ inside each strip (blue squares)

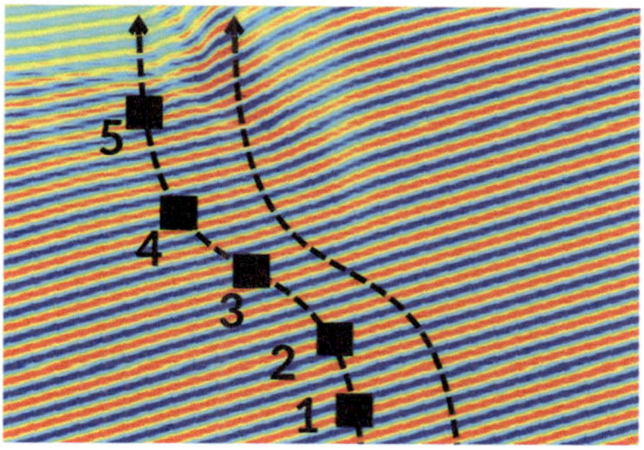

Fig. 12.14 The tsunami wave front (colored strips) propagates from one radar cell (black boxes) to another following the rays predicted by geometrical optics following the bathymetry

tsunami when one signal is time-shifted by the precalculated long wave propagation time. Again, a detection threshold for the correlation peak must be set on empirical grounds.

The detection performance of possible tsunamis events has been quantified in a series of papers [24,25,68,69,97,99,109,127,167,169,178,179] using numerical simulations combined with actual or simulated HFR data. The Japan Tohoku 2011 event triggered a new wave of efforts and progresses and soon a further step was achieved by "replaying" real tsunamis that had been recorded by existing radars but not analyzed in real time [24,66,67,126,167,168,175]. At the same time the radar manufacturers developed real-time detection software and integrated them in their radar systems [71,72,225]. On 14 October 2016, the first real-time tsunami warning was issued automatically by an HFR on the Pacific coast of British Columbia, Canada, in the absence of any seismic alert and a tsunami current pattern was detected about 60 km offshore [72]. This event was later analyzed as a meteo-tsunami triggered by an abrupt low-pressure front that propagated to the west coast in the remains of the tropical storm Tonga [105] and later re-interpreted as a storm surge [60]. Since then, several other atmospheric tsunamis or storm surges have been detected using HFR [70,126,176].

As tsunami waves are transient phenomena with a high propagation speed, they require, in principle, a short integration time. However, this conflicts with the need to resolve small surface currents and the need for a sufficient SNR at large distances. The choice of the integration time is therefore a trade-off between these conflicting requirements. It is usually on the order of 1–2 min. Note, however, that some non-spectral methods using auto-regressive processes [59] can be used to estimate the Doppler spectrum with a higher frequency resolution than that imposed by the usual time-frequency compromise and are thus more adapted to shorter integration times. These methods have proven effective in the context of tsunami detection [60].

Ocean Wave Spectra

A

A.1 Frequency and Wavenumber Spectra

Consider a random wave fields given by the stochastic Fourier integral:

$$\eta(\mathbf{r}, t) = \int a(\mathbf{k}, \omega) \mathrm{e}^{i(\boldsymbol{k}\cdot\boldsymbol{r}-\omega t)} \mathbf{dk} d\omega, \tag{A.1.1}$$

where $a(\mathbf{k}, \omega)$ are independent centered random variables. The associated wavenumber-frequency spectrum $S(\mathbf{k}, \omega)$ is defined as:

$$\langle a(\mathbf{k}, \omega) a^*(\mathbf{k}', \omega') \rangle = S(\mathbf{k}, \omega)\, \delta(\mathbf{k} - \mathbf{k}', \omega - \omega') \tag{A.1.2}$$

It is related to the spatio-temporal autocorrelation function:

$$C(\mathbf{r}, t) = \langle \eta(\mathbf{r}', t') \eta(\mathbf{r} + \mathbf{r}', t + t') \rangle \tag{A.1.3}$$

by a three-dimensional Fourier Transform pair:

$$\begin{aligned} S(\mathbf{k}, \omega) &= \frac{1}{(2\pi)^3} \int \mathrm{e}^{-i(\boldsymbol{k}\cdot\boldsymbol{r}-\omega t)} C(\mathbf{r}, t) d\mathbf{r} dt \\ C(\mathbf{r}, t) &= \int \mathrm{e}^{i(\boldsymbol{k}\cdot\boldsymbol{r}-\omega t)} S(\mathbf{k}, \omega) \mathbf{dk} d\omega \end{aligned} \tag{A.1.4}$$

The wavenumber-frequency spectrum is the most complete description of the field of random elevations, but is difficult to measure in general. In most cases, it can only be partially characterized by measurements in time or space. Using an Eulerian instrument such as a wave gauge or a moored buoy one can only record a time series of water surface elevation at a fixed location, say $\eta(t) = \eta(\mathbf{0}, t)$. The associated autocorrelation function:

$$C(\mathbf{0}, t) = \langle \eta(\mathbf{0}, t' + t) \eta(\mathbf{0}, t') \rangle \tag{A.1.5}$$

C.-A. Guérin, *High-Frequency Radars for the Oceanographer*, Geophysics and Environmental Physics, https://doi.org/10.1007/978-3-031-95642-3

is related to the frequency wave spectrum $S(\omega)$, which is confined to positive frequencies, by a cosine's transform:

$$\begin{aligned} C(\mathbf{0}, t) &= \int_0^{+\infty} S(\omega)\cos(\omega t)d\omega \\ S(\omega) &= \frac{2}{\pi}\int_0^{+\infty} C(\mathbf{0}, t)\cos(\omega t)dt, \ \omega \geq 0. \end{aligned} \tag{A.1.6}$$

By identification with (A.1.4) we have that the frequency spectrum is the integration of the wavenumber spectrum over the **k** space:

$$S(\omega) = 2\int S(\mathbf{k}, \omega)\mathbf{dk}, \ \omega \geq 0, \tag{A.1.7}$$

Some other instruments such as imaging radars or optical systems provide information about the spatial statistical properties of a surface at a given time, say $\eta(\mathbf{r}, 0)$. In this case, the random surface is described by a spatial autocorrelation function:

$$C(\mathbf{r}, 0) = \langle \eta(\mathbf{r}' + \mathbf{r}, 0)\eta(\mathbf{r}', 0)\rangle \tag{A.1.8}$$

which defines a *wavenumber power spectrum* $S_0(\mathbf{k})$ by its Fourier Transform:

$$C(\mathbf{r}, 0) = \int e^{ik\cdot r} S_0(\mathbf{k})\mathbf{dk} \tag{A.1.9}$$

By identification with (A.1.4) again, we see that:

$$S_0(\mathbf{k}) = \int_{-\infty}^{+\infty} S(\mathbf{k}, \omega)d\omega \tag{A.1.10}$$

By construction S_0 is an even function and thus cannot distinguish between ocean waves propagating in opposite directions **k** and $-\mathbf{k}$. To separate them, one can define a directional ocean wavenumber spectrum by restricting the integration to positive frequencies:

$$S(\mathbf{k}) = 2\int_0^{+\infty} S(\mathbf{k}, \omega)d\omega \tag{A.1.11}$$

Unlike S_0, it is not a power spectral density because it is not the Fourier Transform of a correlation function and is not even symmetric ($S(-\mathbf{k}) \neq S(\mathbf{k})$). It appears that the wavenumber power spectrum is a symmetrized version of the directional ocean wavenumber spectrum:

$$S_0(\mathbf{k}) = \frac{1}{2}(S(\mathbf{k}) + S(-\mathbf{k})) \tag{A.1.12}$$

The different types of wave spectra carry different information about the ocean surface. However, they share a common feature in that they integrate to the total wave energy:

$$\langle \eta^2 \rangle = \int S(\mathbf{k}, \omega)\mathbf{dk}d\omega = \int_0^{+\infty} S(\omega)d\omega = \int S(\mathbf{k})\mathbf{dk} = \int S_0(\mathbf{k})\mathbf{dk} \tag{A.1.13}$$

A.2 Polar Decomposition of Ocean Wave Spectra

The directional ocean wavenumber spectrum is often sought in polar coordinates (k, ϕ), where ϕ is the angle with respect to some reference axis. It is usually decomposed into a product of the angular averaged spectral energy and an angular distribution:

$$S(\mathbf{k}) = S(k, \phi) = k^{-1} S_{\text{omni}}(k) D(k, \phi). \tag{A.2.14}$$

The first factor is referred to as the *omni-directional ocean wavenumber spectrum* :

$$S_{\text{omni}}(k) = \int_0^{2\pi} S(k, \phi) k d\phi \tag{A.2.15}$$

while $D(k, \phi)$ is the *angular spreading function*, which is dimensionless and satisfies the normalization condition:

$$\int_{-\pi}^{+\pi} D(k, \phi) d\phi = 1 \tag{A.2.16}$$

The same decomposition holds for the wavenumber power spectrum,

$$S_0(k, \phi) = k^{-1} S_{\text{omni}}(k) D_0(k, \phi), \tag{A.2.17}$$

which shares the same omni-directional part but has a symmetrized spreading function:

$$D_0(k, \phi) = \frac{1}{2}(D(k, \phi) + D(k, \phi + \pi)) \tag{A.2.18}$$

In an analogous manner to the directional ocean wavenumber spectrum, a *directional ocean frequency spectrum* can be defined [61] by:

$$S(\omega, \phi) = 2 \int_0^{+\infty} S(\mathbf{k}, \omega) k dk \tag{A.2.19}$$

together with a spreading function $D(\omega, \phi)$ such that:

$$S(\omega, \phi) = S(\omega) D(\omega, \phi) \tag{A.2.20}$$

The various wave spectra are related by a common integrated value, which is the variance of surface elevations $\eta(\mathbf{r}, t)$:

$$\langle \eta^2 \rangle = \int_{-\pi}^{\pi} \int_0^{\infty} S(\omega, \phi) d\omega d\phi = \int_{-\pi}^{\pi} \int_0^{\infty} S(k, \phi) k dk d\phi = \int_0^{\infty} S_{\text{omni}}(k) dk \tag{A.2.21}$$

In the framework of linear wave theory for deep water gravity surface wave, the spatial and temporal frequencies are related by the dispersion relation $\omega = \sqrt{gk}$. Since ω depends only on the norm of the wave vector and not on its direction, the knowledge of the frequency spectrum is equivalent to knowing the omni-directional ocean spectrum alone:

$$S_{\text{omni}}(k) = S(\omega)\frac{d\omega}{dk}, \ \omega \geq 0. \tag{A.2.22}$$

However, the directional ocean wavenumber and the directional ocean frequency spectra are related one another in any given direction:

$$S(k, \phi) = k^{-1} S(\omega, \phi)\frac{d\omega}{dk}, \ \omega \geq 0, \tag{A.2.23}$$

or, what amounts to the same,

$$D(k, \phi) = D(\omega, \phi) \tag{A.2.24}$$

A.3 Omni-Directional Ocean Spectra

There is a large literature on wave spectra and how there are modelled and measured (see for example the classical monographs by Ochi [207], Goda [93] or Holthuisen [134]). Unfortunately, due to the different ways of defining a sea spectrum as seen above, the terminology is often ambiguous. Depending on the context, the same term "spectrum" can be used to denote either the non-directional ocean frequency spectrum $S(\omega)$, the directional ocean frequency spectrum $S(k, \omega)$, the omni-directional ocean wavenumber spectrum $S_{\text{omni}}(k)$, the directional ocean wavenumber spectrum $S(\mathbf{k})$ or the spatial power spectrum $S_0(\mathbf{k})$. Many models of spectral have been proposed in the literature and we will only mention the most popular and historical ones. The simplest one is the Phillips spectrum [215] which was developed from a dimensional analysis. It is a non-directional ocean frequency spectrum ($S(\omega)$) valid in the equilibrium range:

$$S(\omega) = \alpha g^2 \omega^{-5}, \tag{A.3.25}$$

with a constant that is found to be $\alpha = 7.4\,10^{-3}$ by fitting experimental measurements in a reservoir.

The Pierson-Moskowitz ocean spectrum [218] was derived from measurements in the North Atlantic. It confirms the dimensional form $g^2\omega^{-5}$ of the Phillips spectrum in the equilibrium range with the addition of an exponential cut-off around the peak frequency. It is valid for fully developed sea states. Its expression is given by:

$$S(\omega) = A g^2 \omega^{-5} \mathrm{e}^{-5/4(\omega_p/\omega)^4} \tag{A.3.26}$$

with $A = 8.2\,10^{-3}$ close to the Phillips constant α and $\omega_p = 0.88\, g/U_{19.5}$ the peak frequency, where $U_{19.5}$ is the wind speed at 19.5 meter above sea level. When converted to the omnidirectional ocean wavenumber spectrum it takes the form:

$$S_{\text{omni}}(k) = \frac{1}{2} A k^{-3} \mathrm{e}^{-3/2(k_p/k)^2} \tag{A.3.27}$$

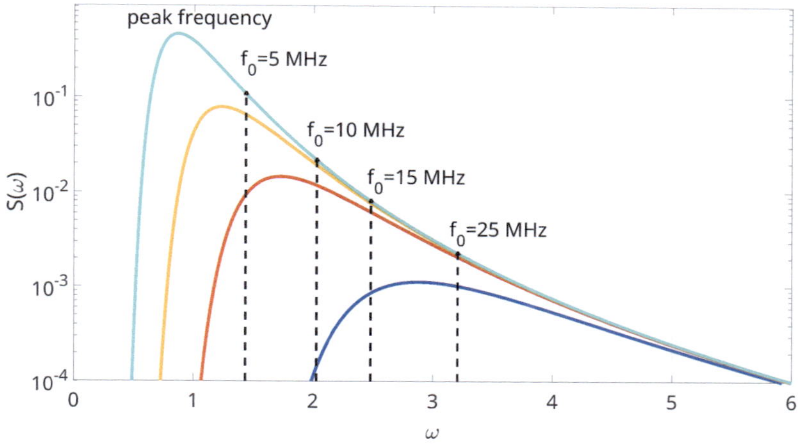

Fig. A.1 The Pierson-Moskowitz frequency spectrum for various wind speeds

where the peak wavenumber is given by $k_p = 0.7g/U_{19.5}^2$. The main sea state parameters can be obtained as simple functions of wind speed. The peak period T_p is simply proportional to wind speed while the significative wave height H_s and the peak wavelength λ_p are proportional to its square:

$$\begin{aligned} T_p &= 2\pi/\omega_p = 0.73\ U_{19.5} \\ H_s &= 4\left(\int_0^\infty S(\omega)d\omega\right)^{1/2} = 0.02\ U_{19.5}^2 \\ \lambda_p &= \frac{2\pi}{k_p} = 0.83\ U_{19.5}^2 \end{aligned} \tag{A.3.28}$$

The Pierson-Moskowitz ocean spectrum is illustrated for several wind speeds on Fig. A.1.

There is a plethora of ocean wave spectra that have been proposed in the literature, which can be found in classic textbooks on ocean waves. HFR are mostly concerned with the long-wave spectra. The most commonly used, besides the Pierson-Moskowitz spectrum, is perhaps the JONSWAP spectrum [117, 122], which is a fetch-dependent non-directional ocean frequency spectrum ($S(\omega)$).

A.4 Angular Spreading Function

The spreading function is less straightforward to measure than the simple frequency spectrum. It can in principle be obtained by forming the cross-correlation of simultaneous time records from arrays of wave gauges (e.g. [184, 207]). Several parametric models are commonly used, the simplest one being the cardioid shaped $\cos^2\phi$ spreading function:

$$D(\phi) = \frac{1}{\pi} \cos^2 \left(\frac{\phi - \phi_w}{2} \right), \tag{A.4.29}$$

where ϕ_w is the direction of wind. It does not depend on frequency or wind speed. A more general model has been proposed by Longuet-Higgins [184] as the $\cos^{2s}$ spreading function:

$$D(\omega, \phi) = \frac{2^{2\,s-1}}{\pi} \frac{\Gamma^2(s+1)}{\Gamma(2\,s+1)} \cos^{2\,s} \left(\frac{\phi - \phi_w}{2} \right), \tag{A.4.30}$$

where Γ is the Gamma function and the frequency and wind speed dependence enters through s. This exponent is referred to as the *spreading parameter* and controls the width of the main lobe. Note that this spreading function vanishes opposite to the wind direction. In order to allow a fraction of the waves to propagate against the wind, as it is often observed, it is common to introduce a small parameter e such that:

$$D(\phi) = \beta(e + (1 - e) \cos^{2\,s} \left(\frac{\phi - \phi_w}{2} \right)) \tag{A.4.31}$$

where β is a numerical normalization constant. Some example are displayed in Fig. A.2 for various values of s and e.

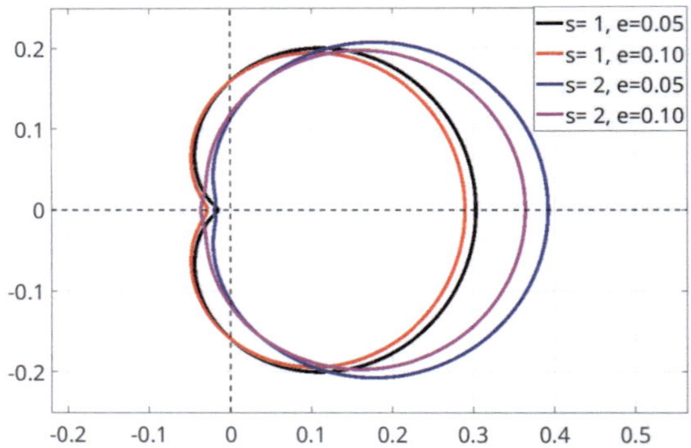

Fig. A.2 Angular spreading function in the $\cos^{2s} \phi$ cardioïd model with $s = 1$ or $s = 2$ and $e = 0.05$ or $e = 0.1$

The Scalar Green's Function

B

The so-called scalar Green's function is a key quantity in scattering from acoustical or EM waves. It represents the elementary solution of the Helmholtz equation for a point source located at the origin in free space:

$$\nabla^2 G(\mathbf{R}) + K_0^2 G(\mathbf{R}) = \delta(\mathbf{R}) \tag{B.0.1}$$

The Laplace operator is isotropic and therefore the solutions of this equation depend only of the radius, $G = G(R)$. One simple technique to find it is to use the Laplacian in spherical coordinates, where it reduces to the radial component:

$$\nabla^2 G = \frac{1}{R}\frac{\partial^2(RG(R))}{\partial R^2} \tag{B.0.2}$$

The one-dimensional function $f(R) = RG(R)$ therefore satisfies the ordinary differential equation:

$$f''(x) + K_0^2 f(x) = 0 \tag{B.0.3}$$

for $x > 0$, with general solutions of the form $f(x) = A\mathrm{e}^{\mathrm{i}K_0 x} + B\mathrm{e}^{-\mathrm{i}K_0 x}$ for some constant A and B. It follows that the scalar Green's function is of the form:

$$G(R) = A\frac{\mathrm{e}^{\mathrm{i}K_0 R}}{R} + B\frac{\mathrm{e}^{-\mathrm{i}K_0 R}}{R} \tag{B.0.4}$$

for $R > 0$. However, only the solution satisfying the outgoing wave radiation condition is physically acceptable as it corresponds to an outwards propagating wave in the temporal representation. With the implicit $e^{-\mathrm{i}\omega_0 t}$ harmonic time dependence, this implies $B = 0$. To find the constant A, we apply the divergence theorem to the vector ∇G over a small sphere of radius R, yielding:

C.-A. Guérin, *High-Frequency Radars for the Oceanographer*, Geophysics and Environmental Physics, https://doi.org/10.1007/978-3-031-95642-3

$$\int_{R'\leq R} \nabla^2 G(\mathbf{R}')d\mathbf{R}' = 4\pi R^2 \frac{\partial G(R)}{\partial R} \tag{B.0.5}$$

In the limit $R \to 0$, the left integral tends to 1 in view of (B.0.1) while the right-hand term tends to $-4\pi A$, yielding in the end:

$$G(R) = -\frac{e^{+\mathrm{i}K_0 R}}{4\pi R} \tag{B.0.6}$$

The 3D scalar Green's function can be written as a superposition of plane waves by means of the so-called *Weyl representation*:

$$G(R) = \frac{1}{8\mathrm{i}\pi^2} \int \frac{\mathrm{e}^{\mathrm{i}(\boldsymbol{k}\cdot\boldsymbol{r}+q|z|)}}{q} d\mathbf{k}, \tag{B.0.7}$$

where $\mathbf{R} = \mathbf{r} + \mathbf{z}$ and $q = \sqrt{K_0^2 - k^2}$ with $\mathrm{Im}[q] \geq 0$. This representation is obtained by taking the partial Fourier Transform of (B.0.1) in the horizontal variables $\mathbf{r} = (x, y)$. The resulting function $G(\mathbf{k}, z)$ satisfies a one-dimensional equation:

$$\frac{\partial^2 G(\mathbf{k}, z)}{\partial z^2} + q^2 G(\mathbf{k}, z) = \delta(z) \tag{B.0.8}$$

The outgoing solution of this last equation is $G(\mathbf{k}, z) = \mathrm{e}^{\mathrm{i}q|z|}/(2\mathrm{i}q)$, which after an inverse Fourier transform yields the Weyl representation (B.0.7). It has a form similar to a Rayleigh expansion with the important difference that the vertical coordinate is taken with an absolute value ($|z|$), so that the components with positive imaginary q (obtained wor $k > K_0$) are evanescent regardless of the space direction. When integrating over the horizontal component $\mathbf{k}$ of the wave vector $\mathbf{K}$, the horizontal areas corresponding to an infinitesimal solid angle $d\Omega$ increase like $d\Omega/q$, explaining this factor in the Weyl integral (see Fig. B.1).

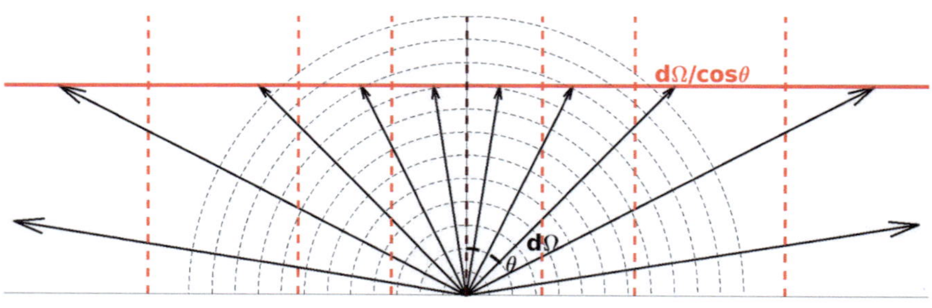

Fig. B.1 In the Weyl representation, plane waves are integrated over the horizontal projection of the wave vector. This induces a weighting factor $1/\cos\theta$ for each solid angle $d\Omega$

The Boundary Conditions at the Interface

C

Solving the electric and magnetic fields above the sea surface requires the knowledge of the boundary conditions at the interface. This is a classical problem in Electromagnetism which can be found in any textbook. We here only recall the main result, that is the continuity of the tangential components of the electric and magnetic fields **E** and **H** across the interface and the continuity of the normal components of the auxiliary fields **B** and **D** across the interface (Fig. C.1). Denoting **n** the local normal vector to an arbitrary interface Σ and $[F]_\Sigma$ the jump of any quantity F across this interface, this can be stated simply by:

$$[\mathbf{E}\times\mathbf{n}]_\Sigma = 0,\ [\mathbf{H}\times\mathbf{n}]_\Sigma = 0,\ [\boldsymbol{D}\cdot\boldsymbol{n}]_\Sigma = 0,\ [\boldsymbol{B}\cdot\boldsymbol{n}]_\Sigma = 0. \tag{C.0.1}$$

The first two relations can be obtained from Maxwell's equations (7.1.2) by taking a small loop across the interface and applying Stokes theorem to transform the curl of **E** and **H** into the circulation of the tangential field; the last two relations follow by taking small flat volume across the interface and applying the divergence theorem to transform the divergence of **B** and **D** into the flux of the normal components.

C.-A. Guérin, *High-Frequency Radars for the Oceanographer*, Geophysics and Environmental Physics, https://doi.org/10.1007/978-3-031-95642-3

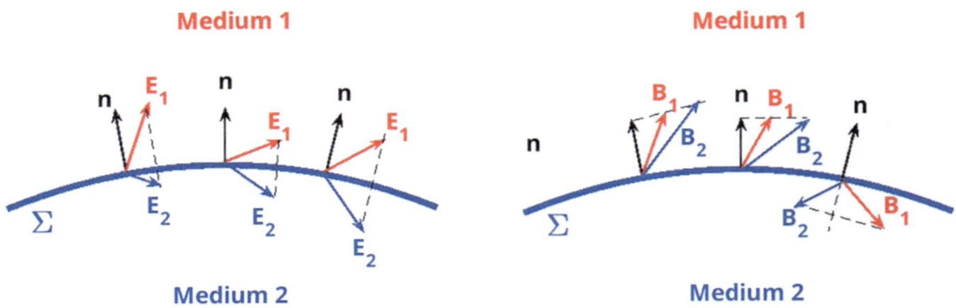

Fig. C.1 The tangential components of the electric and magnetic fields **E** and **H** are continuous at the interface between two homogeneous media. The same holds for the normal components of the auxiliary fields **B** and **D**

At the interface between the air or vacuum and a strongly conductive media, the electric and the magnetic field at the boundary can be related by an *impedance boundary condition* also known as the *Leontovitch boundary condition* [164,233]:

$$\mathbf{n} \times \mathbf{E} = -Z(\mathbf{n} \times \mathbf{H}) \times \mathbf{n}, \tag{C.0.2}$$

where $Z = \sqrt{\mu/\epsilon}$ is the impedance of the medium and $\mathbf{n}$ is the outwards oriented unitary normal to the surface. This relation relates the common tangential components of the electric and magnetic field at the interface and allows for considerable simplifications of the boundary value problem when calculating the electromagnetic wave everywhere in space. It can be derived using the simple following heuristic argument [161]. For a small skin depth $\delta << \lambda_0$, the electric and magnetic field are attenuated over a distance which is much smaller than the wavelength in vacuum, so that their variation along the normal direction to the surface are much larger than their variation along the tangential direction. Just below the surface, it can thus be assimilated to a plane wave propagating along the normal direction $-\mathbf{n}$. It follows that $\nabla \times \mathbf{E} = -\mathrm{i}K\mathbf{n} \times \mathbf{E} = -\mathrm{i}\omega\mu\mathbf{H} =$, where $K = \sqrt{\epsilon_r}K_0$ is the wavenumber in the medium. Since $\mathbf{H} = (\mathbf{n} \times \mathbf{H}) \times \mathbf{n} + (\boldsymbol{H} \cdot \boldsymbol{n})\mathbf{n}$, this implies that $\boldsymbol{H} \cdot \boldsymbol{n} = 0$ and is equivalent to (C.0.2). The impedance Z of a conductive medium is often expressed relative to vacuum through the dimensionless, relative quantity (henceforth simply referred to as the surface impedance):

$$Z_r = \frac{Z}{Z_0} = \frac{1}{\sqrt{\epsilon_r}} \tag{C.0.3}$$

Note that the perfect conductor boundary condition:

$$\mathbf{n} \times \mathbf{E} = \mathbf{0} \tag{C.0.4}$$

is recovered in the limit $Z_r \rightarrow 0$.

The Fresnel Coefficients

D

The Fresnel coefficients play a key role in EM scattering from rough surfaces. They can be found in any textbook but we found useful to recall their derivation by considering them as a particular case of the Rayleigh expansion. Consider a flat interface $z = 0$ separating the air ($z > 0$) from the sea surface ($z < 0$), illuminated from above by an incident linearly polarized plane wave with wave vector $\mathbf{K_0} = \mathbf{k_0} - q_0\mathbf{z}$. Due to the interface, there can be reflected waves propagating upwards or transmitted waves propagating downwards. Even though each first-year student knows that there is only one pair of such waves reflected and transmitted according to Snell-Descartes laws, this assumption is a priori not granted and deserves a rigorous derivation in the framework of EM wave theory. In addition, the simple principles of Optics do not tell which fraction of the wave amplitude is reflected or transmitted, an information which is given by the Fresnel coefficients (Fig. D.1).

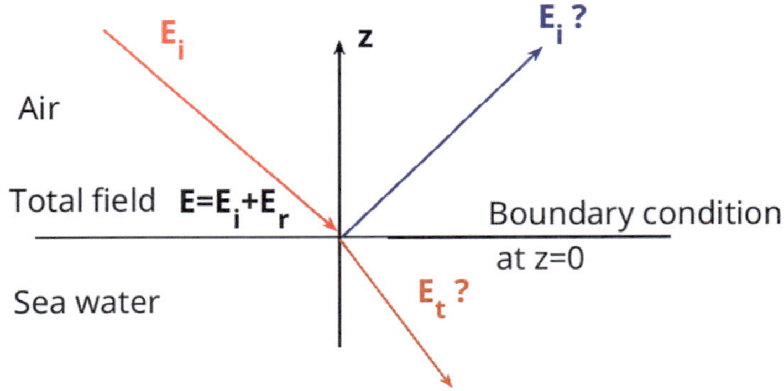

Fig. D.1 The total field above the surface is the sum of the incident and reflected field. It satisfies boundary conditions at the interface $z = 0$, which depend on the incident polarization

C.-A. Guérin, *High-Frequency Radars for the Oceanographer*, Geophysics and Environmental Physics, https://doi.org/10.1007/978-3-031-95642-3

Let us first assume that the incident field is an horizontally polarized EM plane wave propagating downwards with an incident wave vector $\mathbf{K_0} = k_0\mathbf{x} - q_0\mathbf{z}$. The corresponding electric field can be written:

$$\mathbf{E_i}(\mathbf{R}) = \mathbf{E_i}(x, z) = e^{\mathrm{i}(k_0 x - q_0 z)}\mathbf{y} \tag{D.0.1}$$

As explained in Chap. 7, the reflected and transmitted electric fields remain directed along the same direction $\mathbf{y}$ and depend on the same space variables (x, z). Using a Rayleigh expansion and imposing the physical requirement that the reflected field must propagate upwards and the transmitted field must propagate downwards, we can seek the unique component of the electric field in the form:

$$\begin{aligned} E_r(x, z) &= \int_{-\infty}^{+\infty} f(k)e^{\mathrm{i}kx}e^{\mathrm{i}qz}dk, \ z > 0 \\ E_t(x, z) &= \int_{-\infty}^{+\infty} g(k)e^{\mathrm{i}kx}e^{-\mathrm{i}q'z}dk, \ z < 0 \end{aligned} \tag{D.0.2}$$

with the condition $k^2 + q^2 = K_0^2$ in the above medium and $k^2 + (q')^2 = \epsilon_r K_0^2$ in the bottom medium. Now, the total fields $\mathbf{E}$ and $\mathbf{H}$ satisfy the EM boundary conditions at the interface $z = 0$, which in the present case write:

$$\begin{aligned} \mathbf{E}(x, y, 0^+) \times \mathbf{z} &= \mathbf{E}(x, y, 0^-) \times \mathbf{z} \\ \mathbf{H}(x, y, 0^+) \times \mathbf{z} &= \mathbf{H}(x, y, 0^-) \times \mathbf{z}. \end{aligned} \tag{D.0.3}$$

This implies:

$$E_i(x, 0) + E_r(x, 0) = E_t(x, 0) \tag{D.0.4}$$

that is

$$e^{\mathrm{i}k_0 x} + \int_{-\infty}^{+\infty} f(k)e^{\mathrm{i}kx}dk = \int_{-\infty}^{+\infty} g(k)e^{\mathrm{i}kx}dk \tag{D.0.5}$$

Since the complex exponential functions form an independent basis of functions, such a relation can only hold if $f(k) = \mathcal{R}_H(k_0)\delta(k - k_0)$, $g(k) = \mathcal{T}_H(k_0)\delta(k - k_0)$, that is the Rayleigh expansions reduce to a single component:

$$\begin{aligned} E_r(x, z) &= \mathcal{R}_H(k_0)e^{\mathrm{i}(k_0 x + q_0 z)}, \ z > 0 \\ E_t(x, z) &= \mathcal{T}_H(k_0)e^{\mathrm{i}(k_0 x - q_0' z)}, \ z < 0 \end{aligned} \tag{D.0.6}$$

It follows that:

$$1 + \mathcal{R}_H(k_0) = \mathcal{T}_H(k_0) \tag{D.0.7}$$

The quantities $\mathcal{R}_H(k_0)$ and $\mathcal{T}_H(k_0)$ define the reflection and transmission coefficients at the interface, respectively, and they are referred to as the *Fresnel coefficients* . An important consequence is that the reflected wave vector $\mathbf{K_r} = k_0\mathbf{x} + q_0\mathbf{z}$ and the transmitted plane wave vector $\mathbf{K_t} = k_0\mathbf{x} - q_0'\mathbf{z}$ share the same horizontal component

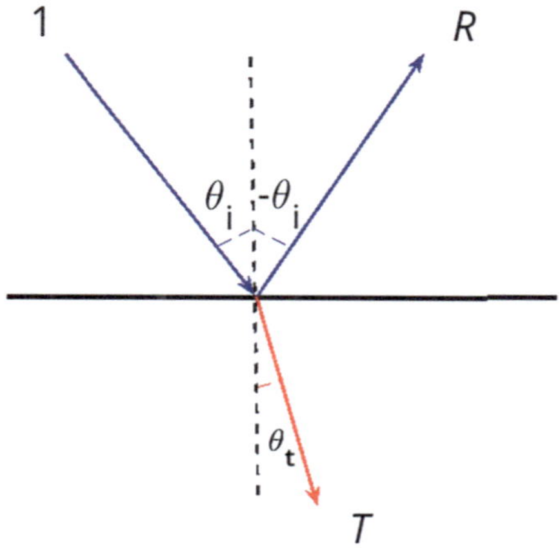

Fig. D.2 Snell-descartes laws of reflection and refraction

(k_0) with the incident wave vector $\mathbf{K_0}$. For a lossless medium for which the *optical index* $n = \sqrt{\epsilon_r}$ is real, this recovers the well-known laws of reflection and refraction derived by Snell and Descartes about 500 years ago. Denoting θ_r and θ_t, respectively, the reflection and transmission angles with respect to the normal direction we have (Fig. D.2):

$$\begin{aligned} \theta_r &= -\theta_i \\ \sin\theta_i &= n\sin\theta_t \end{aligned} \tag{D.0.8}$$

where the optical index is assumed to be one in the above medium. A second relation for the reflection and transmission coefficients can be obtained from the boundary condition for the total magnetic field at the interface (D.0.3). From the harmonic Maxwell-Faraday equation in (7.1.2) we have the continuity of $(\nabla \times \mathbf{E}) \times \mathbf{z}$ at $z = 0$, leading to the continuity of the normal derivative $(\partial_z E)$:

$$\partial_z E_i(x, z)_{|z=0} + \partial_z E_r(x, z)_{|z=0} = \partial_z E_t(x, 0)_{|z=0} \tag{D.0.9}$$

yielding

$$1 - \mathcal{R}_H(k_0) = \frac{q_0'}{q_0}\mathcal{T}_H(k_0) \tag{D.0.10}$$

From the couple of relations (D.0.7)–(D.0.10) we obtain easily:

$$\mathcal{R}_H(k_0) = \frac{q_0 - q_0'}{q_0 + q_0'}, \ \mathcal{T}_H(k_0) = \frac{2q_0}{q_0 + q_0'}. \tag{D.0.11}$$

Note that these coefficients are functions of the incidence angle θ and complex permittivity only and can be rewritten as:

$$\mathcal{R}_H(\theta) = \frac{\cos\theta - \sqrt{\epsilon_r - \sin^2\theta}}{\cos\theta + \sqrt{\epsilon_r - \sin^2\theta}}, \ \mathcal{T}_H(k_0) = \frac{2\cos\theta}{\epsilon_r\cos\theta + \sqrt{\epsilon_r - \sin^2\theta}} \tag{D.0.12}$$

For $\epsilon_r > 1$, which corresponds to the case of nonconducting media, we have $q_0' > q_0$ so that the reflection coefficient is strictly negative and the transmission coefficient is strictly positive, except at the limiting grazing incidence $\theta_i = 90$ deg ($q_0 = 0$).

Next we consider the case of a vertically polarized incident EM plane wave propagating downward with incident wave vector $\mathbf{K_0} = k_0\mathbf{x} - q_0\mathbf{z}$. This time it is the magnetic field $\mathbf{H}$ that has a unique nonzero component along the $\mathbf{y}$ direction and using the same argumentation we may define the Fresnel coefficients on the reflected and transmitted magnetic field:

$$
\begin{aligned}
&H_i(x,z) = e^{\mathrm{i}(k_0 x - q_0 z)} \text{ and } H_r(x,z) = \mathcal{R}_V(k_0) e^{\mathrm{i}(k_0 x + q_0 z)}, \; z > 0 \\
&H_t(x,z) = \mathcal{T}_V(k_0) e^{\mathrm{i}(k_0 x - q_0' z)}, \; z < 0
\end{aligned}
\tag{D.0.13}
$$

The difference with respect to the previous case lies in the boundary conditions. The continuity of the tangential magnetic field yields:

$$
1 + \mathcal{R}_V(k_0) = \mathcal{T}_V(k_0) \tag{D.0.14}
$$

but we now use Maxwell-Ampere equation in (7.1.2) to translate the continuity of the tangential electric field in terms of the continuity of $(1/\epsilon_r)(\nabla \times \mathbf{H}) \times \mathbf{z}$ at $z = 0$. This leads to the continuity of the normal derivative, weighted by the inverse permittivity $((1/\epsilon_r)\partial_z H)$:

$$
1 - \mathcal{R}_V(k_0) = \frac{q_0'}{\epsilon_r q_0} \mathcal{T}_V(k_0) \tag{D.0.15}
$$

Solving the system (D.0.14)–(D.0.15) yields the expression of the Fresnel coefficients in vertical polarization:

$$
\mathcal{R}_V(k_0) = \frac{\epsilon_r q_0 - q_0'}{\epsilon_r q_0 + q_0'}, \quad \mathcal{T}_V(k_0) = \frac{2\epsilon_r q_0}{\epsilon_r q_0 + q_0'}. \tag{D.0.16}
$$

In terms of the incidence angle θ and the complex permittivity, they can be written:

$$
\mathcal{R}_V(\theta) = \frac{\epsilon_r \cos\theta - \sqrt{\epsilon_r - \sin^2\theta}}{\epsilon_r \cos\theta + \sqrt{\epsilon_r - \sin^2\theta}}, \quad \mathcal{T}_V(k_0) = \frac{2\epsilon_r \cos\theta}{\epsilon_r \cos\theta + \sqrt{\epsilon_r - \sin^2\theta}} \tag{D.0.17}
$$

The Zenneck Wave

E

We present here a simple derivation of the surface wave model of Zenneck following the main idea of the original German paper [297] with modern notations. Even though Zenneck's model is not a realistic description of EM ground waves, it helps understanding some key properties of the surface wave, such as its vertical attenuation, its relation to the Fresnel reflection coefficient and its polarization tilt.

We consider the problem of a flat interface $z = 0$ separating the vacuum ($z > 0$) from a conducting medium ($z < 0$) with complex permittivity ϵ_r. We look for an harmonic solution to Maxwell's equation in the form of a monochromatic plane wave propagating along the x-direction, with an electric field $\mathbf{E} = (E_x, 0, E_z)$ in the (x, z) plane and a magnetic field $\mathbf{H} = (0, H_y, 0)$ oriented along the y direction (Fig. E.1). In this particular geometry the Maxwell's equations can be written from (7.2.17)–(7.2.18) as:

$$\begin{aligned} \partial_z E_x - \partial_x E_z &= \mathrm{i}\omega_0 \mu_0 H_y \\ -\partial_z H_y &= \mathrm{i}\omega_0 \epsilon(z) E_x \\ \partial_x H_y &= -\mathrm{i}\omega_0 \epsilon(z) E_z \end{aligned} \tag{E.0.1}$$

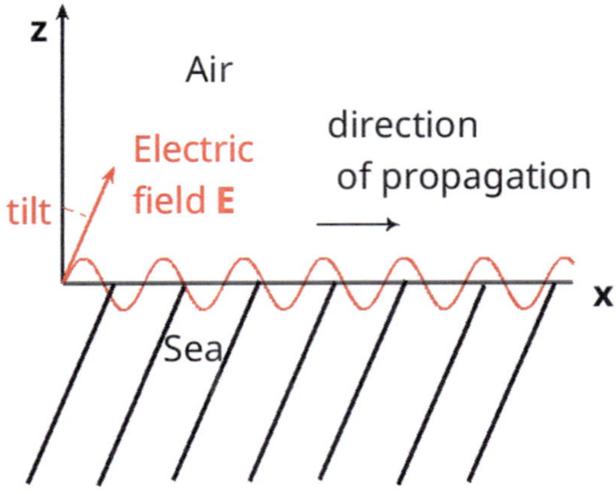

Fig. E.1 The Zenneck wave propagates along the air-sea plane interface and is exponentially attenuated on both sides

C.-A. Guérin, *High-Frequency Radars for the Oceanographer*, Geophysics and Environmental Physics, https://doi.org/10.1007/978-3-031-95642-3

where $\epsilon(z) = \epsilon_0$ if $z > 0$ and $\epsilon(z) = \epsilon_0 \epsilon_r$ if $z < 0$. A consequence of these relations is the Helmholtz equation satisfied by each field component E_x, H_y, E_z in each half-space with the corresponding wavenumber K_0 and $\sqrt{\epsilon_r} K_0$:

$$\frac{\partial^2 E_x}{\partial x^2} + \frac{\partial^2 E_x}{\partial z^2} + \epsilon(z) K_0^2 E_x = 0 \tag{E.0.2}$$

The solution for the three field component (E_x, H_y, E_z) is therefore of the form:

$$\begin{aligned} E_x &= E_{0x} e^{i\alpha x + i\beta z} \\ H_y &= H_{0y} e^{i\alpha x + i\beta z} \\ E_z &= E_{0z} e^{i\alpha x + i\beta z} \end{aligned} \tag{E.0.3}$$

above the interface ($z > 0$) and

$$\begin{aligned} E_x &= E'_{0x} e^{i\alpha x - i\beta' z} \\ H_y &= H'_{0y} e^{i\alpha x - i\beta' z} \\ E_z &= E'_{0z} e^{i\alpha x - i\beta' z} \end{aligned} \tag{E.0.4}$$

where $\alpha > 0$ and

$$\beta = \sqrt{K_0^2 - \alpha^2}, \ \beta' = \sqrt{\epsilon_r^2 K_0^2 - \alpha^2} \tag{E.0.5}$$

are the complex square roots defined with a positive imaginary part ($\mathrm{Im}[\beta] \geq 0$ and $\mathrm{Im}[\beta]' \geq 0$). The sign $\pm i\beta z$ in (E.0.3) and (E.0.4) ensure an exponential decay with $|z|$ on each side of the surface. Now, the boundary conditions (C.0.1) at the interface $z = 0$ imply:

$$E'_{0x} = E_{0x}, \ H'_{0y} = H_{0y}, \ E'_{0x} = Z_r^2 E_{0z}, \tag{E.0.6}$$

where as usual $Z_r = 1/\sqrt{\epsilon_r}$ is the relative surface impendance. Injecting (E.0.6) and (E.0.3)–(E.0.4) in Maxwell's equations (E.0.1) we obtain easily 2 additional relations:

$$\frac{E_{0x}}{E_{0z}} = \frac{\alpha(1 - Z_r^2)}{\beta + \beta'} = -\frac{\beta}{\alpha} \tag{E.0.7}$$

In the strong conductor limit $\left|\sqrt{\epsilon_r}\right| >> 1$, this can be solved easily, leading to:

$$\frac{E_{0x}}{E_{0z}} \simeq Z_r, \ \alpha \simeq K_0 \sqrt{1 - Z_r^2}, \ \beta \simeq -Z_r K_0 \tag{E.0.8}$$

Due to the small positive imaginary part of Z_r, the Zenneck wave is confined to the surface like a guided wave because it decreases exponentially upwards and downwards from the interface $z = 0$. It has also a slow exponential decrease along the positive x direction due to the even smaller negative imaginary part of $(1 - Z_r^2)$. Note the small polarization tilt $\Psi \simeq \mathrm{atan}(E_x/E_z)$ showing that the surface wave does not behave like a transverse free propagating plane wave but has a small longitudinal component. An isotropic version of the Zenneck wave can also be obtained solving Maxwell equations in the form of a cylindrical rather than plane wave. In that case, the asymptotic field is of the form $E(r, z) \sim \mathrm{e}^{\mathrm{i}(\alpha r+\beta z)}/\sqrt{r}$ with the same values α and β.

Not surprinsingly, the complex wavenumber α corresponds to the pole of the Fresnel reflection coefficient $\mathcal{R}_V$ in (7.4.28), which is obtained for $\epsilon_r q_0 = 0$, corresponding to the values $k_0 \simeq K_0\sqrt{1-Z_r^2}$ and $q_0 \simeq -Z_r K_0$. A pole of the reflection coefficient in any surface scattering problem is the mathematical expression of a what is called a *surface wave mode*, a *trapped wave* or a *guided wave* depending on the scientific community. In Optics, it is referred to as a *surface plasmon*. Due to the negative imaginary part of Z_r, the modal value $K_0\sqrt{1-Z_r^2}$ is slighly larger in magnitude than the wavenumber K_0, so that the surface mode cannot be excited by a propagating wave ($|k_0| \leq K_0$) but only by evanescent waves in the near field $|k_0| > K_0$. This might be a reason why a Zenneck wave has never been observed for HF radio waves over the ocean, even though they can be created in numerical simulations with adapted sources [193] or in the laboratory with a metal sheet [209]. Note that taking the average value $Z_r = 0.011 - 0.012i$ for the sea surface impedance leads to $\mathrm{Im}[\alpha] = 1.3\ 10^{-4}\ K_0$ and $\mathrm{Im}[\beta] = 0.012\ K_0$, which implies an unrealistic attenuation distance of the order of a 10000 wavelength along the surface and 100 wavelength above.

F The Weyl Integral in the Sommerfeld Problem

The Sommerfeld problem (Chap. 8) consists in calculating the electric field produced by a dipole antenna above a flat conducting surface, which serves as a paradigm for ground wave propagation over the sea surface. The technical difficulty of this problems lies in the calculation of a specific Weyl integral (see Sect. 8.2), which must be conducted in a cautionous way. Specifically, the calculation of the EM field produced at any position $\mathbf{R} = (\mathbf{r}, z)$ by an infinitesimal vertical dipole current source $\mathbf{j_0}$ located at the position $(\mathbf{0}, H)$ above a conducting surface $(z = 0)$ involves the calculation of the following integral:

$$\mathcal{W}[\mathcal{R}_V - 1](\mathbf{r}, z + H) = -\frac{\mathrm{i}}{8\pi^2} \int_{\mathbb{R}^2} \frac{d\mathbf{k}}{q} \left[\mathcal{R}_V(\mathbf{k}) - 1\right] \mathrm{e}^{\mathrm{i}(\boldsymbol{k}\cdot\boldsymbol{r} + q(z+H))}, \qquad \text{(F.0.1)}$$

where $\mathcal{R}_V$ is the Fresnel reflection coefficient (D.0.16) in vertical polarization. For large value of the optical index $n = \sqrt{\epsilon_r}$ we may approximate:

$$\mathcal{R}_V(k) - 1 = \frac{-2q'}{\epsilon q + q'} = \frac{-2nK_0}{n^2 q + nK_0} + 0(1/n^2) \qquad \text{(F.0.2)}$$

Noting that the imaginary part of the denominator is stricly positive, we can write, using a trick first proposed by Van de Pol [265]:

$$\frac{-2nK_0}{n^2 q + nK_0} = 2\mathrm{i}\, nK_0 \int_0^{\infty} ds\, \mathrm{e}^{\mathrm{i}(n^2 q + nK_0)s} \qquad \text{(F.0.3)}$$

so that the Weyl integral in Eq. (8.2.6) can be rewritten as:

$$\mathcal{W}[\mathcal{R}_V - 1](\mathbf{r}, z + H) = \frac{1}{4\pi^2} nK_0 \int_0^{\infty} ds \left(\mathrm{e}^{\mathrm{i}nK_0 s} \int \frac{d\mathbf{k}}{q}\, \mathrm{e}^{\mathrm{i}\boldsymbol{k}\cdot\boldsymbol{r}} \mathrm{e}^{\mathrm{i}q(z+H+n^2 s)} \right) \qquad \text{(F.0.4)}$$

C.-A. Guérin, *High-Frequency Radars for the Oceanographer*, Geophysics and Environmental Physics, https://doi.org/10.1007/978-3-031-95642-3

The inner integral is identified as the Weyl integral of the free space Green's function at the complex position $\mathbf{R}_s = (\mathbf{r}, z_s)$ where $z_s = z + H + n^2 s$, so that:

$$\mathcal{W}[\mathcal{R}_V - 1](\mathbf{r}, z + H) = -2nK_0 \int_0^\infty ds \frac{e^{iK_0(R_s + ns)}}{4\pi R_s}, \tag{F.0.5}$$

with

$$R_s^2 = r^2 + (z + H + n^2 s)^2 \tag{F.0.6}$$

Due to the large imaginy part of n, the exponential is rapidly attenuated and the main contribution of the integrand arises near $s = 0$. We therefore perform a quadratic expansion:

$$R_s + ns \simeq R_1 + n(1 + n\cos\theta)s + \frac{n^4(1 + \cos^2\theta)}{2R_1} s^2, \tag{F.0.7}$$

where $R_1 = \sqrt{r^2 + (z + H)^2}$ and $\cos\theta = (z + H)/R_1$ and obtain the following approximation:

$$\mathcal{W}[\mathcal{R}_V - 1](\mathbf{r}, z + H) = 2nK_0 G(R_1) \int_0^\infty ds\, e^{(a^2 s^2 + bs)} \tag{F.0.8}$$

with

$$a^2 = iK_0 \frac{(1 + \cos^2\theta)n^4}{2R}, \quad b = iK_0(\cos\theta n^2 + n) \tag{F.0.9}$$

and G the scalar Green's function (B.0.6). The integral (F.0.8) is given by the following identity as long as $\mathrm{Im}[a] \neq 0$:

$$\int_0^\infty ds\, e^{(a^2 s^2 + bs)} = i\sqrt{\frac{\pi}{4a^2}} e^{-b^2/4a^2} \operatorname{erfc}(-i\sqrt{b^2/4a^2}) \tag{F.0.10}$$

Setting

$$w = \frac{b^2}{4a^2} = \frac{iK_0 R}{2n^2} \frac{(1 + n\cos\theta)^2}{1 + \cos^2\theta} \tag{F.0.11}$$

we thus obtain:

$$\mathcal{W}[\mathcal{R}_V - 1](\mathbf{r}, z + H) = \frac{2}{1 + n\cos\theta}(\mathrm{SO}(w) - 1)G(R_1), \tag{F.0.12}$$

where SO is the Sommerfeld attenuation function:

$$\mathrm{SO}(u) = 1 + i\sqrt{\pi u}\, e^{-u} \operatorname{erfc}(-i\sqrt{u}) \tag{F.0.13}$$

The Small Perturbation Method

G

A pertubation expansion makes it possible to obtain tractable analytical approximation of the scattering amplitude from the sea surface when the wave height is small compared to the radar wavelength, an assumption which is in general valid in the context of HFR. We present here an outline of the technique as well as the explicit formulae that have been used to recover Barrick's theory in Chap. 10.

G.1 From Rayleigh to Voronovich

The small perturbation method was first employed by Lord Rayleigh for sound scattering from a sinusoidal surface [220]. His method, now usually referred to as the *Rayleigh method*, can be generalized to any kind of waves and surfaces. In this approach, it is assumed that the Rayleigh expansion describing the scattered field (9.1.3) is valid at the surface itself, although it can only be proven to hold above the maximum vertical excursion of the surface. With this assumption, the boundary conditions satisfied by the electric and magnetic field at the surface provide a closed equation for the scattering amplitude, which can be solved iteratively at different orders in roughness elevations. An example of such derivation is presented in the Appendix G.2 for the simple case of the 2D scalar Dirichlet problem. The Rayleigh method applies in a similar way to the general context of vector EM field above a lossy or dielectric surface but is technically considerably more involved. They imply a set of coupled equations for the scattering kernels derived from the continuity relations (C.0.1) satisfied by the tangential components of the electric and magnetic fields and the continuity of the normal components of the auxiliary fields. The difficulty in that case is that a Rayleigh expansion must be found simultaneously above and below the surface and the perturbation technique must be applied to a system of coupled vector boundary conditions.

In a seminal 1951 paper, Rice generalized Rayleigh's technique to EM scattering from bi-dimensional periodic time-invariant surfaces [222]. He derived the complete

C.-A. Guérin, *High-Frequency Radars for the Oceanographer*, Geophysics and Environmental Physics, https://doi.org/10.1007/978-3-031-95642-3

expression of the second-order field scattered from a perfectly conducting surface at arbitrary incidence angles in horizontal and vertical polarization. He addressed the case of lossy surfaces to show the effect of finite conductivity but dealt with the horizontal polarization only, certainly because of the much greater technical complexity of the vertical polarization case. This last case was treated by Wait [270] for one-dimensional surfaces using the impedance boundary condition, which is valid for media with high conductivity (the case of dielectric surfaces, for which no impedance approximation can be used, was addressed 20 years later by Saillard [230]).

Barrick extended Rice theory to deal with the case of vertically polarized waves propagating over a two-dimensional lossy surface at grazing incidence [11, 13, 14]; by calculating the electric field at the surface in the form of a guided wave, he derived an effective surface impedance, in which the roughness effects come into play at second-order. To simplify the calculations, he used the Leontovich impedance approximation (C.0.2), which allows to uncouple the electric and magnetic field in solving Maxwell's equations and make the calculations more tractable. In these works, Barrick did not take into account the temporal variations of the sea surface. Short after, however, he derived the second-order Doppler cross-section of a moving sea surface and gave explicit formulae for the hydrodynamic and electromagnetic kernels [10, 12, 15, 16]. He did not give the steps of the derivation but it follows from the same methodology that was used to derive the effective impedance surface; he also included the nonlinear hydrodynamical effects, which he described only later in his 1977 companion papers with Weber [27, 283].

The second-order solution for dielectric or lossy two-dimensional surfaces at arbitrary incidences with no further approximation has been derived by Voronovich [266] for the 4 coefficients of the scattering matrix. Their general expression can be also found in [268]. They have also been re-derived using a different approach based on a volume-integral equation of the electric field combined with the half-space Green's function [107]. In the appendix, we have reproduced the expression of the first- and second-order kernels B_1 and B_2 (with some adaptation to our conventions and notations) for the vertical co-polarization state, which is relevant to HFR.

G.2 How Does One Find Explicitely the Perturbation Expansion?

In the perturbation expansion, the scattering amplitude is sought as a functional expansion in surface roughness:

$$\begin{aligned}\mathsf{A}(\mathbf{k},\mathbf{k_0}) &= \mathsf{B}_0(\mathbf{k_0})\delta(\mathbf{k}-\mathbf{k_0}) - i\mathsf{B}_1(\mathbf{k},\mathbf{k_0})\widehat{\eta}(\mathbf{k_B}) \\ &+ \int \mathsf{B}_2(\mathbf{k},\mathbf{k_0},\mathbf{k_0}+\mathbf{k_1})\widehat{\eta}(\mathbf{k_1})\widehat{\eta}(\mathbf{k_B}-\mathbf{k_1})d\mathbf{k_1},\end{aligned} \tag{G.2.1}$$

where $\widehat{\eta}$ is the spatial Fourier transform of elevations and $\mathbf{k_B} = \mathbf{k} - \mathbf{k_0}$ is the bistatic Bragg vector. The kernels B_0, B_1 and B_2 do not depend on roughness but only on the

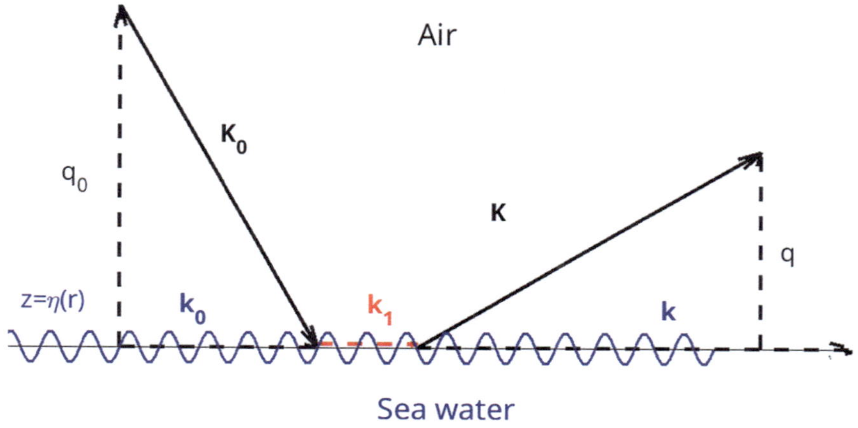

Fig. G.1 Schematic view of the scattering geometry

scattering geometry (Fig. G.1) and the sea surface complex permittivity. Unlike the zero- and first-order kernels B_0 and B_1, the second-order kernel B_2 is not uniquely defined, as several different expressions can lead to the same integral. For example, it can be replaced by its symmetrized expression:

$$\mathsf{B}_2(\mathbf{k}, \mathbf{k_0}, \mathbf{k_0} + \mathbf{k_1}) \rightarrow \frac{1}{2}\left(\mathsf{B}_2(\mathbf{k}, \mathbf{k_0}, \mathbf{k_0} + \mathbf{k_1}) + \mathsf{B}_2(\mathbf{k}, \mathbf{k_0}, \mathbf{k} - \mathbf{k_1})\right), \tag{G.2.2}$$

as can be seen by operating the change of variables $\mathbf{k_1} \rightarrow \mathbf{k_B} - \mathbf{k_1}$ in the expression of A_2. More generally, the second-order integral in Eq. (G.2.1) is left unchanged by operating the following replacement with any arbitrary function J:

$$\mathsf{B}_2(\mathbf{k}, \mathbf{k_0}, \mathbf{k_0} + \mathbf{k_1}) \rightarrow \mathsf{B}_2(\mathbf{k}, \mathbf{k_0}, \mathbf{k_0} + \mathbf{k_1}) + (\mathsf{J}(\mathbf{k}, \mathbf{k_0}, \mathbf{k} - \mathbf{k_1}) - \mathsf{J}(\mathbf{k}, \mathbf{k_0}, \mathbf{k_0} + \mathbf{k_1})) \tag{G.2.3}$$

We will now explain how the perturbation theory works. To highlight the principle of the method without falling into technicalities, we will consider the idealized situation of a scalar two-dimensional problem with Dirichlet boundary condition ($E = 0$ at the surface), which corresponds to the case of a perfectly conducting cylindrical surface illuminated by an horizontally polarized incident plane wave. We will ignore the time dependence of the surface, assuming it is "frozen" during the scattering process. The customary way to find the perturbation expansion is the so-called *Rayleigh method*, which consists in applying the boundary condition to the Rayleigh expansion of the scattered field. Let us consider an incident field $E_i(x, z) = \mathrm{e}^{\mathrm{i}(k_0 x - q_0 z)}$ with incident wave vector $\mathbf{K_0} = (k_0, -q_0)$ in the (x, z) plane impinging on a rough surface $z = \eta(x)$ separating the vacuum (upper medium) from the sea surface (lower medium). The scattered field can be written as:

$$E_s(x, z) = \int_{-\infty}^{+\infty} \mathrm{e}^{\mathrm{i}(kx + qz)} A(k, k_0) dk, \tag{G.2.4}$$

with the usual definition $q = \sqrt{K_0^2 - k^2}$ (which is positive imaginary if $|k| > K_0$). This representation, which is called the Rayleigh expansion, can be obtained in a way very similar to the Weyl's representation of the scalar Green's function. Recall that the scattered field satisfies the Helmholtz equation in the half-plane above the highest surface excursion ($z > z_{max}$):

$$\frac{\partial^2 E(x,z)}{\partial x^2} + \frac{\partial^2 E(x,z)}{\partial z^2} + K_0^2 E(x,z) = 0 \tag{G.2.5}$$

Taking the partial Fourier transform in the horizontal x variable, solving in z at fixed wavenumber k with the outgoing wave condition and back-transforming to the x variable leads to the Rayleigh expansion (G.2.4). This writing is *a priori* valid only above the maximum of surface elevations ($z > z_{max}$). However, one usually resorts to an additional assumption, known as the *Rayleigh hypothesis* (him again !), which postulates that (G.2.4) is also valid at the surface ($z = \eta(x)$). One can then express the boundary condition at the surface:

$$E_i(x, \eta(x)) + E_s(x, \eta(x)) = 0, \tag{G.2.6}$$

as an implicit equation for the scattering amplitude:

$$\mathrm{e}^{\mathrm{i}(k_0 x - q_0 \eta)} + \int_{-\infty}^{+\infty} \mathrm{e}^{\mathrm{i}(kx + q\eta)} A(k, k_0) dk = 0 \tag{G.2.7}$$

The zeroth order scattered amplitude is obtained by setting $\eta = 0$ and is identified as:

$$A_0(k, k_0) = -\delta(k - k_0) \tag{G.2.8}$$

We thus recover the specularly reflected field $E_r(x, y) = -\mathrm{e}^{\mathrm{i}(k_0 x + q_0 z)}$ with the $\mathcal{R} = -1$ Fresnel reflection coefficient of perfectly conducting surfaces. The first-order boundary condition is obtained by linearizing the exponential with respect to the elevations:

$$\mathrm{e}^{\mathrm{i}k_0 x}(1 - \mathrm{i}q_0\eta) + \int_{-\infty}^{+\infty} \mathrm{e}^{\mathrm{i}kx}(1 + \mathrm{i}q\eta)(A_0 + A_1) dk = 0 \tag{G.2.9}$$

After substitution of the zeroth order scattering amplitude and identification of the first-order terms we arrive at:

$$\int_{-\infty}^{+\infty} A_1(k, k_0) \mathrm{e}^{\mathrm{i}kx} dk = 2\mathrm{i}q_0 \mathrm{e}^{\mathrm{i}k_0 x} \eta(x) \tag{G.2.10}$$

The first-order scattering amplitude can be extracted by performing an extra Fourier Transform with respect to the x variable, leading to:

$$A_1(k, k_0) = 2\mathrm{i}q_0 \widehat{\eta}(k_B), \tag{G.2.11}$$

where $\widehat{\eta}$ is the one-dimensional spatial Fourier Transform of elevations:

$$\widehat{\eta}(k) = \frac{1}{2\pi}\int_{-\infty}^{+\infty} \mathrm{e}^{-ikx}\eta(x)dx, \tag{G.2.12}$$

Pursuing the expansion of the boundary condition at second order we obtain:

$$\int_{-\infty}^{+\infty} \mathrm{e}^{ikx}(-\frac{q^2\eta^2}{2}A_0 + \mathrm{i}q\eta A_1 + A_2)dk = \mathrm{e}^{\mathrm{i}k_0x}\frac{q_0^2\eta^2}{2} \tag{G.2.13}$$

which after substitution of the zeroth and first-order solution leads to:

$$\int_{-\infty}^{+\infty} \mathrm{e}^{ikx} A_2(k, k_0)dk = 2qq_0\eta(x)\widehat{\eta}(k_B) \tag{G.2.14}$$

The second-order scattering amplitude can again be extracted via a spatial Fourier Transform yielding:

$$A_2(k, k_0) = -\int_{-\infty}^{+\infty} 2q_0q_1\widehat{\eta}(k-k_1)\widehat{\eta}(k_1-k_0)dk_1 \tag{G.2.15}$$

In this example, the kernels of the scattering amplitude are identified as:

$$B_0(k_0) = -1,\ B_1(k, k_0) = -2q_0,\ B_2(k, k_0, k_1) = -2q_0q_1 \tag{G.2.16}$$

G.3 The Kernels of the Scattering Amplitude

The expression of the kernels have been derived by Voronovich [266] in the most general case of bistatic scattering at arbitrary incidence and polarization from a bidimensional rough surface with complex relative permittivity ϵ_r. We detail here only the vertical co-polarization coefficient, which is relevant for HFR. Note that using Voronovich's original formulae with our convention requires multiplying his first order kernel B_1 by the factor $2\sqrt{qq_0}$ and his second-order kernel by $\sqrt{qq_0}$. The kernel of the first-order scattering amplitude in vertical co-polarization for a dielectric or conducting surface at arbitrary incidence is given by:

$$B_1(\mathbf{k}, \mathbf{k_0}) = \frac{2\sqrt{qq_0}(\epsilon-1)}{(\epsilon q + q')(\epsilon q_0 + q_0')}\left(q'q_0'\widehat{\mathbf{k}}\cdot\widehat{\mathbf{k_0}} - \epsilon kk_0\right) \tag{G.3.17}$$

where we use the notation:

$$\begin{aligned} q_0 &= \sqrt{K_0^2 - k_0^2},\ \ q = \sqrt{K^2 - k^2}, \\ q_0' &= \sqrt{\epsilon K_0^2 - k_0^2},\ q' = \sqrt{\epsilon K_0^2 - k^2} \end{aligned} \tag{G.3.18}$$

and we recall that $\widehat{\mathbf{k}} = \mathbf{k}/k$ is the unit vector along $\mathbf{k}$. The second-order kernel is given by[1]

$$\begin{aligned} B_2(\mathbf{k}, \mathbf{k_0}, \mathbf{k_1}) = & \frac{\sqrt{qq_0}(\epsilon - 1)}{(\epsilon q + q')(\epsilon q_0 + q'_0)} \times \\ & \left[-2\frac{\epsilon - 1}{\epsilon q_1 + q'_1} \left(q'_0 q' (\widehat{\mathbf{k}} \cdot \mathbf{k_1})(\widehat{\mathbf{k_0}} \cdot \mathbf{k_1}) + \epsilon k k_0 k_1^2 \right) \right. \\ & + 2\epsilon \frac{q_1 + q'_1}{\epsilon q_1 + q'_1} \left(k_0 q' \widehat{\mathbf{k}} \cdot \mathbf{k_1} + k q'_0 \widehat{\mathbf{k_0}} \cdot \mathbf{k_1} \right) \\ & \left. - \left(\epsilon K_0^2 (q' + q'_0) + 2 q' q'_0 (q_1 - q'_1) \right) \widehat{\mathbf{k}} \cdot \widehat{\mathbf{k_0}} \right] \end{aligned} \tag{G.3.19}$$

with

$$q_1 = \sqrt{K_0^2 - k_1^2},\ q'_1 = \sqrt{\epsilon K_0^2 - k_1^2} \tag{G.3.20}$$

These expressions can also be recast in terms of Fresnel reflection coefficients (see Appendix B of [107]), which explains the presence of the various factors $1/(\epsilon q + q')$, $1/(\epsilon q_0 + q'_0)$ and $1/(\epsilon q_1 + q'_1)$. For strongly conducting media in the grazing limit ($q'_1 = q' = q'_0 = K_0/Z_r$ and $k = k_0 = K_0$), the above expressions reduce to:

$$B_1(\mathbf{k}, \mathbf{k_0}) = \frac{2\sqrt{qq_0}((\boldsymbol{k} \cdot \boldsymbol{k_0}) - K_0^2)}{(q + K_0 Z_r)(q_0 + K_0 Z_r)} \tag{G.3.21}$$

and

$$\begin{aligned} B_2(\mathbf{k}, \mathbf{k_0}, \mathbf{k_1}) = & \frac{2\sqrt{qq_0}}{(q + K_0 Z_r)(q_0 + K_0 Z_r)(q_1 + K_0 Z_r)} \times \\ & \left(-(\mathbf{k} \cdot \mathbf{k_1})(\mathbf{k_0} \cdot \mathbf{k_1}) + (\boldsymbol{k} \cdot \boldsymbol{k_0} - K_0^2) k_1^2 + K_0^2 (\mathbf{k} + \mathbf{k_0}) \cdot \mathbf{k_1} - K_0^2 \boldsymbol{k} \cdot \boldsymbol{k_0} \right) \end{aligned} \tag{G.3.22}$$

Note the important particular case of forward scattering ($\mathbf{k} = \mathbf{k_0}$)

$$\begin{aligned} B_1(\mathbf{k_0}, \mathbf{k_0}) &= 0 \\ B_2(\mathbf{k_0}, \mathbf{k_0}, \mathbf{k_1}) &= \frac{2q_0 \left[-(\boldsymbol{k_0} \cdot \boldsymbol{k_1})^2 + 2K_0^2 (\boldsymbol{k_0} \cdot \boldsymbol{k_1}) - K_0^4 \right]}{(q_0 + K_0 Z_r)^2 (q_1 + K_0 Z_r)} \end{aligned} \tag{G.3.23}$$

and backscattering ($\mathbf{k} = -\mathbf{k_0}$), which is relevant to monostatic radars:

$$\begin{aligned} B_1(-\mathbf{k_0}, \mathbf{k_0}) &= -\frac{4K_0^2}{(q_0 + K_0 Z_r)} \\ B_2(-\mathbf{k_0}, \mathbf{k_0}, \mathbf{k_1}) &= \frac{2q_0 \left[(\boldsymbol{k_0} \cdot \boldsymbol{k_1})^2 - 2K_0^2 k_1^2 + K_0^4 \right]}{(q_0 + K_0 Z_r)^2 (q_1 + K_0 Z_r)} \end{aligned} \tag{G.3.24}$$

[1] Note that we give the expression of $\mathsf{B}_2(\mathbf{k}, \mathbf{k_0}, \mathbf{k_1})$ as it is customary, rather than $\mathsf{B}_2(\mathbf{k}, \mathbf{k_0}, \mathbf{k_0} + \mathbf{k_1})$, which is the expression involved in the formal perturbation expansion (G.2.1).

The Effective Surface Impedance

H

An effective surface impedance can be attributed to a rough surface above a dielectric or conducting medium by seeking a virtual medium below a flat interface, which would produce the same coherent field in the specular direction. To do this, we consider the coherent scattering amplitude from the rough sea surface, obtained after taking an ensemble average (over surface roughness) of the perturbation expansion (9.2.9):

$$\langle \mathsf{A}(\mathbf{k}, \mathbf{k_0}) \rangle = \mathsf{B}_0(\mathbf{k_0})\delta(\mathbf{k_B}) + \int \mathsf{B}_2(\mathbf{k}, \mathbf{k_0}, \mathbf{k_0} + \mathbf{k_1}) \langle \widehat{\eta}(\mathbf{k_1})\widehat{\eta}(\mathbf{k_B} - \mathbf{k_1}) \rangle d\mathbf{k_1} \tag{H.0.1}$$

Since

$$\langle \widehat{\eta}(\mathbf{k_1})\widehat{\eta}(\mathbf{k_B} - \mathbf{k_1}) \rangle = S_0(\mathbf{k_1})\delta(\mathbf{k_B}), \tag{H.0.2}$$

we have:

$$\langle \mathsf{A}(\mathbf{k}, \mathbf{k_0}) \rangle = \left(\mathsf{B}_0(\mathbf{k_0}) + \int \mathsf{B}_2(\mathbf{k_0}, \mathbf{k_0}, \mathbf{k_0} + \mathbf{k_1}) S_0(\mathbf{k_1}) d\mathbf{k_1} \right) \delta(\mathbf{k_B}), \tag{H.0.3}$$

where $S_0(\mathbf{k}_1)$ is the spatial power spectrum of the surface. This confirms that the mean scattering amplitude is purely specular, as stated in Chap. 9. It follows that the mean scattered field produced by the rough surface is equivalent to the reflected field from a fictitious plane interface (Fig. H.1). By comparing (H.0.3) with the expressions (9.2.9) of the zeroth-order scattering amplitude, we see that the specular field can be associated with an effective Fresnel reflection coefficient $(\mathcal{R}_V)_e$ such that:

$$(\mathcal{R}_V)_e(\mathbf{k_0}) = \mathcal{R}_V(\mathbf{k_0}) + \int B_2(\mathbf{k_0}, \mathbf{k_0}, \mathbf{k_0} + \mathbf{k_1}) S_0(\mathbf{k_1}) d\mathbf{k_1}, \tag{H.0.4}$$

where

$$\mathcal{R}_V = \frac{q_0 - Z_r K_0}{q_0 + Z_r K_0} \tag{H.0.5}$$

C.-A. Guérin, *High-Frequency Radars for the Oceanographer*, Geophysics and Environmental Physics, https://doi.org/10.1007/978-3-031-95642-3

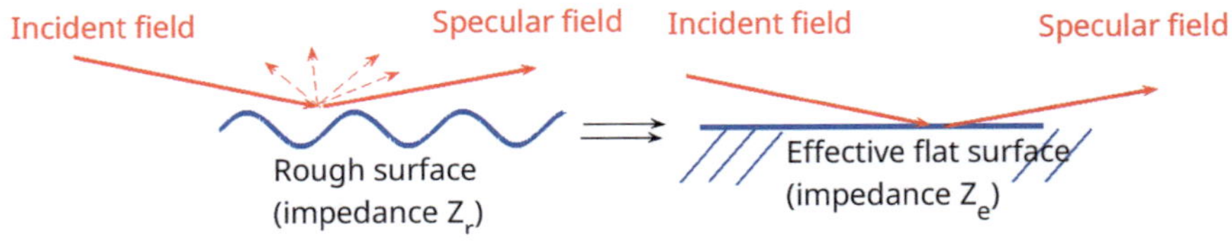

Fig. H.1 The mean field of a rough surface is equivalent to the specular reflection from an effective flat surface with a different impedance

is the Fresnel reflection coefficient in vertical polarization. An effective relative impedance Z_e can therefore be defined from the effective reflection coefficient:

$$(\mathcal{R}_V)_e = \frac{q_0 - Z_e K_0}{q_0 + Z_e K_0}, \tag{H.0.6}$$

Using the expression (G.3.23) given in the Appendix G for the second-order kernel in the forward direction this yields the identity:

$$\frac{q_0 - Z_e K_0}{q_0 + Z_e K_0} = \frac{q_0 - Z_r K_0}{q_0 + Z_r K_0} - \frac{q_0}{(q_0 + K_0 Z_r)^2} \int \frac{2(\boldsymbol{k_0} \cdot \boldsymbol{k_1})^2}{(q_{01} + K_0 Z_r)} S_0(\mathbf{k}_1) d\mathbf{k}_1 \tag{H.0.7}$$

To find the effective impedance we seek a perturbation solution in the form $Z_e = Z_r(1 + \alpha)$ with $\alpha << 1$ and solve the above equation by identifying the first-order term in α. After some algebraic manipulations we obtain:

$$Z_e = Z_r + \int P(\mathbf{k}) S_0(\mathbf{k}) d\mathbf{k}, \tag{H.0.8}$$

with:

$$P(\mathbf{k}) = \frac{(\boldsymbol{k_1} \cdot \boldsymbol{k_0})^2}{K_0 q_{01} + Z_r K_0^2} \tag{H.0.9}$$

with $q_{01} = \sqrt{K_0^2 - \|\mathbf{k_0} + \mathbf{k_1}\|^2}$. Barrick [13] obtained a similar expression more than half a century ago by applying the Zenneck wave model to a slightly rough conducting surface with impedance boundary conditions. He found the same relation (H.0.8) for the effective impedance, with a slightly different kernel in the integral. Adapting Eq. (23) of [13] to our notations,[2] this kernel can be written as:

$$P_{\text{Barrick}}(\mathbf{k_1}) = \frac{(\boldsymbol{k_1} \cdot \boldsymbol{k_0})^2 + Z_r K_0^2 q_{01} (\boldsymbol{k_1} - \boldsymbol{k_0}) \cdot \boldsymbol{k_0}}{K_0 q_{01} + Z_r (K_0^2 + q_{01}^2)} + Z_r \left(\frac{(\boldsymbol{k_1} \cdot \boldsymbol{k_0})^2}{K_0^2} - k_1^2 + \boldsymbol{k_1} \cdot \boldsymbol{k_0} \right) \tag{H.0.10}$$

[2] Beware of the confusion that might appear since Barrick's power spectrum $W(\mathbf{k})$, following the original convention chosen by Rice [222], is four times our wavenumber spectrum $S_0(\mathbf{k})$.

If we expand this last expression in powers of Z_r and neglect the quadratic terms Z_r^2, we can rewrite it in a simpler way after some algebraic manipulation:

$$P_{\text{Barrick}}(\mathbf{k}_1) = \frac{(\boldsymbol{k}_\mathbf{1} \cdot \boldsymbol{k}_\mathbf{0})^2}{K_0 q_{01} + Z_r K_0^2} + \frac{Z_r}{2} k_1^2 \tag{H.0.11}$$

The 2 kernels (H.0.9) and (H.0.11) differ only by the term $\frac{Z_r}{2} k_1^2$. The reason for this small discrepancy is that we dropped the Z_r terms in the strong conductor approximation (G.3.22)–(G.3.23) of the second-order kernels (G.3.19). However, the resulting difference in terms of the effective surface impedance is given by:

$$\int (P(\mathbf{k}) - P_{\text{Barrick}}(\mathbf{k}))\, S_0(\mathbf{k}) d\mathbf{k} = \frac{1}{2} \text{mss}\ Z_r, \tag{H.0.12}$$

where mss is the sea surface mean square slope (12.1.17) and is therefore negligible. Barrick's formulation of the effective impedance based on the Zenneck wave and ours based on the mean specular field are therefore equivalent.

Numerical Evaluation of the Second-Order Doppler Spectrum

I

The numerical evaluation of the integral (10.1.10) describing the second-order Doppler spectrum is non obvious due to the delta function constraint on the ω_1 and ω_2 variables. We will here review the different ways to handle it. For simplicity we will only discuss the monostatic case, the extension to the bistatic case being straightforward.

I.1 A Tricky Integral

Let us introduce the normalized wavenumbers and frequencies:

$$\boldsymbol{\kappa_1} = \mathbf{k}_1/k_B, \ \boldsymbol{\kappa_2} = \mathbf{k}_2/k_B, \ \boldsymbol{\kappa_B} = \mathbf{k}_B/k_B, \ \nu = \omega/\omega_B. \tag{I.1.1}$$

First note that the combined EM and hydrodynamical kernel (10.1.12) depends only on the normalized variables and the Bragg wavenumber:

$$\Gamma(\mathbf{k}_1, \mathbf{k}_2, \omega, \omega_B) = k_B \Gamma(\boldsymbol{\kappa}_1, \boldsymbol{\kappa}_2, \nu, 1). \tag{I.1.2}$$

It follows that the second-order Doppler spectrum can be rewritten in terms of the normalized variables:

$$\sigma_2(\nu) = \mathcal{N} k_B^4 \omega_B^{-1} \sum_{n_1=\pm1, n_2=\pm1} \int S(n_1 k_B \boldsymbol{\kappa}_1) S(n_2 k_B \boldsymbol{\kappa}_2) \left|\Gamma\right|^2 \delta(\nu - n_1\nu_1 - n_2\nu_2) d\boldsymbol{\kappa}_1, \tag{I.1.3}$$

with the relations:

$$\boldsymbol{\kappa}_1 + \boldsymbol{\kappa}_2 = \boldsymbol{\kappa}_B, \ \nu_1 = \sqrt{\kappa_1}, \ \nu_2 = \sqrt{\kappa_2}, \tag{I.1.4}$$

C.-A. Guérin, *High-Frequency Radars for the Oceanographer*, Geophysics and Environmental Physics, https://doi.org/10.1007/978-3-031-95642-3

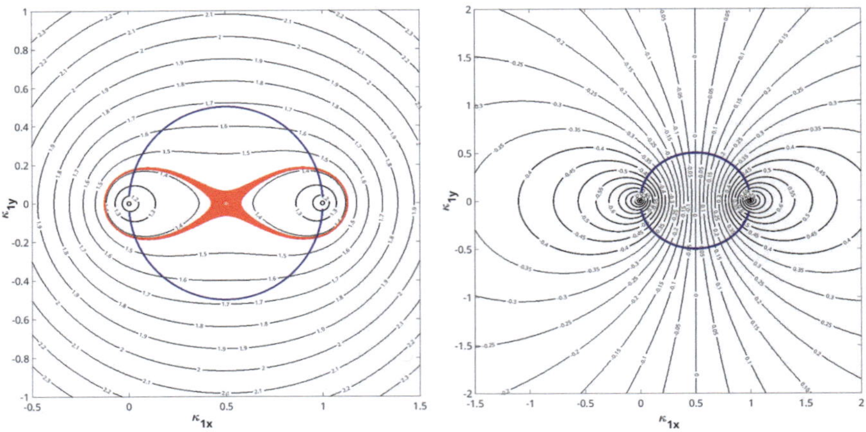

Fig. I.1 Contour lines of the normalized frequency ν in the normalized wavenumber plane for $\nu > 1$ (upper plot) and $|\nu| < 1$ (lower plot). The red region marks the values such that $1.41 < \nu < 1.42$ evidencing the singularity of $\sigma_2(\nu)$ at $\nu = \sqrt{2}$. The blue circle is the locus of the singularities of the EM kernel. For $\nu > 1$, it is tangential to the $2^{3/4} \simeq 1.68$ ν contour line, explaining the second singularity of $\sigma_2(\nu)$ at $\nu = 2^{3/4}$. For $|\nu| < 1$, it is tangential to no contour line

Each of the four components in (I.1.3) addresses a different frequency domain of the Doppler spectrum. The combined geometrical constraints $\nu = n_1\nu_1 + n_2\nu_2$ and $\boldsymbol{\kappa}_1 + \boldsymbol{\kappa}_2 = \boldsymbol{\kappa}_B$ impose $n_1 = n_2 = 1$ if $\nu \geq 1$, $n_1 = n_2 = -1$ if $\nu \leq -1$ and $n_1 = -n_2 = \pm 1$ if $|\nu| \leq 1$. The different combinations of signs thus cover the four distinct regions delimited by the first order Bragg lines. For a prescribed value of ν, the admissible values of $\boldsymbol{\kappa}_1 = (\kappa_{1x}, \kappa_{1y})$ follow curves corresponding to the contour lines shown in Fig. I.1.

The contours have 2 disjoint components for $\nu < \sqrt{2}$ that merge into a single one for $\nu > \sqrt{2}$. The densification of the contour lines at the transition (shown as the red region in the figure) produces a singularity of the function $\sigma_2(\nu)$ at $\nu = \sqrt{2}$. A second singularity is caused by the EM kernel, that has a sharp maximum for $\boldsymbol{\kappa}_1 \cdot \boldsymbol{\kappa}_2 = 0$. The locus of this singularity is shown as the blue circle in Fig. I.1, which is found tangential to the contour line $\nu = 2^{3/4} \simeq 1.68$. This produces a secondary, sharp peak at this value.

Considering the finite integration time and the resulting finite frequency resolution, the theoretical Doppler spectrum is actually smeared by some function Φ, corresponding to the Fourier Transform of the tapering window over the integration time. Hence it is given by:

$$\sigma_2(\nu) = \mathcal{N} k_B^4 \omega_B^{-1} \sum_{n_1=\pm1, n_2=\pm1} \int S(n_1 k_B \boldsymbol{\kappa}_1) S(n_2 k_B \boldsymbol{\kappa}_2) \, |\Gamma|^2 \, \Phi(\nu - n_1\nu_1 - n_2\nu_2) d\boldsymbol{\kappa}_1 \quad \text{(I.1.5)}$$

where the integration domain is now two-dimensional with no further constraint on the integration variable $\mathbf{k}_1$. This allows for a "brute force" computation of σ_2 but requires a tight 2D sampling of the k-space, which is numerically not efficient. An alternative computation can be obtained by reducing the two-dimensional integral

(10.1.10) to a single one by eliminating the delta function through an appropriate change of variables. The standard way to do this (e.g. [173]) is to use polar coordinates (κ_1, θ) for the wave vector $\boldsymbol{\kappa}_1$ and to operate a change a variable $(\kappa_1, \theta) \to (\nu_1 + \nu_2, \theta)$. However, the change of variable is only implicit and requires an extra numerical inversion for each value of ν to calculate the new integrand.

I.2 An Adapted Change of Variables

A less well-known, alternative change of variables was proposed recently by Shahidi and Gill [236–238], leading to a numerically more efficient and fully explicit expression of the second-order Doppler spectrum. We will present here in a slightly different way following the presentation of [101]. Let us detail the calculation of (I.1.3) for the term $n_1 = n_2 = 1$, the other terms being treated in a similar way. This term desribes the right side of the Doppler spectrum with respect to the positive Bragg line ($\nu \geq 1$). We chose to define the x-axis along the Bragg wave vector so that $\kappa_B = (1, 0)$. We then operate the change of variables:

$$(\kappa_{1x}, \kappa_{1y}) \to (\nu_1, \nu_2). \tag{I.2.6}$$

Note that we thus replace the two-dimensional wave vector κ_1 by a set of 2 positive frequencies with the obvious advantage that these appear as the argument of the delta function in the integrand. On the upper half-space $\kappa_{1y} \geq 0$ we can derive a simple one-to-one relationship between old and new integration variables:

$$\begin{aligned} \kappa_{1x}(\nu_1, \nu_2) &= \frac{1}{2}\left(1 + \nu_1^4 - \nu_2^4\right) \\ \kappa_{1y}(\nu_1, \nu_2) &= \left|\nu_1^4 - \frac{1}{4}\left(1 + \nu_1^4 - \nu_2^4\right)^2\right|^{1/2} \end{aligned} \tag{I.2.7}$$

and the same holds on the lower half-space by changing the sign of κ_{1y} in (I.2.7). The Jacobian of the transformation is found to be:

$$J(\nu_1, \nu_2) = \left|\frac{4\nu_1^3\nu_2^3}{\kappa_{1y}(\nu_1, \nu_2)}\right|. \tag{I.2.8}$$

We now evaluate the second-order Doppler spectrum in this new set of variables by treating separately the frequency intervals between and outside the Bragg lines.

I.3 Outside the Bragg Lines

We start with the case($\nu > 1$ ($n_1 = n_2 = 1$) or $\nu < -1$ ($n_1 = n_2 = -1$). By separating the integrals over the two half-spaces and performing the integration over the ν_2 variable to eliminate the delta function we obtain:

$$\sigma_2(\nu) = \mathcal{N} k_B^4 \omega_B^{-1} \int_{I(\nu)} \mathcal{K}(\nu_1)\gamma(\nu_1) J(\nu_1, |\nu| - \nu_1)\, d\nu_1 \tag{I.3.9}$$

where γ is the normalized kernel:

$$\gamma(\nu_1) = \left|\Gamma(\boldsymbol{\kappa}_1^+, \boldsymbol{\kappa}_2^+, \nu_1, 1)\right|^2 = \left|\Gamma(\boldsymbol{\kappa}_1^-, \boldsymbol{\kappa}_2^-, \nu_1, 1)\right|^2, \tag{I.3.10}$$

$\boldsymbol{\kappa}_1^\pm, \boldsymbol{\kappa}_2^\pm$ are the normalized wave vectors in the upper/lower half-space:

$$\begin{aligned} \boldsymbol{\kappa}_1^\pm &= (\kappa_{1x}(\nu_1, |\nu| - \nu_1), \pm\kappa_{1y}(\nu_1, |\nu| - \nu_1)) \\ \boldsymbol{\kappa}_2^\pm &= (1 - \kappa_{1x}(\nu_1, |\nu| - \nu_1), \mp\kappa_{1y}(\nu_1, |\nu| - \nu_1)) \end{aligned} \tag{I.3.11}$$

and $\mathcal{K}$ is given by:

$$\mathcal{K}(\nu_1) = S(n_1 k_B \boldsymbol{\kappa}_1^+) S(n_1 k_B \boldsymbol{\kappa}_2^+) + S(n_1 k_B \boldsymbol{\kappa}_1^-) S(n_1 k_B \boldsymbol{\kappa}_2^-). \tag{I.3.12}$$

The integration domain $I(\nu)$ is a subset of the positive axis. It is limited to the values of ν_1 for which $\nu_2 \geq 0$ and for which the argument of the square root in (I.2.7) is positive, that is:

$$\nu_1^4 - \frac{1}{4}\left(1 + \nu_1^4 - (|\nu| - \nu_1)^4\right)^2 \geq 0. \tag{I.3.13}$$

The limits of the domain can easily be identified analytically with the roots of the above equation. The integration domain is composed of one or two disjoint intervals depending on the value of ν

$$I(\nu) = \begin{cases} \left[\frac{\nu^2-1}{2|\nu|}, \frac{|\nu|-\sqrt{2-\nu^2}}{2}\right] \cup \left[\frac{|\nu|+\sqrt{2-\nu^2}}{2}, \frac{1+\nu^2}{2|\nu|}\right], & 1 \leq |\nu| \leq \sqrt{2}, \\ \left[\frac{\nu^2-1}{2|\nu|}, \frac{\nu^2+1}{2|\nu|}\right], & |\nu| > \sqrt{2}. \end{cases} \tag{I.3.14}$$

Note that the integration domain is aligned in two disjoints intervals for $1 < |\nu| < \sqrt{2}$. However, these two intervals are interchanged by the transformation $\nu_1 \mapsto \nu_1 - \nu$. Since this transformation amounts to swap ν_1 and ν_2 in the integrand, this leaves γ, J and $\mathcal{K}$ unchanged. It follows that the 2 integrals on the lower and upper interval are identical so that it suffices to calculate one of them and multiply the result by two.

I.4 Inside the Bragg Lines

We now address the case $|\nu| < 1$ ($n_1 = -n_2 = \pm 1$). Similar calculations lead to the following expression for the second-order Doppler spectrum inside the Bragg lines:

$$\sigma_2(\nu) = \mathcal{N} k_B^4 \omega_B^{-1} \left\{ \int_{I_{1,-1}(\nu)} + \int_{I_{-1,1}(\nu)} \right\} \mathcal{K}(\nu_1)\gamma(\nu_1) J(\nu_1, |\nu| - \nu_1)\, d\nu_1 \tag{I.4.15}$$

with the following adaptation of the integrand. On the interval $I_{1,-1}(\nu)$, the variable $|\nu| - \nu_1$ should be replaced by $\nu_1 - \nu$ in (I.3.9) and (I.3.11), and (I.3.12) should be replaced by:

$$\mathcal{K}(\nu_1) = S(k_B \kappa_1^+) S(-k_B \kappa_2^+) + S(k_B \kappa_1^-) S(-k_B \kappa_2^-). \tag{I.4.16}$$

On the interval $I_{-1,1}(\nu)$, the variable $|\nu| - \nu_1$ should be replaced by $\nu_1 + \nu$ in (I.3.9) and in (I.3.11), and (I.3.12) should be replaced by

$$\mathcal{K}(\nu_1) = S(-k_B \kappa_1^+) S(k_B \kappa_2^+) + S(-k_B \kappa_1^-) S(k_B \kappa_2^-). \tag{I.4.17}$$

The integration domain $I_{1,-1}(\nu)$ is constrained by the conditions

$$\nu_1^4 - \frac{1}{4}\left(1 + \nu_1^4 - (\nu - \nu_1)^4\right)^2 \geq 0, \;\; \nu_1 \geq \nu \tag{I.4.18}$$

leading to:

$$I_{1,-1}(\nu) = \left[\frac{\nu + \sqrt{2 - \nu^2}}{2}, \frac{1 + \mathrm{sign}(\nu)\nu^2}{2\,|\nu|}\right]. \tag{I.4.19}$$

Likewise the integration domain $I_{-1,1}(\nu)$ is constrained by the conditions

$$\nu_1^4 - \frac{1}{4}\left(1 + \nu_1^4 - (\nu + \nu_1)^4\right)^2 \geq 0, \;\; \nu_1 \geq -\nu \tag{I.4.20}$$

leading to:

$$I_{-1,1}(\nu) = \left[\frac{-\nu + \sqrt{2 - \nu^2}}{2}, \frac{1 - \mathrm{sign}(\nu)\nu^2}{2\,|\nu|}\right]. \tag{I.4.21}$$

Note that the transformation $\nu_1 \mapsto \nu_1 - \nu$ maps the interval $I_{1,-1}(\nu)$ to $I_{-1,1}(\nu)$. Again, since this transformation amounts to swap ν_1 and ν_2 in the integrand, this leaves γ and J unchanged in the integral and transforms (I.4.16)–(I.4.17). It follows that the 2 integrals in (I.4.15) are identical so that it suffices to calculate one of them

$$\sigma_2(\omega) = 2N k_B^4 \omega_B^{-1} \int_{I_{-1,1}(\nu)} \mathcal{K}(\nu_1)\gamma(\nu_1) J(\nu_1, |\nu| - \nu_1)\, d\nu_1. \tag{I.4.22}$$

When ν_1 approaches the border of the domain, corresponding to one end of the intervals $I(\nu)$, $I_{1,-1}(\nu)$ or $I_{-1,1}(\nu)$, the Jacobian term $J(\nu_1, |\nu| - \nu_1)$ has an integrable $1/\sqrt{x}$ singularity, except at the particular values $\nu = \pm\sqrt{2}$, where its denominator has a double root causing an non-integrable $1/x$ singularity. This recovers the result established in [142] that the Doppler spectrum has a logarithmic singularity at $\nu = \pm\sqrt{2}$. Apart from these 2 specific points, the singularity of the Jacobian at the borders of the domain can be handled properly by using the Gauss-Jabobi quadrature rule for numerical integration. This allows the calculation of the second-order Doppler spectrum (I.3.9) at the cost of an ordinary one-dimensional integral.

The Weighting Function J

As was shown in Chap. 12, the main wave parameters can be obtained through a direct integration of the second-order Doppler spectrum, providing an approximation that consists in replacing the coupling coefficient $|\Gamma|^2$ by its average value inside the (10.1.10). This last quantity has been introduced by Barrick [17, 18] and is referred to as the *weighting function*. It is defined as:

$$\frac{K_0^2}{8}\mathcal{W}(\nu) = \overline{|\Gamma|^2} \tag{J.0.1}$$

where $\nu = \omega/\omega_B$ is the normalized Doppler frequency and the right-hand side is the angular average of the coupling coefficient $|\Gamma|^2$ in the $\mathbf{k_1}$ variable. It has been recently shown [101] that other choices of weighting functions are possible, depending on the integration variables. With the double frequency variables as proposed in the Appendix I, one obtains a slighly different weighting function, which has been shown on the basis of numerical simulations to be a bit more accurate in estimating the main wave parameters. An approximate analytical fit of this weighting function has been found as:

$$\mathcal{W}(\nu) = \begin{cases} \exp(-129.1\nu^3 + 143.4\nu^2 - 63.4\nu + 12.7) & \text{if } \nu < 0.48 \\ 3\exp(-3.66(\nu - 0.48)) & \text{if } 0.48 < \nu < 0.64 \\ 1.67\exp(1.23(\nu - 0.64)) & \text{if } 0.64 < \nu < 1 \\ 5.43 & \text{if } 1 < \nu < 2^{1/2} \\ 5.43\exp(-10.24(\nu - 1.41)) & \text{if } 2^{1/2} < \nu < 1.51 \\ 1.95\exp(22.11(\nu - 1.51)) & \text{if } 1.51 < \nu < 2^{3/4} \\ 83.6\exp(-28.17(\nu - 1.68)) & \text{if } 2^{3/4} < \nu < 1.78 \\ 5\exp(2.1(\nu - 1.78)) & \text{if } \nu > 1.78 \end{cases} \tag{J.0.2}$$

C.-A. Guérin, *High-Frequency Radars for the Oceanographer*, Geophysics and Environmental Physics, https://doi.org/10.1007/978-3-031-95642-3

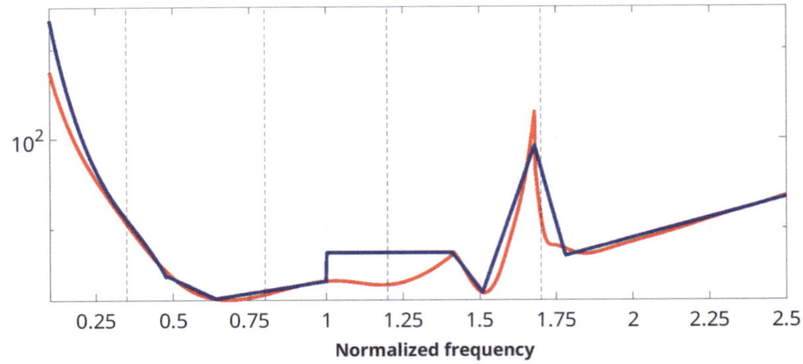

Fig. J.1 Barrick's weighting function (red solid line) and the alternative weighting function (J.0.2) (blue solid line)

Figure J.1 shows the fits of the two weighting functions. Barrick's weighting function is difficult to implement from the original publication. However, a numerical routine has been made available in the free RadarWIC library [38], that uses a digitized version of the Fig. 3 of [18].

Notations

K

We used the following typographic convention that a bold character **a** denotes a vector, the same plain character denotes its norm, $a = \|\mathbf{a}\|$, and the same bold character with a "hat" symbol denotes the corresponding unit vector, $\widehat{\mathbf{a}} = \mathbf{a}/a$. In addition, the following notations have been used consistently throughout this book.

1. Geometry of the sea surface

$\mathbf{R} = (x, y, z)$	3-dimensional position vector
$\mathbf{r} = (x, y)$	2-dimensional horizontal vector
R	Distance in 3-dimension
r	Horizontal distance
ϕ	Azimuth or bearing
	$(x = r\cos\phi,\ y = r\sin\phi)$
$\eta(\mathbf{r}, t)$	Sea surfaces elevations with respect to their mean plane
$\eta(\mathbf{r})$	Sea surfaces elevations ignoring the time dependence
$\widehat{s}(\omega) = \int_{-\infty}^{+\infty} e^{-i\omega t} s(t) dt$	Fourier transform of a time signal $s(t)$
$\widehat{\eta}(\mathbf{k}) = (2\pi)^{-2} \int e^{-i\mathbf{k}\cdot\mathbf{r}} \eta(\mathbf{r}) d\mathbf{r}$	Spatial Fourier transform of elevations
$\mathbf{k}, \mathbf{k_1}, \mathbf{k_2}$	2-dimensional wavevector of ocean waves (or sea surface elevations)
(k, θ)	2-dimensional wavevector in polar coordinates
ω	Ocean wave circular frequency
λ	Ocean wave wavelength
$S(\mathbf{k})$ or $S(k, \theta)$	Directional wavenumber spectrum
$S(\omega, \theta)$	Directional frequency spectrum
$S(\omega)$	... if no angular dependence
$S_{omni}(k)$	Omnidirectional wavenumber spectrum
$S_0(\mathbf{k})$	Wavenumber power spectrum
$D(k, \theta)$	Angular spreading function
$D(\theta)$	... if no wavenumber dependence

C.-A. Guérin, *High-Frequency Radars for the Oceanographer*, Geophysics and Environmental Physics, https://doi.org/10.1007/978-3-031-95642-3

2. Electromagnetic quantities

f_0	Radar carrier frequency
$\omega_0 = 2\pi f_0$	Radar carrier circular frequency
λ_0	Electromagnetic wavelength in vacuum
K_0	EM vacuum wavenumber at the radar carrier frequency
$\mathbf{K_0}$	Incident electromagnetic wave vector
$\mathbf{K}$	Scattered electromagnetic wave vector
θ_i, θ	Polar angle giving the incidence of the EM wave vector with respect to the z axis
λ_B	Bragg wavelength ($= \lambda_0/2$ in the monostatic case)
$\mathbf{k_0}, \mathbf{k}$	Horizontal projection of the incident and scattered EM wave vector
q_0, q	Vertical component of the incident and scattered EM wave vector ($q_0 = K_0 \cos\theta_i,\ q = K_0 \cos\theta$)
q_0', q'	Vertical component of EM wave vector in sea water
φ	Bistatic angle (= half angle between $\mathbf{k}$ and $\mathbf{k_0}$)
$\mathbf{k_B} = -2\mathbf{k_0}$ or $\mathbf{k} - \mathbf{k_0}$	Monostatic or bistatic Bragg wave vector
$k_B = 2k_0$ or $2k_0 \cos\varphi$	Monostatic or bistatic Bragg wave vector
$\mathsf{A}(\mathbf{k}, \mathbf{k_0})$	Scattering amplitude
$\mathsf{B}_0, \mathsf{B}_1, \mathsf{B}_2$	Kernels of the perturbation expansion
ϵ_r	Complex relative permittivity
$Z_0 = \sqrt{\mu_0/\epsilon_0}$	Vacuum impedance
$Z = \sqrt{\mu_0/\epsilon}$	Medium impedance
$Z_r = 1/\sqrt{\epsilon_r}$	Surface relative impedance
$n = \sqrt{\epsilon_r}$	Surface optical index

3. Doppler spectrum

f or f_D	Doppler frequency
$\omega = 2\pi f$	Radian Doppler frequency
$g = 9.81\ m.s^{-2}$	Gravitational acceleration at the Earth's surface
$\omega(\mathbf{k})$ or ω	Radian wave frequency from gravity waves dispersion relation ($= \sqrt{gk}$ in absence of current)
f_B	Monostatic or bistatic Bragg frequency
$\omega_B = 2\pi f_B = \sqrt{gk_B}$	Monostatic or bistatic radian Bragg frequency
$\Delta f, \Delta\omega$	Ordinar and circular frequency resolution (depends on integration time)
R	Range of a radar cell
ϕ	Azimuth of a radar cell
ΔR	Range resolution
$\Delta\phi$	Azimuthal resolution
$\sigma(\omega)$	Normalized Doppler cross-section, often simply called "Doppler spectrum"
$\sigma_1(\omega)$	First-order Doppler spectrum
$\sigma_2(\omega)$	Second-order Doppler spectrum
Γ_{EM}	Elecromagnetic kernel
Γ_H	Hydrodynamic kernel
Γ	Total kernel
$\mathcal{N} = 2^5\pi K_0^4$	Normalization coefficient of the Doppler spectrum
$\nu = f/f_B$	Normalized Doppler frequency
$\kappa = k/k_B$	Normalized wavenumber

References

1. J.L. Ahearn, S.R. Curley, J.M. Headrick, D.B. Trizna, Tests of remote skywave measurement of ocean surface conditions. Proc. IEEE **62**(6), 681–687 (1974)
2. Z.R. Alattabi, D. Cahl, G. Voulgaris, Swell and wind wave inversion using a single very high frequency (VHF) radar. J. Atmos. Oceanic Tech. **36**(6), 987–1013 (2019)
3. Z.R. Alattabi, G. Voulgaris, D.C. Conley, Evaluation and validation of HF radar swell and wind-wave inversion method. J. Atmos. Oceanic Tech. **38**(10), 1747–1775 (2021)
4. G. Apaydin, L. Sevgi, Numerical investigations of and path loss predictions for surface wave propagation over sea paths including hilly island transitions. IEEE Trans. Antennas Propag. **58**(4), 1302–1314 (2010). IEEE
5. M.R. Archer, L.K. Shay, J. Martinez-Pedraja, Evaluation of WERA HF radar observations: currents, winds and waves, in *2015 IEEE/OES Eleventh Current, Waves and Turbulence Measurement (CWTM)* (IEEE, 2015), pp. 1–9
6. F. Ardhuin, L. Marié, N. Rascle, P. Forget, A. Roland, Observation and estimation of Lagrangian, Stokes, and Eulerian currents induced by wind and waves at the sea surface. J. Phys. Ocean. **39**(11), 2820–2838 (2009)
7. J.N. Atanga, L.R. Wyatt, Comparison of inversion algorithms for HF radar wave measurements. IEEE J. Ocean. Eng. **22**(4), 593–603 (1997)
8. C.A. Balanis, *Antenna Theory: Analysis and Design* (Wiley, 2016)
9. N.F. Barber, F. Ursell, J. Darbyshire, M.J. Tucker, A frequency analyser used in the study of ocean waves. Nature **158**(4010), 329–332 (1946)
10. D.E. Barrick, Remote sensing of sea state by radar, in *Remote Sensing of the Troposphere*, vol. 12, ed. by V.E. Derr (US Government, 1972), pp. 1–46
11. D.E. Barrick, The interaction of HF/VHF radio waves with the sea surface and its implications, in *AGARD Electromagnetics of the sea state* (1970)
12. D.E. Barrick, Dependence of second-order Doppler sidebands in HF sea echo upon sea state, in *1971 Antennas and Propagation Society International Symposium*, vol. 9 (IEEE, 1971), pp. 194–197
13. D.E. Barrick, Theory of HF and VHF propagation across the rough sea, 1, The effective surface impedance for a slightly rough highly conducting medium at grazing incidence. Radio Sci. **6**(5), 517–526 (1971)
14. D.E. Barrick, Theory of HF and VHF propagation across the rough sea, 2, Application to HF and VHF propagation above the sea. Radio Sci. **6**(5), 527–533 (1971)
15. D.E. Barrick, First-order theory and analysis of MF/HF/VHF scatter from the sea. IEEE Trans. Antennas Propag. **20**(1), 2–10 (1972)

C.-A. Guérin, *High-Frequency Radars for the Oceanographer*, Geophysics and Environmental Physics, https://doi.org/10.1007/978-3-031-95642-3

16. D.E. Barrick, Remote sensing of sea state by radar, in *Ocean 72-IEEE International Conference On Engineering in the Ocean Environment* (IEEE, 1972), pp. 186–192
17. D.E. Barrick, Extraction of wave parameters from measured HF radar sea-echo Doppler spectra. Radio Sci. **12**(3), 415–424 (1977)
18. D.E. Barrick, The ocean waveheight nondirectional spectrum from inversion of the HF sea-echo Doppler spectrum. Remote Sens. Environ. **6**(3), 201–227 (1977)
19. D.E. Barrick, HF radio oceanography: a review. Bound.-Layer Meteorol. **13**(1–4), 23–43 (1978)
20. D.E. Barrick, A coastal radar system for tsunami warning. Remote Sens. Environ. **8**(4), 353–358 (1979)
21. D.E. Barrick, Accuracy of parameter extraction from sample-averaged sea-echo Doppler spectra. IEEE Trans. Antennas Propag. **28**(1), 1–11 (1980)
22. D.E. Barrick, M.W. Evans, B.L. Weber, Ocean surface currents mapped by Radar: mobile coastal units can map variable surface currents in real time to 70 km, using ocean wave scatter. Science **198**(4313), 138–144 (1977)
23. D.E. Barrick, J.M. Headrick, R.W. Bogle, D.D. Crombie, Sea backscatter at HF: interpretation and utilization of the Echo. Proc. IEEE **62**(6), 673–680 (1974)
24. D.E. Barrick, B. Lipa, Japan tsunami detected by HF radars on two continents, in *OCEANS 2011* (IEEE, 2011), pp. 1–5
25. D.E. Barrick, B. Lipa, C. Whelan, Bathymetric influence on tsunami spatial/temporal patterns in HF radar currents, in *OCEANS, 2012-Yeosu* (IEEE, 2012), pp. 1–6
26. D.E. Barrick, B.J. Lipa, Evolution of bearing determination in HF current mapping radars. Oceanography **10**(2), 72–75 (1997)
27. D.E. Barrick, B.L. Weber, On the nonlinear theory for gravity waves on the ocean's surface. Part II: interpretation and applications. J. Phys. Ocean. **7**(1), 11–21 (1977)
28. M.S. Bartlett, Multivariate analysis. Suppl. J. R. Stat. Soc. **9**(2), 176–197 (1947)
29. M.S. Bartlett, Smoothing periodograms from time-series with continuous spectra. Nature **161**(4096), 686–687 (1948)
30. M.S. Bartlett, Periodogram analysis and continuous spectra. Biometrika **37**(1/2), 1–16 (1950)
31. L. Bellomo et al., Toward an integrated HF radar network in the Mediterranean Sea to improve search and rescue and oil spill response: the TOSCA project experience. J. Oper. Ocean. **8**(2), 95–107 (2015)
32. D.J. Benney, Non-linear gravity wave interactions. J. Fluid Mech. **14**(4), 577–584 (1962)
33. G. Bienvenu, L. Kopp, Optimality of high resolution array processing using the eigensystem approach. IEEE Trans. Acoust. Speech Signal Process. **31**(5), 1235–1248 (1983)
34. C. Bourlier, G. Kubicke, Ground wave propagation along an inhomogeneous rough surface in the HF band: Millington effect for a flat Earth. IEEE Trans. Geosci. Remote Sens. **49**(4), 1374–1382 (2010). IEEE
35. C. Bourlier, G. Kubicke, HF ground wave propagation over a curved rough sea surface in the presence of islands. Waves Random Complex Media **21**(3), 469–484 (2011). Taylor & Francis
36. H. Bremmer, The extension of Sommerfeld's formula for the propagation of radio waves over a flat earth, to different conductivities of the soil. Physica **20**(1–6), 441–460 (1954). Elsevier
37. P. Broche, J.C. Demaistre, P. Forget, Mesure par radar décamétrique cohérent des courants superficiels engendrés par le vent. Oceanol. Acta **6**(1), 43–53 (1983)
38. D. Cahl, G. Voulgaris, Z. Alattabi, Wave radar inversion code (WaveRIC) V1.0.1 (version V1.0.1) (2019)
39. F. Capodici, S. Cosoli, G. Ciraolo, C. Nasello, A. Maltese, P.-M. Poulainc, A. Drago, J. Azzopardi, A. Gauci, Validation of HF radar sea surface currents in the Malta-Sicily channel. Remote Sens. Environ. **225**, 65–76 (2019)
40. R.D. Chapman, H.C. Graber, Validation of HF radar measurements. Oceanography **10**(2), 76–79 (1997)
41. R.D. Chapman, L.K. Shay, H.C. Graber, J.B. Edson, A. Karachintsev, C.L. Trump, D.B. Ross, On the accuracy of HF radar surface current measurements: intercomparisons with ship-based sensors. J. Geophys. Res. C: Oceans **102**(C8), 18737–18748 (1997)

42. C. Chavanne, Do high-frequency radars measure the wave-induced Stokes drift? J. Atmos. Oceanic Tech. **35**(5), 1023–1031 (2018)
43. Z. Chen, C. Zezong, J. Yanni, F. Lingang, Z. Gengfei, Exploration and validation of wave-height measurement using multifrequency HF radar. J. Atmos. Oceanic Tech. **30**(9), 2189–2202 (2013)
44. Z. Chen, L. Zhang, C. Zhao, J. Li, Calibration and evaluation of a circular antenna array for HF radar based on AIS information. IEEE Geosci. Remote Sens. Lett. **17**(6), 988–992 (2019)
45. L.Z.-H. Chuang, Y.R. Chen, Y.J. Chung, L.C. Wu, T.M. Tien, Bragg-region recognition of high-frequency radar spectra based on deep learning and image fusion processing. Int. J. Remote Sens. **43**(18), 6766–6782 (2022)
46. P.C. Clemmow, Ground-wave propagation across a land/sea boundary. Nature **165**(4186), 107–108 (1950). Nature Publishing Group UK London
47. J.H. Cloete, Is B or H the fundamental magnetic field? in *Proceedings of IEEE AFRICON'96*, vol. 1 (IEEE, 1996), pp. 354–361
48. R.E. Collin, Hertzian dipole radiating over a lossy earth or sea: some early and late 20th-century controversies. IEEE Antennas Propag. Mag. **46**(2), 64–79 (2004)
49. S. Cosoli, A. Mazzoldi, M. Gačić, Validation of surface current measurements in the northern Adriatic sea from high-frequency radars. J. Atmos. Oceanic Tech. **27**(5), 908–919 (2010)
50. C. Cox, W. Munk, Measurement of the roughness of the sea surface from photographs of the sun's glitter. J. Opt. Soc. Am. **44**, 838–850 (1954)
51. G.D. Crapper, *Introduction to Water Waves* (ELLIS HORWOOD LIMITED, 1984)
52. D.B. Creamer, F. Henyey, R. Schult, J. Wright, Improved linear representation of ocean surface waves. J. Fluid Mech. **205**, 135–161 (1989)
53. D.D. Crombie, Resonant backscatter from the sea and its application to physical oceanography, in *Ocean 72-IEEE International Conference on Engineering in the Ocean Environment* (IEEE, 1972), pp. 174–179
54. D.D. Crombie, Doppler spectrum of sea echo at 13.56 Mc./s. Nature **175**(4459), 681–682 (1955)
55. D.D. Crombie, Backscatter of HF radio waves from the sea. Electromagn. Probing Geophys. **79**, 131 (1971)
56. D.D. Crombie, K. Hasselmann, W. Sell, High-frequency radar observations of sea waves travelling in opposition to the wind. Bound.-Layer Meteorol. **13**, 45–54 (1978)
57. D.D. Crombie, J.M. Watts, Observations of coherent backscatter of 2–10 MHz radio surface waves from the sea. Deep-Sea Res. Oceanogr. Abstr. **15**(1), 81–87 (1968)
58. D.D. Crombie, J.M. Watts, W.M. Beery, Spectral characteristics of HF ground wave signals backscattered from the sea, in *AGARD Conference Proceedings of the Electromagnetics of the Sea (No. 77)* (1970)
59. B. Domps, D. Dumas, C.-A. Guérin, J. Marmain, High-frequency radar ocean current mapping at rapid scale with autoregressive modeling. IEEE J. Oceanic Eng. **46**(3), 891–899 (2021)
60. B. Domps, J. Marmain, C.-A. Guérin, *A Reanalysis of the October 2016 "MeteotsunamiâŁž in British Columbia with help of high-frequency radars and autoregressive modeling* (IEEE Geosci. Remote Sens, Lett, 2021)
61. M.A. Donelan, J. Hamilton, W.H. Hui, Directional spectra of wind-generated waves. Philos. Trans. R. Soc. London, Ser. A **315**(1534), 509–562 (1985)
62. D. Dumas, A.-C. Bennis, C.-A. Guérin, G. Lopez, L. Benoit, High-frequency radar observation of strong and contrasted currents: The Alderney race paradigm. Ocean Dyn. **75**(45) (2025)
63. D. Dumas, A. Gramoullé, C.-A. Guérin, A. Molcard, Y. Ourmieres, B. Zakardjian, Multistatic estimation of high-frequency radar surface currents in the region of Toulon. Ocean Dyn. **70**(12), 1485–1503 (2020)
64. D. Dumas, C.-A. Guérin, New signal processing techniques for phased-array oceanographic radars: self-calibration, antenna grouping, and de noising. J. Atmos. Oceanic Tech. **40**(6), 753–769 (2023)
65. A. Dussol, C. Chavanne, Estimation of the wind field with a single high-frequency radar. Remote Sens. **16**(13), 2258 (2024)

66. A. Dzvonkovskaya, Ocean surface current measurements using HF radar during the 2011 Japan tsunami hitting Chilean coast, in *Geoscience and Remote Sensing Symposium (IGARSS), 2012 IEEE International* (IEEE, 2012), pp. 7605–7608
67. A. Dzvonkovskaya, D. Figueroa, K.-W. Gurgel, H. Rohling, T. Schlick, HF radar observation of a tsunami near Chile after the recent great earthquake in Japan, in *Radar Symposium (IRS), 2011 Proceedings International* (IEEE, 2011), pp 125–130
68. A. Dzvonkovskaya, K.-W. Gurgel, T. Pohlmann, T. Schlick, J. Xu, Simulation of tsunami signatures in ocean surface current maps measured by HF radar, in *OCEANS 2009-EUROPE* (IEEE, 2009), pp. 1–6
69. A. Dzvonkovskaya, K.-W. Gurgel, T. Pohlmann, T. Schlick, J. Xu, Tsunami detection using HF radar WERA: a simulation approach, in *Radar Conference-Surveillance for a Safer World, 2009. RADAR. International* (IEEE, 2009), pp. 1–6
70. A. Dzvonkovskaya, D. Nikolic, V. Orlic, M.V. Peric, N. Tosic, Remote observation of a small meteotsunami in the bight of Benin using HF radar operating in lower HF band. IEEE Access: Pract. Innov., Open Solut. **7**, 88601–88608 (2019)
71. A. Dzvonkovskaya, L. Petersen, T. Helzel, M. Kniephoff, High-frequency ocean radar support for Tsunami early warning systems. Geosci. Lett. **5**(29) (2018)
72. A. Dzvonkovskaya, L. Petersen, T.L. Insua, *Real-time capability of meteotsunami detection by WERA ocean radar system, in International Radar Symposium* (Czech Republic, Prague, 2017)
73. T.M. Elfouhaily, C.-A. Guérin, A critical survey of approximate scattering wave theories from random rough surfaces. Waves Random Media **14**(4), R1 (2004)
74. S.A.A. Ellingsen, Y. Li, Approximate dispersion relations for waves on arbitrary shear flows. J. Geophys. Res. C: Oceans **122**(12), 9889–9905 (2017)
75. B.M. Emery, A. Kirincich, HF radar observation of nearshore winds, in *Ocean Remote Sensing Technologies-High-Frequency, Marine and GNSS-based Radar*. ed. by W. Huang, E.W. Gill (SciTech Publishing Raleigh, NC, USA, 2021), pp.191–216
76. B.M. Emery, L. Washburn, C. Whelan, D.E. Barrick, J. Harlan, Measuring antenna patterns for ocean surface current HF radars with ships of opportunity. J. Atmos. Oceanic Tech. **31**(7), 1564–1582 (2014)
77. H.-H. Essen, K.-W. Gurgel, T. Schlick, Measurement of ocean wave height and direction by means of HF radar: an empirical approach. Deutsche Hydrografische Zeitschrift **51**(4), 369–383 (1999)
78. H.-H. Essen, K.-W. Gurgel, T. Schlick, On the accuracy of current measurements by means of HF radar. IEEE J. Oceanic Eng. **25**(4), 472–480 (2000)
79. R.M. Fano, R.B. Adler, L.J. Chu, *Electromagnetic Fields, Energy, and Forces* (Taylor & Francis, 1960)
80. D.M. Fernandez, J.F. Vesecky, C.C. Teague, Measurements of upper ocean surface current shear with high-frequency radar. J. Geophys. Res. C: Oceans **101**(C12), 28615–28625 (1996)
81. D.M. Fernandez, J.F. Vesecky, C.C. Teague, Calibration of HF radar systems with ships of opportunity, in *IGARSS 2003. 2003 IEEE International Geoscience and Remote Sensing Symposium. Proceedings (IEEE Cat. No. 03CH37477)*, vol. 7 (IEEE, 2003), pp. 4271–4273
82. D.M. Fernandez, J.F. Vesecky, C.C. Teague, Phase corrections of small-loop HF radar system receive arrays with ships of opportunity. IEEE J. Oceanic Eng. **31**(4), 919–921 (2006)
83. X. Flores-Vidal, P. Flament, R. Durazo, C. Chavanne, K.-W. Gurgel, High-frequency radars: beamforming calibrations using ships as reflectors. J. Atmos. Oceanic Tech. **30**(3), 638–648 (2013)
84. P. Forget, Noise properties of HF radar measurement of ocean surface currents. Radio Sci. **50**(8), 764–777 (2015)
85. P. Forget, P. Broche, A study of VHF radio wave propagation over a water surface of variable conductivity. Radio Sci. **26**(5), 1229–1237 (1991)
86. P. Forget, P. Broche, J.C. De Maistre, Attenuation with distance and wind speed of HF surface waves over the ocean. Radio Sci. **17**(3), 599–610 (1982)

87. P. Forget, P. Broche, J.C. De Maistre, A. Fontanel, Sea state frequency features observed by ground wave HF doppler radar. Radio Sci. **16**(05), 917–925 (1981)
88. S. Fujii, M.L. Heron, K. Kim, J.-W. Lai, S.-H. Lee, X. Wu, X. Wu, L.R. Wyatt, W.-C. Yang, An overview of developments and applications of oceanographic radar networks in Asia and Oceania countries. Ocean Sci. Journ. **48**, 69–97 (2013)
89. E.W. Gill, W. Huang, J. Walsh, On the development of a second-order bistatic radar cross section of the ocean surface: a high-frequency result for a finite scatering patch. IEEE J. Ocean. Eng. **31**, 740–750 (2006)
90. E.W. Gill, W.W., Huang, HF radar, in *a maritime environment, in Ocean Remote Sensing Technologies. High-Frequency, Marine and GNSS-based Radar*. ed. by W. Huang, E.W. Gill (SciTech Publishing Raleigh, NC, USA, 2021), pp.1–39
91. E.W. Gill, W.W. Huang, Oceanographic applications of high-frequency (HF) radar backscatter, in *Ocean Remote Sensing Technologies-High-Frequency, Marine and GNSS-Based Radar*. ed. by W. Huang, E.W. Gill (SciTech Publishing Raleigh, NC, USA, 2021), pp.41–58
92. E.W. Gill, J. Walsh, High-frequency bistatic cross sections of the ocean surface. Radio Sci. **36**(6), 1459–1475 (2001)
93. Y. Goda, *Random Seas and Design of Maritime Structures*, vol. 33 (World Scientific Publishing Company, 2010)
94. T.H. Gomez, R. Tran, A. Ramdhani, R. Triyono, HF radar validation and accuracy analysis using baseline comparison approach in the Sunda strait, in *Global Oceans 2020: Singapore-US Gulf Coast* (IEEE, 2020), pp. 1–5
95. H.C. Graber, B.K. Haus, R.D. Chapman, L.K. Shay, HF radar comparisons with moored estimates of current speed and direction: expected differences and implications. J. Geophys. Res. C: Oceans **102**(C8), 18749–18766 (1997)
96. H.C. Graber, M.L. Heron, Wave height measurements from HF radar. Oceanography **10**(2), 90–92 (1997)
97. S.T. Grilli, C.A. Guérin, M. Shelby, A.R. Grilli, P. Moran, S. Grosdidier, T.L. Insua, Tsunami detection by high frequency radar beyond the continental shelf: II. Extension of algorithms and validation on realistic case studies. Pure and Appl. Geophys. **174**(1), 3003–3028 (2017)
98. S.T. Grilli, S. Grosdidier, C.-A. Guérin, Tsunami detection by high-frequency radar beyond the continental shelf, in *Global Tsunami Sci.: Past and Future*, vol. I (Springer, Berlin, 2015), pp. 3895–3934
99. S.T. Grilli, S. Grosdidier, C.-A. Guérin, Tsunami detection by high frequency radar beyond the continental shelf. I. Algorithms and validation on idealized case studies. Pure Appl. Geophys. **173**(12), 3,895–3,934 (2016)
100. S. Grosdidier, P. Forget, Y. Barbin, C.-A. Guérin, HF bistatic ocean doppler spectra: simulation versus experimentation. IEEE Trans. Geosci. and Remote Sens. **52**(4), 2138–2148 (2014)
101. C.-A. Guérin, Improved calculation of the second-order ocean Doppler spectrum for sea state inversion. IEEE J. Ocean Eng. **50**(3), 1895–1905 (2025)
102. C.-A. Guérin, D. Dumas, A. Gramoullé, C. Quentin, M. Saillard, A. Molcard, The multistatic HF radar network in Toulon, in *IEEE Radar 2019 Conference* (IEEE, 2019)
103. C.-A. Guérin, D. Dumas, A. Molcard, C. Quentin, B. Zakardjian, A. Gramoullé, M. Berta, High-frequency radar measurements with CODAR in the region of nice: improved calibration and performance. J. Atmos. Oceanic Tech. **38**(11), 2003–2016 (2021)
104. C.-A. Guérin, S.T. Grilli, A probabilistic method for the estimation of ocean surface currents from short time series of HF radar data. Ocean Model. **121**, 105–116 (2018)
105. C.-A. Guérin, S.T. Grilli, P. Moran, A.R. Grilli, T.L. Insua, Tsunami detection by high-frequency radar in British Columbia: performance assessment of the time-correlation algorithm for synthetic and real events. Ocean Dyn. **68**(4–5), 423–438 (2018)
106. C.-A. Guérin, M. Rozel, D. Dumas, M. Menelle, V. Morales-Marquez, L. Perus, High-frequency surface wave radars in the Mediterranean Sea: civil and environmental applications (IEEE RADAR 2024, Rennes, 2024)
107. C.-A. Guérin, A. Sentenac, Second-order perturbation theory for scattering from heterogeneous rough surfaces. J. Opt. Soc. Am. A, Opt. Image Sci. **21**(7), 1251–1260 (2004)

108. K.-W. Gurgel, G. Antonischki, H.-H. Essen, T. Schlick, Wellen Radar (WERA): a new ground-wave HF radar for ocean remote sensing. Coast. Eng. **37**(3), 219–234 (1999)
109. K.-W. Gurgel, A. Dzvonkovskaya, T. Pohlmann, T. Schlick, E. Gill, Simulation and detection of tsunami signatures in ocean surface currents measured by HF radar. Ocean Dyn. **61**(10), 1495–1507 (2011)
110. K.-W. Gurgel, H.-H. Essen, S.P. Kingsley, High-frequency radars: physical limitations and recent developments. Coast. Eng. **37**(3–4), 201–218 (1999)
111. K.W. Gurgel, H.H. Essen, T. Schlick, An empirical method to derive ocean waves from second-order Bragg scattering: prospects and limitations. IEEE J. Oceanic Eng. **31**(4), 804–811 (2006)
112. T.M. Hammond, C.B. Pattiaratchi, D. Eccles, M.J. Osborne, L.A. Nash, M.B. Collins, Ocean surface current radar (OSCR) vector measurements on the inner continental shelf. Cont. Shelf Res. **7**(4), 411–431 (1987)
113. P. Hansen, Measurements of basic transmission loss for HF ground wave propagation over seawater. Radio Sci. **12**(3), 397–404 (1977)
114. R.L. Hardman, L.R. Wyatt, C.C. Engleback, Measuring the directional ocean spectrum from simulated bistatic HF radar data. Remote Sens. **12**(2), 313 (2020)
115. R.L. Hardman, *Remote Sensing of Ocean Winds and Waves with Bistatic HF Radar*. Ph.D. thesis, University of Sheffield (2019)
116. J.A. Harlan, T.M. Georges, An empirical relation between ocean-surface wind direction and the Bragg line ratio of HF radar sea echo spectra. J. Geophys. Res. C: Oceans **99**(C4), 7971–7978 (1994)
117. D.E. Hasselmann, M. Dunckel, J.A. Ewing, Directional wave spectra observed during JONSWAP 1973. J. Phys. Ocean. **10**(8), 1264–1280 (1980)
118. K. Hasselmann, On the non-linear energy transfer in a gravity-wave spectrum Part 1. Gen. Theory. J. Fluid Mech. **12**(4), 481–500 (1962)
119. K. Hasselmann, On the non-linear energy transfer in a gravity wave spectrum Part 2. Conservation theorems; wave-particle analogy; irrevesibility. J. Fluid Mech. **15**(2), 273–281 (1963)
120. K. Hasselmann, On the non-linear energy transfer in a gravity-wave spectrum. Part 3. Evaluation of the energy flux and swell-sea interaction for a Neumann spectrum. J. Fluid Mech. **15**(3), 385–398 (1963)
121. K. Hasselmann, Determination of ocean wave spectra from Doppler radio return from the sea surface. Nature **229**(1), 16–17 (1971)
122. K. Hasselmann, et al., Measurements of wind-wave growth and swell decay during the joint north sea wave project (JONSWAP). Ergänzungsheft zur Deutschen Hydrographischen Zeitschrift Reihe (1973)
123. I. Hernández-Carrasco, L. Solabarrieta, A. Rubio, G. Esnaola, E. Reyes, A. Orfila, Impact of HF radar current gap-filling methodologies on the Lagrangian assessment of coastal dynamics. Ocean Sci. **14**(4), 827–847 (2018)
124. M.L. Heron, Y. Barbin, Bistatic HF ocean radar: errors and limitations. Oceans **2009**, 1–5 (2009)
125. M.L. Heron, Ocean wave measurement, in *Ocean Remote Sensing Technologies: High-Frequency, Marine and GNSS-based Radar*. ed. by W. Huang, E.W. Gill (SciTech Publishing Raleigh, NC, USA, 2021), pp.217–237
126. M.L. Heron, B. Domps, C.-A. Guérin, M. Wang, L. Petersen, HF radar real-time alert to a tsunami-like disturbance at Tofino on January 5, 2020: surge or tsunami? IEEE J. Oceanic Eng. **49**(1), 105–113 (2024)
127. M.L. Heron, A. Prytz, S.F. Heron, T. Helzel, T. Schlick, D.J.M. Greenslade, E. Schulz, W.J. Skirving, Tsunami observations by coastal ocean radar. Int. J. Remote Sens. **29**(21), 6347–6359 (2008)
128. M.L. Heron, R. Rose, On the application of HF ocean radar to the observation of temporal and spatial changes in wind direction. IEEE J. Oceanic Eng. **11**(2), 210–218 (1986)
129. S.F. Heron, M.L. Heron, A comparison of algorithms for extracting significant wave height from HF radar ocean backscatter spectra. J. Atmos. Oceanic Tech. **15**(5), 1157–1163 (1998)

130. Y. Hisaki, Nonlinear inversion of the integral equation to estimate ocean wave spectra from HF radar. Radio Sci. **31**(1), 25–39 (1996)
131. Y. Hisaki, Inter-comparison of wave data obtained from single high-frequency radar, in situ observation, and model prediction. Int. J. Remote Sens. **35**(10), 3459–3481 (2014)
132. Y. Hisaki, Development of HF radar inversion algorithm for spectrum estimation (HIAS). J. Geophys. Res. C: Oceans **120**(3), 1725–1740 (2015)
133. Y. Hisaki, Development of HF radar inversion algorithm for spectrum estimation (HIAS). J. Geophys. Res. Oceans **120**(3), 1725–1740 (2015)
134. L.H. Holthuijsen. *Waves in Oceanic and Coastal Waters* (Cambridge University Press, 2010)
135. J. Horstmann, A. Dzvonkovskaya, High frequency radar, in *Springer Handbook of Atmospheric Measurements*, ed. by T. Foken (Springer, 2021)
136. J.-C. Hou, S.-C. Wu, Z.-J. Yang, Remote sensing of ocean surface currents by HF radar. Oceanogr. Lit. Rev. **11**(44), 1384 (1997)
137. R. Howell, J. Walsh, Measurement of ocean wave spectra using narrow-beam HE radar. IEEE J. Oceanic Eng. **18**(3), 296–305 (1993)
138. N.E. Huang, C.-C. Tung, The dispersion relation for a nonlinear random gravity wave field. J. Fluid Mech. **75**(2), 337–345 (1976)
139. W. Huang, E. Gill, X. Wu, L. Li, Measurement of sea surface wind direction using bistatic high-frequency radar. IEEE Trans. Geosci. Remote Sens. **50**(10), 4117–4122 (2012)
140. W. Huang, E.W. Gill, HF surface wave radar. *Wiley Encyclopedia of Electrical and Electronics Engineering* (1999), pp. 1–11
141. D.V. Ivonin, P. Broche, J.-L. Devenon, V.I. Shrira, Validation of HF radar probing of the vertical shear of surface currents by acoustic Doppler current profiler measurements. J. Geophys. Res. C: Oceans **109**(C4) (2004)
142. D.V. Ivonin, V.I. Shrira, P. Broche, On the singular nature of the second-order peaks in HF radar sea echo. IEEE J. Oceanic Eng. **31**(4), 751–767 (2006)
143. J.D. Jackson, *Classical Electrodynamics*, 3rd ed. (Wiley, 2021)
144. D.H. Johnson, The application of spectral estimation methods to bearing estimation problems. Proc. IEEE **70**(9), 1018–1028 (1982)
145. E.C. Jordan, *Electromagnetic Waves and Radiating Systems* (Prentice-Hall, 1958)
146. J.W. Maresca, T.M. Georges, Measuring RMS wave height and the scalar ocean wave spectrum with HF skywave radar. J. Geophys. Res. C: Oceans **85**(C5), 2759–2771 (1980)
147. A. Kalampokis, M. Uttieri, P.-M. Poulain, E. Zambianchi, Validation of HF radar-derived currents in the gulf of Naples with Lagrangian data. IEEE Geosci. Remote Sens. Lett. **13**(10), 1452–1456 (2016)
148. S.Y. Kim, E.J. Terrill, B.D. Cornuelle, Mapping surface currents from HF radar radial velocity measurements using optimal interpolation. J. Geophys. Res. C: Oceans **113**(C10) (2008)
149. R.J. King, Electromagnetic field of a vertical dipole over an imperfectly conducting half-space. Radio Sci. **25**(2), 149–160 (1990)
150. B. Kinsman, *Wind Waves: Their Generation and Propagation on the Ocean Surface* (Courier Corporation, 1984)
151. J.T. Kirby, T.-M. Chen, Surface waves on vertically sheared flows: approximate dispersion relations. J. Geophys. Res. Oceans **94**(C1), 1013–1027 (1989)
152. A. Kirincich, Toward real-time, remote observations of the coastal wind resource using high-frequency radar. Mar. Technol. Soc. J. **47**(4), 206–217 (2013)
153. A. Kirincich, Remote sensing of the surface wind field over the coastal ocean via direct calibration of HF radar backscatter power. J. Atmos. Oceanic Tech. **33**(7), 1377–1392 (2016)
154. A. Kirincich, Improved detection of the first-order region for direction-finding HF radars using image processing techniques. J. Atmos. Oceanic Tech. **34**(8), 1679–1691 (2017)
155. A. Kirincich, B. Emery, Revisiting HF ground wave propagation losses over the ocean: a comparison of long-term observations and models. Radio Sci. **58**(4), 1–24 (2023)
156. A. Kirincich, B. Emery, L. Washburn, P. Flament, Improving surface current resolution using direction finding algorithms for multiantenna high-frequency radars. J. Atmos. Oceanic Tech. **36**(10), 1997–2014 (2019)

157. J.R. Klauder, A.C. Price, S. Darlington, W.J. Albersheim, The theory and design of chirp radars. Bell Syst. Tech. J. **39**(4), 745–808 (1960)
158. H. Krim, M. Viberg, *Two decades of array signal processing research* (IEEE Signal Process, Mag, 1996)
159. H. Kuroiwa, Y. Oono, Y. Hisaki, A. Nadai, HF ocean radar system at Okinawa radio observatory, in *Proceedings of the PORSEC 1992 in Okinawa: Conference for Pacific Ocean Environments and Probing*, vol. 1 (1992), pp. 47–52
160. Y. Lai, Y. Wang, H. Zhou, First-order peaks determination for direction-finding high-frequency radar. J. Marine Sci. and Eng. **9**(1), 8 (2020)
161. L.D. Landau, E.M. Lifshits, *Electrodynamics of Continuous Media* (Pergamon Press, Oxford, 1946)
162. D. Lawrence, P. Smith, Evaluation of HF ground-wave radar on the east coast of Canada. IEEE J. Ocean. Eng. **11**(2), 246–250 (1986)
163. K. Laws, J.D. Paduan, J. Vesecky, Estimation and assessment of errors related to antenna pattern distortion in CODAR SeaSonde high-frequency radar ocean current measurements. J. Atmos. Oceanic Tech. **27**(6), 1029–1043 (2010)
164. M.A. Leontovich, V.A. Fock, Solution of the problem of propagation of electromagnetic waves along the earth's surface by the method of parabolic equation. J. Phys. Ussr **10**(1), 13–23 (1946)
165. C. Levis, J.T. Johnson, F.L. Teixeira, *Radiowave Propagation: Physics and Applications* (Wiley, 2010)
166. B.J. Lipa, D.E. Barrick, Least-squares methods for the extraction of surface currents from CODAR crossed-loop data: application at ARSLOE. IEEE J. Oceanic Eng. **8**(4), 226–253 (1983)
167. B.J. Lipa, D.E. Barrick, S. Diposaptono, J. Isaacson, B.K. Jena, B. Nyden, K. Rajesh, T.S. Kumar, High Frequency (HF) Radar detection of the weak 2012 Indonesian tsunamis. Remote Sens. **4**(10), 2944–2956 (2012)
168. B. Lipa, D.E. Barrick, S.-I. Saitoh, Y. Ishikawa, T. Awaji, J. Largier, N. Garfield, Japan tsunami current flows observed by HF radars on two continents. Remote Sens. **3**(8), 1663–1679 (2011)
169. B.J. Lipa, D.E. Barrick, J. Bourg, B.B. Nyden, HF radar detection of tsunamis. J. Ocean. **62**(5), 705–716 (2006)
170. B.J. Lipa, Derivation of directional ocean-wave spectra by integral inversion of second-order radar echoes. Radio Sci. **12**(3), 425–434 (1977)
171. B.J. Lipa, Inversion of second-order radar echoes from the sea. J. Geophys. Res. C: Oceans **83**(C2), 959–962 (1978)
172. B.J. Lipa, D.E. Barrick, Methods for the extraction of long-period ocean wave parameters from narrow beam HF radar sea echo. Radio Sci. **15**(4), 843–853 (1980)
173. B.J. Lipa, D.E. Barrick, Extraction of sea state from HF radar sea echo: mathematical theory and modeling. Radio Sci. **21**(1), 81–100 (1986)
174. B.J. Lipa, D.E. Barrick, J. Isaacson, Section: Chapter 5: Coastal tsunami coastal tsunami warning with deployed HF radar systems, in *Tsunami, Environmental Science* (InTech, 2016), pp. 73–112
175. B.J. Lipa, D.E. Barrick, J. Isaacson, Evaluating HF coastal radar site performance for tsunami warning. Remote Sens. **11**(23), 2773 (2019)
176. B.J. Lipa, D.E. Barrick, C. Whelan, M. Losekoot, H. Parikh, First open-coast HF radar observations of a 2-phase volcanic tsunami, Tonga 2022. Remote Sens. **15**(9), 2325 (2023)
177. B.J. Lipa, B. Nyden, Directional wave information from the SeaSonde. IEEE J. Oceanic Eng. **30**(1), 221–231 (2005)
178. B.J. Lipa, H. Parikh, D.E. Barrick, H. Roarty, S. Glenn, High frequency radar observations of the June 2013 US East Coast meteotsunami. Nat. Hazards **74**, 109–122 (2014)
179. B.J. Lipa, C. Whelan, B. Rector, B. Nyden, HF radar bistatic measurement of surface current velocities: drifter comparisons and radar consistency checks. Remote Sens. **1**(4), 1190–1211 (2009)

180. Y. Liu, R.H. Weisberg, C.R. Merz, Assessment of CODAR SeaSonde and WERA HF radars in mapping surface currents on the west Florida Shelf. J. Atmos. Oceanic Tech. **31**(6), 1363–1382 (2014)
181. A. Long, D. Trizna, Mapping of North Atlantic winds by HF radar sea backscatter interpretation. IEEE Trans. Antennas Propag. **21**(5), 680–685 (1973)
182. M.S. Longuet-Higgins, The directional spectrum of ocean waves, and processes of wave generation. Proc. R. Soc. London. Ser. A. Math. Phys. Sci. **265**(1322), 286–315 (1962)
183. M.S. Longuet-Higgins, The effect of non-linearities on statistical distributions in the theory of sea waves. J. Fluid Mech. **17**(03), 459–480 (1963)
184. M.S. Longuet-Higgins, *Observation of the directional spectrum of sea waves using the motions of a floating buoy* (Oc, Wave Spectra, 1963)
185. G. Lopez, A.-C. Bennis, Y. Barbin, A. Sentchev, L. Benoit, L. Marié, Surface hydrodynamics in the Alderney Race from HF Radar measurements and 3D modelling. Phil. Trans. R. Soc. A **378**, 20190494 (2020)
186. G. Lopez, D.C. Conley, D. Greaves, Calibration, validation, and analysis of an empirical algorithm for the retrieval of wave spectra from HF radar sea echo. J. Atmos. Oceanic Tech. **33**(2), 245–261 (2016)
187. P. Lorente, et al., On the performance of high frequency radar in the western Mediterranean during the record-breaking storm Gloria. Front. Mar. Sci. **8** (2021)
188. P. Lorente et al., Coastal high-frequency radars in the Mediterranean - Part 1: status of operations and a framework for future development. Ocean Sci. **18**, 761–795 (2022)
189. P. Lorente, S. Piedracoba, J. Soto-Navarro, E. Alvarez-Fanjul, Evaluating the surface circulation in the Ebro delta (northeastern Spain) with quality-controlled high-frequency radar measurements. Ocean Sci. **11**(6), 921–935 (2015)
190. J. Maclean, G. Wu, *Radiowave Propagation over Ground* (Springer, US, 1993)
191. C. Mantovani et al., Best practices on high frequency radar deployment and operation for ocean current measurement. Front. Mar. Sci. **7**, 210 (2020)
192. C.C. Mei, M.A. Stiassnie, D.K.-P. Yue, *Theory and Applications of Ocean Surface Waves: Part 1: Linear Aspects* (World Scientific, 2005)
193. F. Mesa, D.R. Jackson, Excitation of the Zenneck wave by a tapered line source above the earth or ocean. IEEE Trans. Ant. Prop. **68**(6), 4848–4859 (2020)
194. G. Millington, Ground-wave propagation over an inhomogeneous smooth earth. Proc. IEE-Part III: Radio Commun. Eng. **96**(39), 53–64 (1949)
195. G. Millington, G.A. Isted, Ground-wave propagation over an inhomogeneous smooth earth. Part 2: experimental evidence and practical implications. Proc. IEE-Part III: Radio Commun. Eng. **97**(48), 209–221 (1950)
196. A. Molcard, P.M. Poulain, P. Forget, A.L. Griffa, Y. Barbin, J. Gaggelli, J.C. Maistre, M. Rixen, Comparison between VHF radar observations and data from drifter clusters in the gulf of la Spezia (Mediterranean sea). J. Mar. Syst. **78**, S79–S89 (2009)
197. V. Morales-Márquez, D. Dumas, C.-A. Guérin, HF radar estimation of ocean wave parameters: second-order Doppler spectrum versus Bragg wave modulation approach. Coastal Eng. **198**, 104719 (2025)
198. R. Mucci, A comparison of efficient beamforming algorithms. IEEE Trans. Acous. Speech Signal Process. **32**(3), 548–558 (1984)
199. K.F. Niessen, Zur Entscheidung zwischen den beiden Sommerfeldschen Formeln für die Fortpflanzung von drahtlosen Wellen. Ann. Phys. **42**(17), 585–596 (1937)
200. K.A. Norton, The propagation of radio waves over the surface of the earth and in the upper atmosphere. Part I. Proc. Inst. Radio Eng. **24**(10), 1367–1387 (1936)
201. K.A. Norton, The physical reality of space and surface waves in the radiation field of radio antennas. Proc. Inst. Radio Eng. **25**(9), 1192–1202 (1937)
202. K.A. Norton, The propagation of radio waves over the surface of the earth and in the upper atmosphere. Part II. Proc. Inst. Radio Eng. **25**(9), 1203–1236 (1937)
203. K.A. Norton, The calculation of ground-wave field intensity over a finitely conducting spherical earth. Proc. IRE **29**(12), 623–639 (1941)

204. K.A. Norton, Transmission loss in radio propagation. Proc. IRE **41**(1), 146–152 (1953)
205. K.A. Norton, System loss in radio wave propagation. Proc. IRE **47** (1959)
206. F. Nouguier, C.-A. Guérin, and B. Chapron. "Choppy wave" model for nonlinear gravity waves. *J. Geophys. Res.: Oceans*, 114(C9), 2009
207. M.K. Ochi, *Ocean Waves* (Cambridge University Press, 1998)
208. A. Orasi et al., HF radar for wind waves measurements in the Malta-Sicily channel. Measurement **128**, 446–454 (2018)
209. S.K. Oruganti et al., Experimental realization of Zenneck type wave-based non-radiative, non-coupled wireless power transmission. Sci. Rep. **10**(1), 925 (2020)
210. E.A. Fanjul, P. Lorente, J. Soto-Navarro, S. Piedracoba, Accuracy assessment of high frequency radar current measurements in the Strait of Gibraltar. J. Oper. Ocean. **7**(2), 59–73 (2014)
211. J.D. Paduan, R. Delgado, J.F. Vesecky, Y. Fernandez, J. Daida, C. Teague, Mapping coastal winds with HF radar, in *Proceedings of the IEEE Sixth Working Conference on Current Measurement (Cat. No. 99CH36331)* (IEEE, 1999), pp. 28–32
212. J.D. Paduan, K.C. Kim, M.S. Cook, F.P. Chavez, Calibration and validation of direction-finding high-frequency radar ocean surface current observations. IEEE J. Ocean. Eng. **31**(4), 862–875 (2006)
213. J.D. Paduan, L. Washburn, High-frequency radar observations of ocean surface currents. Ann. Rev. Mar. Sci. **5**, 115–136 (2013)
214. A. Papoulis, S.U. Pillai, *Probability, Random Variables and Stochastic Processes* (McGraw-Hill, 2002)
215. O.M. Phillips, The equilibrium range in the spectrum of wind-generated waves. J. Fluid Mech. **4**(4), 426–434 (1958)
216. O.M. Phillips, On the dynamics of unsteady gravity waves of finite amplitude Part 1: the elementary interactions. J. Fluid Mech. **9**(2), 193–217 (1960)
217. O.M. Phillips, *The Dynamics of the Upper Ocean* (Cambridge University Press, 1966)
218. W.J. Pierson, L. Moskowitz, A proposed spectral form for fully developed wind seas based on the similarity theory of SA Kitaigorodskii. J. Geophys. Res. **69**(24), 5181–5190 (1964)
219. R.J. Ramos, H.C. Graber, B.K. Haus, Observation of wave energy evolution in coastal areas using HF radar. J. Atmos. Oceanic Tech. **26**(9), 1891–1909 (2009)
220. J.W.S. Rayleigh, *The Theory of Sound*, vol. 2 (Macmillan, 1896)
221. E. Reyes et al., Coastal high-frequency radars in the Mediterranean - Part 2: applications in support of science priorities and societal needs. Ocean Sci. **18**, 797–837 (2022)
222. S.O. Rice, Reflection of electromagnetic waves from slightly rough surfaces. Commun. Pure Appl. Math. **4**(2–3), 351–378 (1951)
223. H. Roarty et al., The global high frequency radar network. Front. Mar. Sci. **6**, 164 (2019)
224. H. Roarty, C. Evans, S. Glenn, H. Zhou, Evaluation of algorithms for wave height measurements with high frequency radar, in *2015 IEEE/OES Eleveth Current, Waves and Turbulence Measurement (CWTM)* (IEEE, 2015), pp. 1–4
225. H. Roarty, G. Garcia, T. Nichols, Evaluation of the CODAR tsunami detection algorithm and software, in *Oceans 2019 MTS/IEEE Seattle* (IEEE, 2019), pp. 1–6
226. A.M. Robinson, L.R. Wyatt, M.J. Howarth, A two year comparison between HF radar and ADCP current measurements in Liverpool Bay. J. Oper. Ocean. **4**(1), 33–45 (2011)
227. R.E. Robson, Simplified theory of first-and second-order scattering of HF radio waves from the sea. Radio Sci. **19**(6), 1499–1504 (1984)
228. J. Röhrs, A.K. Sperrevik, K.H. Christensen, G. Broström, Ø. Breivik, Comparison of HF radar measurements with Eulerian and Lagrangian surface currents. Ocean Dyn. **65**(5), 679–690 (2015)
229. A. Rubio et al., HF radar activity in European coastal seas: next steps toward a pan-European HF radar network. Front. Mar. Sci. **4**, 8 (2017)
230. M. Saillard, A characterization tool for dielectric random rough surfaces: Brewster's phenomenon. Waves Random Media **2**(1), 67 (1992). (January)

231. S. Saviano, F. De Leo, G. Besio, E. Zambianchi, M. Uttieri, HF radar measurements of surface waves in the gulf of Naples (Southeastern Tyrrhenian sea): comparison with Hindcast results at different scales. Front. Mar. Sci. **7** (2020)
232. R. Schmidt, Multiple emitter location and signal parameter estimation. IEEE Trans. Antennas Propag. **34**(3), 276–280 (1986)
233. T.B.A. Senior, Impedance boundary conditions for imperfectly conducting surfaces. Appl. Sci. Res., Sect. B **8**(1), 418–436 (1960)
234. A. Sentchev, P. Forget, P. Fraunié, Surface current dynamics under sea breeze conditions observed by simultaneous HF radar. ADCP and drifter measurements. Ocean Dyn. **67**, 499–512 (2017)
235. L. Sevgi, Groundwave modeling and simulation strategies and path loss prediction virtual tools. IEEE Trans. Antennas Propag. **55**(6), 1591–1598 (2007)
236. R. Shahidi, E.W. Gill, A new method for ocean spectrum extraction from high-frequency second-order doppler radar data, in *OCEANS 2016 MTS/IEEE Monterey* (IEEE, 2016), pp. 1–4
237. R. Shahidi, E.W. Gill, An efficient and accurate solution for the extraction of non-directional ocean wave spectra from second-order high-frequency radar Doppler spectra, in *OCEANS 2017-Aberdeen* (IEEE, 2017), pp. 1–6
238. R. Shahidi, E.W. Gill, A new automatic nonlinear optimization-based method for directional ocean wave spectrum extraction from monostatic HF-radar data. IEEE J. Ocean. Eng. **46**(3), 900–918 (2020)
239. R. Shahidi, E.W. Gill, Two new methods for the extraction of significant wave heights from received HF-radar time series. IEEE Geosci. Remote Sens. Lett. **17**(12), 2070–2074 (2020). https://doi.org/10.1109/LGRS.2019.2961917
240. R. Shahidi, M.U.I. Loya, E.W. Gill, Simplified extraction of peak wave period from HF-radar data. IEEE Geosci. Remote Sens. Lett. **21**, 1–3 (2024)
241. E.D.R. Shearman, Propagation and scattering in MF/HF groundwave radar, in *IEE Proceedings F (Communications, Radar and Signal Processing)*, vol. 130 (IET, 1983), pp. 579–590
242. E.D.R. Shearman, Radio science and oceanography. Radio Sci. **18**(3), 299–320 (1983)
243. W. Shen, K.-W. Gurgel, Wind direction inversion from narrow-beam HF radar backscatter signals in low and high wind conditions at different radar frequencies. Remote Sens. **10**(9), 1480 (2018)
244. W. Shen, K.-W. Gurgel, G. Voulgaris, T. Schlick, D. Stammer, Wind-speed inversion from HF radar first-order backscatter signal. Ocean Dyn. **62**, 105–121 (2012)
245. V.I. Shrira, D.V. Ivonin, P. Broche, J.C. de Maistre, On remote sensing of vertical shear of ocean surface currents by means of a Single-frequency VHF radar. Geophys. Res. Lett. **28**(20), 3955–3958 (2001)
246. M.T. Silva, W. Huang, E.W. Gill, Bistatic high-frequency radar cross-section of the ocean surface with arbitrary wave heights. Remote Sens. **12**(4), 667 (2020)
247. M.T. Silva, R. Shahidi, E.W. Gill, W. Huang, Nonlinear extraction of directional ocean wave spectrum from synthetic bistatic high-frequency surface wave radar data. IEEE J. Ocean. Eng. **45**(3), 1004–1021 (2019)
248. A. Sommerfeld, Über die Ausbreitung der Wellen inder drahtlosen Telegraphie. Ann. Phys. **333**, 665–736 (1909)
249. R.M. Sorensen, *Basic Wave Mechanics for Coastal and Ocean Engineers* (Wiley, Inc., 1993)
250. R. Stewart, Higher order scattering of radio waves from the sea, in *1971 Antennas and Propagation Society International Symposium*, vol. 9 (IEEE, 1971), pp. 190–193
251. R.H. Stewart, J.W. Joy, HF radio measurements of surface currents, in *Deep Sea Research and Oceanographic Abstracts*, vol. 21 (Elsevier, 1974), pp. 1039–1049
252. R.H. Stewart, J.R. Barnum, Radio measurements of oceanic winds at long ranges: an evaluation. Radio Sci. **10**(10), 853–857 (1975)
253. V.I. Tatarskii, M.I. Charnotskii, On the universal behavior of scattering from a rough surface for small grazing angles. IEEE Trans. Antennas Propag. **46**(1), 67–72 (1998)

254. C.C. Teague, Multifrequency HF radar observations of currents and current shears. IEEE J. Ocean. Eng. **11**(2), 258–269 (1986)
255. C.C. Teague, J.F. Vesecky, Z.R. Hallock, A comparison of multifrequency HF radar and ADCP measurements of near-surface currents during COPE-3. IEEE J. Ocean. Eng. **26**(3), 399–405 (2001)
256. Y. Tian, Z. Tian, J. Zhao, B. Wen, W. Huang, Wave height field extraction from first-order doppler spectra of a dual-frequency wide-beam high-frequency surface wave radar. IEEE Trans. Geosci. Remote Sens. **58**(2), 1017–1029 (2019)
257. Y. Tian, Z. Tian, J. Zhao, B. Wen, W. Huang, Wave height field extraction from first-order doppler spectra of a dual-frequency wide-beam high-frequency surface wave radar. IEEE Trans. Geosci. Remote Sens. **58**(2), 1017–1029 (2020)
258. Y. Tian, B. Wen, Z. Li, Y. Yin, W. Huang, Analysis and validation of an improved method for measuring HF surface wave radar antenna pattern. IEEE Antennas Wirel. Propag. Lett. **18**(4), 659–663 (2019)
259. Y. Tian, B. Wen, H. Zhou, C. Wang, J. Yang, W. Huang, Wave height estimation from first-order backscatter of a dual-frequency high frequency radar. Remote Sens. **9**(11), 1186 (2017)
260. Z. Tian, Y. Tian, B. Wen, S. Wang, J. Zhao, W. Huang, E.W. Gill, Wave-height mapping from second-order harmonic peaks of wide-beam HF radar backscatter spectra. IEEE Trans. Geosci. Remote Sens. **58**(2), 925–937 (2019)
261. L.J. Tick, A non-linear random model of gravity waves I. J. Math. Mech. 643–651 (1959)
262. D. Trizna, L. Xu, Target classification and remote sensing of ocean current shear using a dual-use multifrequency HF radar. IEEE J. Oceanic Eng. **31**(4), 904–918 (2006)
263. D.B. Trizna, A bistatic HF radar for current mapping and robust ship tracking. Oceans **2008**, 1–6 (2008)
264. D.B. Trizna, L. Xu, Dual-use bistatic multi-frequency HF radar for ship classification and mapping of oceanic currents and vector winds, in *2005 IEEE Antennas and Propagation Society International Symposium*, vol. 1A (2005), pp. 392–395
265. B. Van Der Pol, Theory of the reflection of the light from a point source by a finitely conducting flat mirror, with an application to radiotelegraphy. Physica **2**(1–12), 843–853 (1935)
266. A. Voronovich, Small-slope approximation for electromagnetic wave scattering at a rough interface of two dielectric half-spaces. Waves Random Media **4**(3), 337 (1994)
267. A.G. Voronovich, *Wave Scattering from Rough Surfaces*, vol. 17 (Springer Science & Business Media, 2013)
268. A.G. Voronovich, V.U. Zavorotny, Theoretical model for scattering of radar signals in K u-and C-bands from a rough sea surface with breaking waves. Waves Random Media **11**(3), 247 (2001)
269. J.R. Wait, *Electromagnetic Probing in Geophysics*, vol. 79 (The Golem Press, 1971)
270. J.R. Wait, Perturbation analysis for reflection from two-dimensional periodic sea waves. Radio Sci. **6**(3), 387–391 (1971)
271. J.R. Wait, *Electromagnetic Wave Theory* (Harper and Row, New-York, 1985)
272. J.R. Wait, The ancient and modern history of EM ground-wave propagation. IEEE Antennas Propag. Mag. **40**(5), 7–24 (1998)
273. J. Walsh, R. Donnelly, Consolidated approach to two-body electromagnetic scattering. Phys. Rev. A **36**(9), 4474 (1987)
274. J. Walsh, R. Donnelly, A new technique for studying propagation and scattering for mixed paths with discontinuities. Proc. R. Soc. London. A. Math. Phys. Sci. **412**(1842), 125–167 (1987)
275. J. Walsh, E.W. Gill, An analysis of the scattering of high-frequency electromagnetic radiation from rough surfaces with application to pulse radar operating in backscatter mode. Radio Sci. **35**(6), 1337–1359 (2000)
276. J. Walsh, S.K. Srivastava, Rough surface propagation and scatter 1. General formulation and solution for periodic surfaces. Radio Sci. **22**(02), 193–208 (1987)
277. J. Walsh, J. Zhang, E.W. Gill, High-frequency radar cross section of the Ocean surface for an FMCW waveform. IEEE J. Oceanic Eng. **36**(4), 615–626 (2011)

278. W. Wang, P. Forget, C. Guan, Inversion of swell frequency from a 1-year HF radar dataset collected in Brittany (France). Ocean Dyn. **64**, 1447–1456 (2014)
279. W. Wang, P. Forget, C. Guan, Inversion and assessment of swell waveheights from HF radar spectra in the Iroise Sea. Ocean Dyn. **66**, 527–538 (2016)
280. W. Wang, E.W. Gill, Evaluation of beamforming and direction finding for a phased array HF ocean current radar. J. Atmos. Ocean. Tech. **33**(12), 2599–2613 (2016)
281. J.F. Ward, Power spectra from ocean movements measured remotely by ionospheric radio backscatter. Nature **223**(5213), 1325–1330 (1969)
282. L. Washburn, E. Romero, C. Johnson, B. Emery, C. Gotschalk, Measurement of antenna patterns for oceanographic radars using aerial drones. J. Atmos. Ocean. Tech. **34**(5), 971–981 (2017)
283. B.L. Weber, D.E. Barrick, On the nonlinear theory for gravity waves on the Ocean's surface. J. Phys. Ocean. **7**(1) (1977)
284. H. Weyl, Ausbreitung elektromagnetischer Wellen über einem ebenen Leiter. Ann. Phys. **365**(21), 481–500 (1919)
285. W.H. Wise, The physical reality of Zenneck's surface wave. Bell Syst. Tech. J. **16**(1), 35–44 (1937)
286. L.R. Wyatt, On Barrick's inversion procedure. Technical report, University of Birmingham (1981)
287. L.R. Wyatt, The measurement of the ocean wave directional spectrum from HF radar Doppler spectra. Radio Sci. **21**(3), 473–485 (1986). (May)
288. L.R. Wyatt, A relaxation method for integral inversion applied to HF radar measurement of the ocean wave directional spectrum. Int. J. Remote Sens. **11**(8), 1481–1494 (1990)
289. L.R. Wyatt, Limits to the inversion of HF radar backscatter for ocean wave measurement. J. Atmos. Ocean. Tech. **17**(12), 1651–1666 (2000)
290. L.R. Wyatt, An evaluation of wave parameters measured using a single HF radar system. Can. J. Remote. Sens. **28**(2), 205–218 (2002)
291. L.R. Wyatt, Shortwave direction and spreading measured with HF radar. J. Atmos. Ocean. Technol. **29**(2), 286–299 (2012)
292. L.R. Wyatt, Measuring the ocean wave directional spectrum 'First Five' with HF radar. Ocean Dyn. **69**(1), 123–144 (2019)
293. L.R. Wyatt, Ocean wave measurement, in *Ocean Remote Sensing Technologies-High-Frequency, Marine and GNSS-based Radar*. ed. by W. Huang, E.W. Gill (SciTech Publishing Raleigh, NC, USA, 2021), pp.145–178
294. L.R. Wyatt, et al.: HF radar: applications in coastal monitoring, planning and engineering. in: *Coasts and Ports 2013: 21st Australasian Coastal and Ocean Eng. Conference and the 14th Australasian Port and Harbour Conference* (Engineers Australia, 2013), pp. 878
295. L.R. Wyatt, J.J. Green, A. Middleditch, HF radar data quality requirements for wave measurement. Coastal Eng. (2011)
296. L.R. Wyatt, L.J. Ledgard, C.W. Anderson, Maximum-likelihood estimation of the directional distribution of 0.53-Hz ocean waves. J. Atmos. Oceanic. Tech. **14**(3), 591–603 (1997)
297. J. Zenneck, Über die Fortpflanzung ebener elektromagnetischer Wellen längs einer ebenen Leiterfläche und ihre Beziehung zur drahtlosen Telegraphie. Ann. Phys. **328**(10), 846–866 (1907)
298. J. Zhang, E.W. Gill, Extraction of ocean wave spectra from simulated noisy bistatic high-frequency radar data. IEEE J. Ocean. Eng. **31**(4), 779–796 (2006)
299. C. Zhao, Z. Chen, J. Li, F. Ding, W. Huang, L. Fan, Validation and evaluation of a ship echo-based array phase manifold calibration method for HF surface wave radar DOA estimation and current measurement. Remote Sens. **12**(17), 2761 (2020)
300. H. Zhou, B. Wen, Wave height extraction from the first-order Bragg peaks in high-frequency radars. IEEE Geosci. Remote Sens. Lett. **12**(11), 2296–2300 (2015)

Index

C.-A. Guérin, *High-Frequency Radars for the Oceanographer*, Geophysics and Environmental Physics, https://doi.org/10.1007/978-3-031-95642-3

MIX
Papier aus verantwortungsvollen Quellen
Paper from responsible sources
FSC® C105338

If you have any concerns about our products,
you can contact us on
ProductSafety@springernature.com

In case Publisher is established outside the EU,
the EU authorized representative is:
Springer Nature Customer Service Center GmbH
Europaplatz 3, 69115 Heidelberg, Germany

Printed by Libri Plureos GmbH
in Hamburg, Germany